CULTURE AND GLOBAL CHANGE

Investigations into cultural aspects of ways of life can be crucial in making valuable assessments of processes of change. Understanding culture in a broad conceptual framework helps us interpret what things mean to different people in different cultures and places.

Culture and Global Change presents twenty-five chapters which each offer their own particular take on 'culture', provoking new debate, opening up new areas of research and questioning conventional assumptions. Multiple meanings of culture – culture as representation, as discourse, as practice, as product, as action, as explanation – are represented here in this unique collection of contributions from leading writers in the field.

The book is divided into eight parts: Culture and Development; Questioning Cultural Assumptions; Representations and Cultural Commodification; Culture as Explanation; Culture and Resistance; Culture and Human Rights; Religion, Culture and Politics; Culture as Product: Culture as Pleasure. The editors guide readers through the debates, themes and places covered within individual chapters, ranging across topics which include: cricket in the Caribbean; Indian cinema and Hollywood; sex tourism in Thailand and the Philippines; London's representation through the *Evening Standard*; the Japanese economic miracle and workplace; street children in Brazil; Mexican–US and Latin America–UK migration; female circumcision in Sudan; women and New Hinduism; and film in Soviet and Russian society.

Exploring the significance and meaning of cultural issues for different people in different parts of the contemporary world, this book highlights the dangers of using reductive models of culture and of adopting culturally deterministic positions. The book stresses that we have to understand social action within the context of cultural change and both recognise and respect the diversity, dynamism and relevance of cultures. With chapters dealing with the importance of 'Third World' cultures but also with change in Russia, Japan, the USA and the UK, this book considers the relationship between culture and development at a sophisticated level and within a truly global context.

Tracey Skelton is a lecturer in the Department of International Studies, Nottingham Trent University and **Tim Allen** is a lecturer at the Development Studies Institute, London School of Economics.

CULTURE AND GLOBAL CHANGE

TRACEY SKELTON AND TIM ALLEN

London and New York

First published 1999
by Routledge
11 New Fetter Lane, London EC4P 4EE

Simultaneously published in the USA and Canada
by Routledge
29 West 35th Street, New York, NY 10001

Reprinted 2000, 2002

Routledge is an imprint of the Taylor & Francis Group

Typeset in Sabon by Routledge
Printed and bound in Great Britain by TJ International Ltd, Padstow, Cornwall

British Library Cataloguing in Publication Data
A catalogue record for this book is available from the British Library

Library of Congress Cataloguing in Publication Data
A catalogue record for this book has been requested

ISBN 0–415–13916–3 (hbk)
ISBN 0–415–13917–1 (pbk)

CONTENTS

List of illustrations ix
Notes on contributors xi
Acknowledgements xv

Culture and global change: an introduction 1
TRACEY SKELTON AND TIM ALLEN

I
CULTURE AND DEVELOPMENT 11

1 Classic conceptions of culture 13
PETER WORSLEY

2 Globalised culture: the triumph of the West? 22
JOHN TOMLINSON

3 Culture and development theory 30
PETER WORSLEY

II
QUESTIONING CULTURAL ASSUMPTIONS 43

4 Modernisation versus the environment? Shifting objectives of progress 45
ALAN THOMAS

5 Local knowledges and changing technologies 58
GORDON WILSON

6 Understanding health 70
MURRAY LAST

III
REPRESENTATIONS AND CULTURAL COMMODIFICATION 85

7 Finding the right image: British development NGOs and the regulation of imagery 87
HENRIETTA LIDCHI

8 Representations of conflict in the Western media: the manufacture of a barbaric periphery 102
PHILIPPA ATKINSON

9 Sex tourism: the complexities of power 109
JAN JINDY PETTMAN

10 The city and identity: news frames and the representation of London and Londoners in the *Evening Standard* 117
JENNY OWEN

IV
CULTURE AS EXPLANATION 125

11 Culture as ideology: explanations for the development of the Japanese economic miracle 127
ROGER GOODMAN

12 Cultural disease and British industrial decline: Weber in reverse 137
MIKE HICKOX

13 Ethnicity 145
JOHN EADE AND TIM ALLEN

V
CULTURE AND RESISTANCE 157

14 Local forms of resistance: weapons of the weak 159
HAZEL JOHNSON

15 'The people's radio' of Vila Nossa Senhora Aparecida: alternative communication and cultures of resistance in Brazil 167
VIVIAN SCHELLING

16 The new migrants: 'flexible workers' in a global economy 180
CHRIS J. MARTIN

VI
CULTURE AND HUMAN RIGHTS 191

17 The West, its Other and human rights 193
ROLANDO GAETE

18 Female circumcision and cultures of sexuality 201
MELISSA PARKER

19 Street lives and family lives in Brazil 212
TOM HEWITT AND INES SMYTH

VII
RELIGION, CULTURE AND POLITICS 221

20 Religion and political transformation 223
JEFF HAYNES

21 Religion and development 232
PARVATI RAGHURAM

22 Paying the price of femininity: women and the New Hinduism 240
DINA ABBOTT

VIII

CULTURE AS PRODUCT: CULTURE AS PLEASURE 249

23 Whose game is it anyway? West Indies cricket and post-colonial cultural globalism 251
HILARY McD. BECKLES

24 Bollywood versus Hollywood: battle of the dream factories 260
HEATHER TYRRELL

25 Mimicking Mammon? What future for the post-communist Russian film industry? 274
KATE HUDSON

References 282
Index 308

ILLUSTRATIONS

PLATES

3.1	Uganda/Sudanese repatriation. These are new arrivals from Nimule, Sudan, en route between Bibia and Arinyapia, where a large new reception centre has been built.	39
5.1	Cheziya Bakery, in Masvingo, Zimbabwe. Labour-intensive, Fordist production styles are subtly adapted, resulting in a highly flexible and effective work-force.	62
5.2	Metal-working in Mbare suburb of Harare, Zimbabwe. The 'informal sector' workshops of Mbara cover a large area with significant clusters of metal-working, wood-working and other trades – learning and copying from, sharing and competing with, one another.	65
6.1	Laropi, Uganda, 1991. A young woman is possessed by spirits while being trained as an *Ojo*, spirit medium/healer. The importance of indigenous religious practice remains strong, even among the young.	78
6.2	A *nyanga* in Tonga, Zimbabwe. *Nyangas* are formally registered local healers who still form an important part of Zimbabwe's health care system.	79
7.1	Disasters Emergency Committee East African Appeal, press advertisement, June 1980.	88
7.2	Christian Aid, poster advertisement, April–May 1990.	89
7.3	Christian Aid, press advertisement, April–May 1990.	95
7.4	Christian Aid advertising schedule, 1989–90.	96
13.1	A refugee from Bosnia-Herzegovina, now in Croatia, holds a newspaper with the headline, 'Through Terror to Great Serbia'. She has lived through some of that terror.	146
15.1	A street in Vila Aparecida.	170
15.2	Arriving in Vila Aparecida, the bus stops by a football field and circus tent.	171
15.3.	Photographs on the church wall record events and themes in the history of Vila Aparecida as a Basic Community. These photographs are images of the 'Day of the Migrant', when the church, 'committed to justice, the land and brotherhood', was inaugurated.	171
15.4.	The church built by the inhabitants of Vila Aparecida with loudspeakers on the roof.	172
15.5	A cloth banner behind the stage where the Festival was being held: 'Blacks: A Cry for Justice'.	172
21.1	Growing up in the desert offers few distractions for Algerian /Saharan refugee children of Smara Camp in the Tindouff region. This Koran teacher keeps a group of youngsters busy during the school holidays.	233

22.1 An RSS poster of Bharat Mata (Mother India) spreading luminously across and beyond the South Asian continent, erasing national boundaries from Afghanistan to China and Russia. 241
22.2 Celebrating the space of the Sati Sthal, where Roop Kanwar was burned to death on her husband's funeral pyre on 4 September 1987, in Deorala, Sikar district, Rajasthan. She was eighteen years old. 246
22.3 Photo collage sold in Deorala of a beatific Roop Kanwar on the pyre, glorifying her Sati. 246
24.1 Front cover of *Movie International* magazine, February 1996 issue, showing (clockwise from top right) Shah Rukh Khan, Madhuri Dixit, Shammi Kapoor, Sharmila Tagore, Rekha, Raj Kapoor, Amitabh Bachchan, Dilip Kumar, and (centre) Madhubala. 262
24.2 Drawing of Amitabh Bachchan by Deepa Mashru, age nine, from Leicester. 264
24.3 Publicity poster for *Cliffhanger* in Hindi. 265
24.4 Cassette cover of soundtrack to Mani Ratnam's *Bombay*. 265
24.5 Still from Mani Ratnam's *Bombay*. 267
24.6 Sanjeev Baskar, presenter of BBC 2's *Bollywood or Bust!* 269

FIGURES

11.1 Short-termism in the west: what managers cite as their top priority. 129
11.2 Typical Japanese office set-up. 130
19.1 Diagram to show the complexities of children's lives on and off the street in Brazil. 213

TABLES

9.1 Arrivals of tourists from abroad and receipts from international tourism. 111
9.2 Tourist arrivals and income in Thailand, 1985–90. 112
11.1 Opposing values said to underlie 'Japanese' and 'western' capitalism. 128
11.2 Total days lost by annual labour disputes. 129
14.1 Domination and disguised resistance. 163

CONTRIBUTORS

Dina Abbott is a Lecturer in Development Studies at the University of Derby and an associate lecturer for the Open University. She has carried out extensive fieldwork in Maharastra, India and published on women's poverty and income generation, slum housing and living conditions, and religiously motivated conflict in Bombay. She is currently collaborating on a major research project on slum dynamics with the University of Bombay.

Tim Allen is a Lecturer in the Development Studies Institute at the London School of Economics and senior visiting lecturer at the Open University. He presents the radio programme series 'Developing World' and is the author/editor of numerous books and articles on aspects of development. He has carried out long-term fieldwork in Uganda and Sudan. His most recent publications include *The Media in Conflict* (with Jean Seaton) and *Divided Europeans: Understanding Ethnicities in Conflict* (with John Eade).

Philippa Atkinson is based at the Development Studies Institute of the LSE and has done local level research over many years in war zones in Africa.

Hilary McD. Beckles is Professor of History at the University of the West Indies, Cave Hill campus in Barbados. He is the author and editor of several books on Caribbean slavery, emancipation and independence. He is also Director of the Centre for Cricket Research.

John Eade is a Reader at the Roehampton Institute in London.

Rolando Gaete is a Reader in Socio-Legal Studies at South Bank University and a lawyer. He took part in the agrarian reform process in Chile from 1970 to 1973 and defended political prisoners from 1973 to 1976, during the early period of Pinochet's dictatorship. He got a Ph.D at the LSE and settled in the UK. He is the author of *Human Rights and the Limits of Critical Reason* and articles in various academic journals.

Roger Goodman is University Lecturer in the Social Anthropology of Japan and a Fellow of St Anthony's College, University of Oxford. He is the author of *Japan's 'International Youth': The Emergence of a New Class of Ideology and Practice in Modern Japan* (Routledge, 1992), and author and co-editor of *Ideology and Practice in Modern Japan* (Routledge, 1992), *Case Studies on Human Rights in Japan* (Curzon, 1996) and the *East Asian Welfare State: Welfare Orientalism and the State* (Routledge, 1998).

Jeff Haynes is a Reader in the Department of Politics and Modern History, London Guildhall University, where he teaches courses on Third World, African and international politics. His most recent books are: *Religion, Globalization and Political Culture in the Third World* (editor, 1999); *Religion in Global Politics* (1998) and *Democracy and Civil Society in the Third World* (1997).

Tom Hewitt is Senior Lecturer in Development Studies in the Technology Faculty of the Open University. He has been director of the Open University's M.Sc. in Development Management

(1995–8) and most recently has published, as co-editor, *Managing Inter-Organisational Relationships* (1999). His research interests are in social policy, technology policy and industry in Latin America and Africa.

Mike Hickox is Principal Lecturer in Politics at South Bank University. Research interests include the sociology of educational vocationalism, cultural Englishness and art history, especially in relation to the Pre-Raphaelites. He has lectured and published in all these areas, and is currently researching the life and work of the Pre-Raphaelite artist, John Brett.

Kate Hudson is Senior Lecturer in Russian and East European Politics at South Bank University, London. Before entering academic life, Kate worked in the film industry, both as an archivist specialising in Soviet documentaries and newsreels, and as a film critic. She is editor of the international journal *Contemporary Politics* and has a book forthcoming in autumn 1999 entitled *European Communism since 1989: Survival and Revival?* (Macmillan).

Hazel Johnson is a Lecturer in the Faculty of Technology at the Open University. She is the co-editor of *Internationalisation and Development* and *Rural Livelihoods: Crises and Responses.*

Murray Last is Professor of Anthropology at University College, London. One of the founders of the British Medical Anthropology Society, he started the teaching of medical anthropology in the University of London in 1976. Since 1987 he has been the editor of the International African Institute's journal *Africa*; he has edited *The Professionalisation of African Medicine* (1986) and *Healing the Social Wounds of War* (in press).

Henrietta Lidchi is a Curator at the Department of Ethnography, British Museum, London, working on North American Collections. She has completed research and published in the fields of development studies and the politics of representation while at the Open University and the British Museum. In the last few years she has undertaken fieldwork in the Southwest United States. She is co-editor of *Imaging The Arctic.*

Chris J. Martin is a Lecturer at South Bank University and also works for the British Council in Mexico.

Jenny Owen is a Senior Lecturer in Media at South Bank University, London. She has written on the importance of journalism and documentary for contemporary society and is currently working on a book about women, crime and the media.

Melissa Parker is Director of the International Medical Anthropology Programme, Brunel University. She has undertaken fieldwork in Northern Sudan on female circumcision as well as tropical diseases, health and well-being. More recently she has investigated sexual networks and the transmission of HIV in London.

Jan Jindy Pettman is Reader in International Politics and Director of the Centre for Women's Studies at the Australian National University. Her most recent book is *Worlding Women: A Feminist International Politics* (Routledge, 1996). She is currently working on a project entitled 'A Feminist Perspective on "Australia in Asia"'. She is co-editor of the new Routledge *International Feminist Journal of Politics.*

Parvati Raghuram is a Lecturer in the Department of International Studies at Nottingham Trent University where she teaches geography at undergraduate level and research methods to postgraduate students. Her research interests include gender, domestic work and migration in India. More recently she has worked on a project on minority ethnic enterprise in the UK and on the new international middle class. She is a co-author of *Feminist Geographies* and a member of the Editorial Board for the CRICC/Sage Series *Global Power/Cultural Spaces.*

Vivian Schelling was born and brought up in South America where she lived until she came to England to complete her university education. She is a Senior Lecturer at the University of East

London where she has been teaching Third World and Development Studies. Her particular field of interest is Latin American, and in particular Brazilian, culture on which she has several publications.

Tracey Skelton lectures in geography, development and cultural studies. She is the co-editor of *Cool Places: Geographies of Youth Cultures* and co-author of *Feminist Geographies*. Her research is conducted mostly in the Caribbean and especially on the island of Montserrat. She is currently the chair of the Women and Geography Study Group and is the co-editor of a series with Routledge called *Critical Geographies*.

Ines Smyth is a Policy Advisor at Oxfam. She has been a Lecturer in Development Studies at the Institute of Social Studies and at the Development Studies Institute, LSE. She has carried out extensive fieldwork in Indonesia, and shorter term research in Uganda, South Sudan and East Timor. She has published articles and books on gender and development. The most recent is *Gender Works: Oxfam's Experience in Policy and Practice* (with Fenella Porter and Caroline Sweetman).

Alan Thomas is Senior Lecturer in Systems and Co-Chair of the Development Studies Subject Group at the Open University, and team member of the GECOU (Global Environmental Change – Open University) interdisciplinary research project on NGO environmental advocacy, funded by the UK Economic and Social Research Council under its Global Environmental Change initiative. Most of the examples mentioned in his paper were researched as part of the GECOU project, and the financial support of the ESRC and the Open University is acknowledged.

John Tomlinson is Director of the Centre for Research in International Communication and Culture (CRICC) at Nottingham Trent University. He has published widely on issues of globalisation, media and culture. His first book, *Cultural Imperialism* was published in 1991 and has since been widely translated. His latest book, *Globalization and Culture* was published by Polity Press in March 1999

Heather Tyrrell has been researching a Ph.D at the Nottingham Trent University on 'Bollywood in Britain' since 1995. She began studying Bollywood during a Masters degree at Glasgow's John Logie Baird Centre. She has lived two years in Sri Lanka, been a teacher and a journalist, and was the Media Education Officer at Broadway, Nottingham's Media Centre, for three years. She now works as a television researcher.

Gordon Wilson is a Senior Lecturer in the Technology Faculty of the Open University where he chairs the undergraduate course in Development Studies and the postgraduate course 'Development: Context and Practice'. His research interests include the processes by which small-scale producers in 'developing' countries develop technological capabilities and of how partnership and participation arrangements may be negotiated in development interventions in ways that lead to institutional sustainability.

Peter Worsley is Emeritus Professor at Manchester University. He is the author of *The Three Worlds* (1966), *The Third World* (1984) and *Knowledges* (1997).

ACKNOWLEDGEMENTS

We would like to thank the following for their kind permission to reproduce their photographs and images: UNHCR for Plates 3.1, 13.1 and 21.1; the East African Emergency Committee for Plate 7.1; Christian Aid for Plates 7.2 through to 7.4; *Manushi: A Journal About Women and Society* for Plates 22.2 and 22.3; *Movie International* magazine for Plate 24.1; Colombia Tristar/*Film Information* for Plate 24.3; Spark International for Plates 24.4 and 24.5; and Ken Green/BBC for Plate 24.6. We thank the *Observer* Newspaper for permission to reprint Figure 11.2.

We would also like to thank the following authors for providing their own photographs for Plates 5.1 and 5.3 (Gordon Wilson); 6.1 and 6.2 (Tim Allen); 15.1 through to 15.5 (Vivian Schelling); and to Deepa Mashru for Plate 24.2.

Every attempt has been made to obtain permission to reproduce copyright material. If any proper acknowledgement has not been made, we would invite copyright holders to inform us of the oversight.

The final production of the book owes much to work by the Routledge team, which has changed considerably from the time of contract through to completion. Tristan Palmer was the original enthusiast for the proposal and Sarah Lloyd was the very patient and supportive editor who saw it through to the time of manuscript submission. Sarah Carty, John Dixon and Casey Mein all played important parts in the final stages of the production of the book. We also thank Deb Booler and Patricia Skelton for their help during the final proofreading.

The editors would like to thank the group in Development Policy and Practice, of the Faculty of Technology at the Open University, who invited them to put together the initial book proposal, and provided a small subsidy for the initial work on the project. They would also like to acknowledge the patience and enthusiasm of the book's contributors during what turned out to be a very long writing and editing process.

Tracey Skelton would also like to thank Tim for being such a positive work companion, even though he could never remember where she was when. He was always witty! He is the only person who calls her 'babe' and is still living. Tracey's deepest thanks go to S. who made cups of tea, cooked dinners and made her laugh throughout the whole business.

Tim Allen would like to say that it was impossible to know where Tracey was and that he has an entire page in his phone book allocated to her peregrinations. Nevertheless it was a joy to try to keep up with her. He would also like to thank Melissa who rarely made him cups of tea or cooked, but was lovely to live with.

CULTURE AND GLOBAL CHANGE

An introduction

Tracey Skelton and Tim Allen

This is a book with a very grand title. We need to begin by explaining what it means, and, equally importantly, what it does not. This is a little easier with respect to 'global change' than it is for 'culture', so let us start with the former term.

WHAT IS GLOBAL CHANGE?

Global has become a popular adjective in recent years. There are three main reasons for this. The first is simply the lack of an alternative way of indicating the whole world. The expression 'worldly change' would carry quite different connotations. Indeed, it might suggest that this book is about the promotion of moral values.

The second reason is that there is a generally recognised need for terms to describe world-wide events and processes. A variety of factors, including the end of the Cold War and the rapid economic growth of several East Asian countries, has rendered conventional ways of categorising the world into parts redundant, or at least much more problematic. If there is still usefulness, for example, in a label like 'Third World', it is to point to shared histories rather than to regions which can be lumped together as a clearly delineated category. Moreover application of the expression 'Third World' has to be tempered with awareness of the growing *integration* of aspects of social life in Africa, Latin America, the Caribbean, Asia and Oceania with aspects of social life in the 'First' and 'Second' Worlds, particularly (though not exclusively) for élite groups.

A third reason for the choice of the word 'global' relates to this integration. Several chapters presented here specifically focus on it and comment on some of its consequences (e.g. Tomlinson, Atkinson, Lidchi and Beckles). Since the mid-1980s a fascination with connections across the world has resulted in a large literature on 'globalisation'. Some of the analysis has focused specifically on matters like the radical changes in financial transactions or the spread of CNN-style real-time television. Other scholars have viewed globalisation as a more pervasive phenomenon which is transforming, or may be about to transform, virtually everything. From this perspective, globalisation is understood as *the* key quality of the contemporary situation. In numerous publications, a great deal of space has been allocated to the elaboration of a sophisticated, diffuse, and often rather obscure, range of general models and hypotheses, all of which attempt to come to terms with the enormity of what is occurring.

This book has determinedly avoided such an approach. It is likely to be of most interest to students of globalisation not for its abstractions, but for its investigation of examples (of which there is a striking paucity in most publications on the subject). Several of our contributors are in fact very sceptical of the far-reaching assumptions and vague assertions made in the globalisation discourse, and suspect that there is a considerable element of 'old wine in new bottles'. Certainly for millions of people modernity seems to be characterised more by systematic exclusion and marginality rather than interconnectedness and the formation of new hybrid identities.

Thus we have opted for the adjective 'global' rather than the noun 'globalisation' because it is a less loaded word. We employ it to refer both to the important world-wide networks of integration (some of which are new and some of which are less so), and simply to important transformations going on in different places around the world.

We might have called these transformations 'development', but again we have preferred something more neutral for the title, so we refer instead to 'change', which has the distinct advantage, in this context, of not carrying the complex and wide-ranging associations of 'development'. Our intention is also to interrogate the notion of 'development' as an historically formulated concept, imbued with relations of power.

Clearly there is common ground here with the so-called 'post-development' school. This group of scholars have suggested that development is a dangerous, or even pernicious, concept in that it implies something positive but offers nothing more than a kind of mirage of progress, based on rigorous homogenisation of ways of being human. Most of the authors of this book share such concerns about traditional development theory and policy making, with its Western or Eurocentric biases. Nevertheless, this is not really a post-development book. The majority of the authors still work within what might be termed a 'development studies' framework in that they are unwilling to euphemise the experience of poverty by analysing it as a form of discourse, and remain committed both to a structural linking of poverty with affluence, and to the need to engage in practical action to alleviate it.

WHAT IS CULTURE?

So our use of the expression 'global change' is consciously straightforward. We would like the same to be the case for our use of the word 'culture', but that is wishful thinking. Raymond Williams famously described 'culture' as one of the most complicated words in the English language (1958). Yet most of the time its use is not so much complicated as just confused and confusing. It can refer to matters as diverse as a Beethoven symphony, a code of conduct or the quality of being human. Often it is clear that an individual uses it in different ways in different circumstances, and its application as an analytical category can be singularly unhelpful in that it is held to mean a variety of things, some of which might be contradictory. What the term means to a reader may be quite at variance with what was meant by the author (assuming she or he knew what they meant in the first place). Part of the problem relates to the etymology of the word.

Its origins are linked to 'cultivate' and 'cultivation', and a list of definitions given in *The Concise Oxford Dictionary* begins with: 'Tillage of the soil; rearing, production, (of bees, oysters, fish, silk, bacteria); quality of bacteria thus produced'. This is why 'culture' for natural scientists is still commonly something found in a petri-dish. During the seventeenth century, the word also began to be used in a metaphorical sense, to refer to the growth of individuals or of human society. Particularly in Germany, by the late eighteenth century, 'culture' (or *Cultur/Kultur*) was being used in scholarly works on historical progress, and was also being employed in the plural, to refer to distinct social groups. By the turn of the twentieth century, 'culture' had already become a concept with a complex of overlapping, but potentially different meanings. Three of these have proved particularly significant in the social sciences.

The broadest meaning related to debates about the theory of evolution – debates which continue up to the present. This is the argument about the extent to which human behaviour is determined by biology. For those who maintain that humans are not just another type of primate, the crucial difference between humans and animals is culture. Here culture refers to 'learned, adapted symbolic behaviour, based on a full-fledged language, associated with technical inventiveness, a complex of skills that in turn depends on a capacity to organise exchange relationships between communities' (Kuper 1994: 90).

A second meaning of culture was more overtly value-laden. Culture was again conceptualised as singular, but it was viewed as less pervasive. It was what 'a person *ought* to acquire in order to become a fully worthwhile moral agent' (Barnard and Spencer 1996: 136). Some people (i.e. well-educated English gentlemen) and some human products (i.e. classical music), were understood as having more culture than others.

In contrast, a third meaning was plural and relativistic. The world is divided into many cultures, each of which is valuable. 'Any particular person is a product of the particular culture in which he or she has lived, and differences between human beings are to be explained (but not judged) by differences in their cultures (rather than their race)' (Barnard and Spencer 1996: 136).

Although they are very different, these three diverse conceptions have rarely been kept completely separate. In popular discourse about culture, and in academic writing there has usually been a shifting between them. As Peter Worsely points out in Chapter 1, the academic discipline which has, until relatively recently, dominated discussion of culture is anthropology. Anthropologists are usually associated with the third meaning. Indeed Barnard and Spencer call this 'the anthropological sense' in the *Encyclopaedia of Social and Cultural Anthropology* (1996: 136). However, social and cultural anthropologists have implicitly or explicitly located their in-depth studies of particular cultures within the context of the first meaning, and will revert to culture in the singular if responding to suggestions of biological determinism.

More subtly, the second meaning has also informed their work too, though usually at an unconscious level. For a long time anthropologists persisted in describing their subject matter as 'primitive', 'native', 'tribal' or 'savage' society. Later, other terms were employed, notably 'the local'. However, as James Ferguson notes:

> Even if it is true that all social processes are in some sense 'local', it is also clear that, in normal anthropological practice, some problems, some research settings, even some people, seem to be more 'local' than others. Unsurprisingly, it is the least 'developed' who are generally understood to be the most 'local'.
>
> (Ferguson 1996: 159)

Moreover, while some anthropologists now work 'at home', the discipline remains obsessed with 'the other', which Said (1978) reminds us means 'those who are not us'. Those who are not us must be different to us, and, in a rather crude adaptation of much more sophisticated debates, the next line of argument is about who is better or worse, and a hierarchy of cultural development, cultural sophistication and concepts of superiority can filter through.

Anthropologists have tried to avoid this danger by emphasising their relativism, and by replacing earlier models of how behaviour is determined by the qualities of a culture with less functionalist theories which stress social agency and view culture as actively negotiated, not just inherited. However, the power, practice and production of 'Western-centric' development has a culture all of its own, and it is hegemonic enough to appropriate what fits from anthropological studies, and leave aside what does not. It represents 'others' as passive objects of development, trapped in their own poverty and lack of knowledge and understanding, constrained by traditions and cultural practices, fixed in time and needy of 'help' to develop and to emulate the successful 'West'.

In recent years the situation has been further complicated within academia by the rise of cultural studies, the 'cultural turn' in geography and the resurgence of cultural anthropology (see for example the overviews in Jackson 1989; Jenks 1993; Nugent and Shore 1997; During 1993; and Crang 1998). Elsewhere (Allen 1995; Skelton 1996a and 1996b) we have discussed the ways in which 'culture' is being included in academic discourse on development, globalisation and global political economy. More and more books and articles elaborate insights into what culture might mean, with debates even *within* particular disciplines, let alone *between* them, sometimes indulging 'in purely definitional argy-bargy' (Bauman 1996: 11). In addition there has been an increased tendency for economists and political scientists to invoke culture as an explanation for everything from 'transaction costs', to civil war, to the rapid industrialisation of China (e.g. North 1990; Fukuyama 1995; Huntington 1996). While others working from intellectually more relativist and locally grounded starting points call for a 'cultural critique of economics as a foundation structure of modernity' and assert the need for 'the formation of a culture-based political economy' (Escobar 1995: vii).

One exasperated senior British anthropologist, Adam Kuper, now takes the view that the term 'culture' covers too much and in too vague a manner to be of use, and should usually be set aside as a category in social science (personal communication). There have been moments in the course of working on this book when we have thought this might be good advice. Trying to pull together threads which have a commonalty of colour ('culture') but combine to form different and complex patterns (geography, content, focus, emphases, conclusions and suggestions) is in many ways an editor's nightmare. But obviously we have resisted going so far as to reject the concept of 'culture' – or our book would have been called something else.

Culture as a concept is everywhere, and we cannot just wish it away because it is a difficult thing to define and write about. There are common-sense understandings of the term and it is important that we engage and debate with the ways in which people use it. 'Culture' remains a significant part of people's lives. Understanding culture in a broad conceptual framework can help us interpret what things mean to people and, as many chapters in this collection demonstrate, nuanced and sophisticated investigations into cultural aspects of ways of life can be very significant in making assessments of processes of change.

There is also another reason why we feel we cannot reject an engagement with the concept of culture altogether. The huge body of knowledge and writing related to culture across and beyond the social sciences may be rather mind-boggling, but some of this material is extremely valuable. It may not be consistent, but its very diversity, complexity and variance are part of what make it interesting. The fact that in these many discourses the word 'culture' carries different meanings can be unsettling, but it does not necessarily mean that one discourse is correct and another one is wrong.

Our own position, and that of all the authors of this book, is that what has to be avoided is the construction of models in which culture is conceived in reified form. It may be that some people experience aspects of culture as if they are natural, but analysis must explain how these experiences are historically or socially constructed. This does not lead us to sacrifice 'a good, if complicated word altogether' (Baumann 1996: 11), but rather to engage with it critically.

Thus we reject conceptions of culture as fixed, coherent or 'natural', and instead view it as dynamically changing over time and space – the product of ongoing human interaction. This means that we accept the term as ambiguous and suggestive rather than as analytically precise. It reflects or encapsulates the muddles of living. We recognise that there are ideas and practices which may be maintained over long periods of time, from generation to generation, but culture is always contingent upon historical processes (extremely important in the context of development debates and the historical legacies of Empire). It is also influenced by, influences and generally interacts with, contemporary social, economic and political factors. Geography too is significant. It is not just about where you are on the world map, for example, but about the ways in which space and place interact with understandings about being a person. Moreover, any one individual's experience of culture will be affected by the multiple aspects of their identity – 'race', gender, age, sexuality, class, caste position, religion, geography and so forth – and it is likely to alter in various circumstances.

Multiple meanings of culture debated at length in the wider literature are represented here – culture as representation, as discourse, as practice, as product, as action, as explanation. Within each chapter though there is a general understanding that culture is not just about high and low, not about product and audience, not just about ways of doing things. It is of course all of these things.

WHAT IS CULTURE AND GLOBAL CHANGE?

Some scholars would argue that culture has always had an important place in the study of global change, even if this was not always consciously recognised. It is worth recalling that the origins of the notion of 'development' in the sense of promoting social progress lie in the same set of nineteenth century ideas which are associated with the origins of the concept of culture. Indeed culture and development sometimes meant almost the same thing until discourses of development planning and cultural specificity diverged in the early decades of the present century. As Ferguson points out this is one reason why the disciplines of anthropology and development studies have such a fraught relationship:

> Like an unwanted ghost, or an uninvited relative, development haunts the house of anthropology. Fundamentally disliked by a discipline that at heart loves all those things that development intends to destroy. Anthropology's evil twin remains too close a relative to be simply kicked out.
>
> (Ferguson 1996: 160)

However, as suggested above, what traditional development thinking has considered to be 'culture' often draws less upon local specificities than the overtly value-laden meaning of culture associated with a hierarchy of ways of life. It stresses the need to move away from 'primitive' (or 'local') traditions and adopt more 'Western', 'progressive' cultural practices. Another perception is that culture is fixed in time. For example, in the eyes of the United States Agency for International Development, agricultural practices along the Nile are fossilised, unable to change and therefore in need of technological injections to bring them into the twentieth century (Mitchell 1995). Similarly, aid workers in Somalia have characterised the country's population as trapped in a rigid and fatalistic value system (Allen 1992a: 339). When development swoops in on its objects, they are stripped of their history, placed in typologies defining what they are now, what they once were and where development will carry them (Crush 1995b). What this means is that the 'cultures' of people who are subject to development are still too often viewed as frozen in static traditions, and thereby simplistically (and often incorrectly) perceived as obstacles.

It is an important call within this book that, when development approaches take culture into account, this is done with thought and care. Pitfalls of western centrism must be avoided. There must be a recognition of the dynamism of people's world views. Diversity must be respected and reflected, and indigenous knowledges must be investigated and valued but not reified and romanticised.

There needs to be an inclusive and representative understanding of culture which becomes part of development discourse. The ways in which the 'development machine' (Ferguson 1990) analyses the world have to be fundamentally challenged and transformed.

The relationship between culture and development has to go further and deeper than just recognising the importance of 'Third World cultures'. As many of the authors in this volume demonstrate the very 'culture of development' has to be acknowledged and changed. This is a reason why we have included several chapters which do not deal with the 'Third World', but with change in Russia, Japan, the USA and the UK, or which consider issues of representation, migrations and philosophical debates that transcend national boundaries.

Finally, it is our contention that 'culture' must be understood as the over-arching context in which development and all forms of social change occur. The great strength of the term, for its many limitations, is that it directs our attention to the inter-connectedness between conventionally divided-up parts of social life. It should warn us not to believe too firmly in models that divorce economic transactions from religious beliefs, or kinship systems from the invigoration of democratic institutions. All the chapters in this book contribute to this end. The point is not to end up with almost unreadable and unusable thick description, but to provoke new questions, to research new areas and to question our conventional assumptions.

STRUCTURE OF THE BOOK

There is an enormous diversity contained within this collection. We present twenty-five chapters divided into eight parts. Each part provides a particular take on 'culture', but even within the parts there are further nuances of interpretations discussed in the constituent chapters. Consequently it is important to remember that each part is far from discrete or bounded. There are many chapters which could have been placed under different headings. An example is Hilary McD. Beckles' chapter on cricket and the Caribbean which could have been placed within the part entitled 'Representations and Cultural Commodification' because he discusses the ways in which cricket has become a globalised and commoditised form. However the chapter is actually placed in the section called 'Culture as Product: Culture as Pleasure' because cricket is also a sporting product, a product of national identities and is still viewed and played as a pleasure. Heather Tyrrell's chapter on Indian cinema is placed in 'Culture as Product: Culture as Pleasure' but could also have been placed in the part on 'Culture and Resistance'. Through film, India resists cultural hegemony from Hollywood and hence the USA. Rolando Gaete's chapter, which questions Western notions of human rights, is placed under the part heading of 'Culture and Human Rights' but could also have found a home within 'Questioning Cultural Assumptions'.

Hence it is critical that, just as we recognise the multiple interpretations of 'culture' and 'development', we recognise that the structure of the book could be as fluid and dynamic as the very concepts we draw upon. Our structure of eight parts is artificial and yet has purpose and meaning at the same time. Our aim is to guide you, the reader, through a range of debates, focuses and places. The parts could have followed a different order and while you will find many connections as you read, you will also be able to dip into chapters and form your own pathways through the text. Now to an outline of the parts of the book and a brief guide to the chapters.

Part I is entitled 'Culture and Development'. Here the chapters focus explicitly on debates surrounding questions of culture and its connections with development, both over time and in the contemporary context. They establish themes which are picked up through the rest of the text by other authors. Peter Worsley, one of the leading figures of modern sociology, presents two chapters, placed as Chapters 1 and 3. The first, 'Classic Conceptions of Culture', takes the reader through some of the key debates on culture within different academic disciplines (both in the UK and the USA), namely anthropology and sociology. It also describes the significant influences Gramsci and Williams have had in British studies of culture. John Tomlinson's chapter focuses on arguments, anxieties and assumptions surrounding the concept of a 'globalised culture'. He engages with the debates around the totalising impact of the West and stresses that while some aspects are embraced others are firmly rejected (a theme in Tyrrell's chapter). He states that if a globalised culture is emerging (and he presents considerable evidence which argues against this) then it is very much a hybrid culture. We cannot deny the power of Western culture but it might also be that the processes which consolidate this power – globalisation – might also lead to a decline of the West; it might lose its 'edge' against the 'rest'. In Chapter 3, 'Culture and Development Theory', Worsley, in his own distinctive style, considers the predominance of economic analyses within development theory, a legacy of modernisation theory. He shows that even alternatives to this theory did not take culture adequately into account. He argues that only when expected outcomes of development failed to materialise, or were highly successful, did questions begin to be asked about the role of culture in development. These three chapters, along with our introduction, combine to form a framework of debate about, and contextualisation of, 'culture' as a central concept.

Part II, 'Questioning Cultural Assumptions', develops more specific examples of some of the issues raised by Tomlinson. Each chapter forces the reader to engage with a reconsideration of what is 'appropriate' in non-Western contexts and cultures. Alan Thomas questions the ways in which modernisation as a concept underpins development theory. He argues that to date there is no real alternative to such an approach but suggests that a challenge may come from global environmentalism. Combinations of very different groups of people express and campaign around global environmental concerns and have forced the issue onto development agendas. However, Thomas urges considerable caution because global environmentalism is far from a single movement. There are very different cultures contained within the approach to environment and this can often lead to groups being in direct conflict. He reminds us that there are considerable differences, many of them revolving around culture and politics, between environmental groups of the South and the North. Gordon Wilson considers 'Local Knowledges and Changing Technologies'. He emphasises the fact that technology is far from culturally neutral as much development theory and process assumes. Technology is connected to knowledge and the latter is very much a cultural construction. As Wilson clearly demonstrates the cultural embeddedness of technology is evident in the gendered impacts its introduction has. If technology is going to have beneficial impacts then it has to be part of the cultural experiences and knowledges of the people who use it. Murray Last's chapter, 'Understanding Health' is the final one in this section. Last admires the aims of development health organisations of achieving 'health for all' but states that they are unrealistic. His central argument is that there should be a recognition of the value of local knowledge so as to allow people to have senses of self-worth and empowerment in relation to their own health. Echoing Tomlinson in examining what globalisation actually means, Last shows that globalisation has direct effects on people's health. It creates the means of rapid dissemination of diseases and because of incorporation into the global economy exacerbates illnesses which are linked to poverty. He examines the assumptions around biomedicine, which is assumed to be culture-free, when in fact it is tightly bound by existing power structures (see Parker's chapter). Last calls for a definition of health which emphasises well-being, recognises the role of the community in definitions of health/illness and the provision of health care, and works within existing cultures of health.

In Part III we have four chapters which on the surface appear very different. However, what all of them do is illustrate the ways in which 'Representations and Cultural Commodification' are part of our ways of understanding the world around us and are also pivotal to power relations between those who construct and those who are constructed. However, as illustrated in more detail in Part V, there is always and everywhere resistance. Henrietta Lidchi's chapter, 'Finding the Right Image: British Development NGOs and the Regulation of Imagery', focuses upon the ways in which representations construct meaning in and through images of development. She demonstrates that NGOs (Non-Governmental Organisations) are cultural organisations and they have to respond to external forces – not only from the 'Third World' but also from within the 'First World'. She considers the growth of concern about the 'image of development' and the increased recognition of the power of the image. She offers detailed reading of the images of an NGO campaign which tried to balance the need to raise funds with providing positive representations of their partners in development. Philippa Atkinson considers the ways in which the media represent other places to us – places we have not been ourselves. In 'Representations of Conflict in the Western Media: The Manufacture of a Barbaric Periphery' she argues that the media inevitably influence public opinion and consequently have to bear responsibility for the ways in which places and people are represented. She outlines the thesis of 'new barbarism', which represents Africa as a continent riddled with conflicts of its own making. Such persistent misrepresentation of Africa means that it is consistently misunderstood. Conflicts are represented without history, meaning or analysis. She then discusses the role aid agencies can and do play in monitoring and countering such representations and demonstrates the role that positive media transmission can have in conflict resolution (the importance of radio is taken up by Schelling's chapter). Turning our focus to a different part of the world and to a different form of commodification Jan Jindy Pettman writes about 'Sex Tourism: The Complexities of Power'. She takes us through the historical development of prostitution and thence sex tourism in Thailand and the Philippines; the involvement of the USA is clearly shown. The chapter considers the ways in which women, and increasingly children, are represented as sexual commodities and how this forms an international cultural politics, constructed, in this case, through tourism. She closes by looking at the ways in which women internationally are resisting exploitative representations as well as economic, political and social relations. For the fourth chapter in this section we focus on London as a global city and consider the way in which the city is represented through its 'local' paper, the *London Evening Standard*. Jenny Owen interrogates the context of 'The City and Identity' and shows the many ways in which the newspaper represents the city and the multiplicity of cultures within it. She argues that the *Standard* has to tread a fine balance between rendering London distinct and yet to provide spaces for difference. She shows how the paper wants to reclaim London from an agenda of crime, decay and loss of community and represent it as a living cultural ecology.

Part IV 'Culture as Explanation' offers chapters which critique three different examples of culture being used as an explanatory context for certain conditions. Roger Goodman, in 'Culture as Ideology', looks at debates which argue that there is a particular culture which has stimulated the Japanese economic miracle. He states that the 'West' was fascinated with what the Japanese did to achieve success and was hungry for texts which described corporate culture and history so that the approach might be copied. He shows that there are serious problems with such representations – they are functionalist, essentialist and static. It also allows no room for agency among the Japanese; they appear locked into a particular way of working. Goodman takes us into the reality of the Japanese workplace and shows that there is enormous diversity and the picture is much more complex than the 'how to' manuals imply. Perhaps, he suggests, Japan's economic success is more linked to the fact it has one of the most highly educated workforces in the world. In a similar vein Mike Hickox's chapter, 'Cultural Disease and British Industrial Decline: Weber in Reverse', considers the argument that Britain faces a 'cultural malaise'. He shows that while there are differing explanations as to where the 'malaise' originates and what might be done about it, there is a consensus about the fact that it exists. He critiques approaches which foster culture as explanation and clearly shows that there are flaws and over-statements in many of the arguments about Britain's apparent cultural malaise, complacency and chauvinism. He concludes that the so-called cultural traditionalism of Britain may in fact provide a distinctive edge in a rapidly globalising world. In their chapter 'Ethnicity', Tim Allen and John Eade engage with the problems of adequately defining ethnicity and of not conflating it with 'race'. They consider three different approaches to ethnicity within the discipline of anthropology and show the importance of relevant and meaningful understandings of the concept of ethnicity – not only for anthropology but also within every other context of social and cultural life. The chapter argues against those following a culturally functionalist perspective which treats 'ethnicity', 'culture' and 'society' as interchangeable concepts. It critiques the view that ethnicity should be used as a euphemism for the cultural specificities of social groups and highlights the common tendency to use ethnicity as a polite alternative term for 'race'. It provocatively points out that much discussion of ethnicity is in fact racist.

The fifth section of the book focuses on 'Culture and Resistance' and covers debates about resistance among the poor of the 'Third World' and ways in which local communities resist and self-empower. It also recognises that making choices and demonstrating creativity are aspects of resistance. Hazel Johnson considers 'Local Forms of Resistance: Weapons of the Weak' and interrogates the complexities of defining what is resistance. She asks whether there are genuine weapons which the weak can use to bring about long-term change and improvement. The chapter considers the ways in which power and weakness are conceptualised and how strongly poverty links to weakness. Nevertheless, as the chapter shows, the weapons that people do have are culturally embedded and there are several ways of interpreting successful outcomes of different forms of resistance. The people's radio of Vila Nossa Senhora Aparecida is the focus of Vivian Schelling's chapter about alternative communications and cultures of resistance in Brazil. A community of north-eastern migrants, settled on the edge of São Paulo, have developed their own radio station which is both informed by, and informs, their relatively 'new' urban experiences. It is also a means of resisting the erosion of their cultural identities as they face a particular form of modernity which invalidates their skills and patterns of life. As Schelling illustrates, the people's radio is in fact a significant counter-hegemonic cultural form. Chris J. Martin's chapter, 'The New Migrants: "Flexible Workers" in a Global Economy', is not ostensibly about cultures of resistance, and yet this is an implicit factor in his discussion of contemporary voluntary migrants. He places an emphasis on the fact that the migrants are forging new destinies and are being creative in the ways in which they take opportunities. He provides a different representation of migrants to the one of downtrodden people subject to forces beyond their control. He focuses on Mexican–US and Latin America – UK migration and shows the ways in which workers are creating new niches for themselves and have revitalised whole areas of cities. He concludes that migrants have more capacity for success than existing migration theories have given them credit for.

Closely connected with the chapters on resistance are those which focus on issues of human rights and which form Part VI of the book, 'Culture and Human Rights'. The three chapters force us to question conceptions of human rights and to think hard about the ways in which the rights discourse is one imbued with 'Western' conceptions of cultural and moral norms. Rolando Gaete's chapter, 'The West, Its Other and Human Rights' considers the historical antecedents of contemporary rights discourses in which the West constructed itself as the civilising force of the world. The West established itself as being what the rest of the world, in particular the colonised world now the 'Third World', was not. Gaete argues that the West currently assumes that with globalisation has come the end of history, the end of particularities. For the West, human rights are 'cultural artefacts grounded in universal consensus'. He shows that the West allows for aspects of cultural relativism in internal decisions about rights but refuses the same flexibility in its analysis of human rights situations in non-Western countries. Gaete calls for greater subtlety in the interpretation of human rights considerations and for the West to recognise that there are important aspects of cultural relativity. Melissa Parker tackles the complex and difficult matter of female circumcision in Sudan. Her chapter is both critical of existing material on the subject and also thoughtfully reflective about her own witnessing of a circumcision ceremony. She shows that the subject area is one fraught with problems of definitions and interpretations, and recognises that even using the term 'circumcision' will condemn her to severe criticism from some quarters. The intense emotions revolving around this practice, which is often done to girls of a very young age, has meant that the usual criteria for rigour and clarity in academic writing within the disciplines of biomedicine and anthropology are often abandoned. She carefully shows this through her reading of the existing literature. She concludes that Western preoccupations with female circumcision and the subsequent focus on cultures of sexuality has a great deal to do with our own concerns about sexuality. The final chapter of this section, by Tom Hewitt and Ines Smyth, focuses on children and raises questions about the ways in which so-called 'street children' in Brazil are defined and conceptualised. The complexities around the meanings of 'street children' are explored and there is a call for a move away from the idealised and often 'Western' notions of family life.

The penultimate section, Part VII, 'Religion, Culture and Politics', has three chapters which consider the ways in which culture, religion, secularisation, modernisation, development and political identities interact and intersect. In the first chapter Jeff Haynes investigates the relationship between 'Religion and Political Transformation'. He discusses the way in which through principles of modernisation there have been attempts to secularise nations and separate politics from religion and shows how this has failed. Haynes demonstrates the importance of identity within the context of religion and how this interacts with processes of development. With rapid change comes a loss of identity, distrust of governments and a turning to other sources which validate and give meaning to individuals and communities, a significant one being different forms of religion. Parvati Raghuram writes about 'Religion and Development' and shows how the Abrahamic religions (Islam, Christianity and Judaism) have been strongly interconnected with modernisation theory, in particular through missionaries and evangelism. However, there have been notable shifts in the relationship between development and religion. As Raghuram shows, one of the most significant has been evident in Latin America where Catholic Liberation Theology has emphasised equity and empowerment approaches in development. Our attention is then drawn to India and the rise of Hindu fundamentalism and nationalism. Dina Abbott in her chapter 'Paying the Price of Femininity: Women and the New Hinduism' asks questions about the processes of redefining national identities, especially those which place emphasis on religion and ethnicity. Abbott demonstrates the ways in which the Pariwar (a group of fascistic parties) are galvanising Hindus around a redefinition of nationhood, often with extremely violent results. Her focus is then on what this means for women and the lower castes and she analyses the growing trend for a resurrection of the ceremony of Suti where a widow immolates herself on her husband's funeral pyre.

The final section, Part VIII, considers 'Culture as Product: Culture as Pleasure'. The chapters consider particular forms of cultural production and entertainment which provide not only pleasure, but also form arenas in which global change and power are played out. Hilary McD. Beckles investigates, often with wry humour, the complexities of power around cricket, colonial and post-colonial identities. His chapter is called 'Whose Game is it Anyway? West Indies Cricket and Post-Colonial Cultural Globalism' and considers the cultural and ideological baggage the West Indies and England carry to a cricket match. He focuses on a particular match played in Barbados where local capital and cultural 'guardians' conspired with global capital (specially arranged tourist trips) to bring the English in and keep the Bajans out. Through this he shows the inequalities of access and pleasure experienced by those on different sides of the post-colonial global process. From the Caribbean and sport we travel to India, the USA, and film. Heather Tyrrell introduces us to the fascinating world of Indian cinema (film, video and television, within India and the Indian diaspora) in 'Bollywood versus Hollywood: Battle of the Dream Factories'. She engages with the theoretical debates about Third Cinema and the ambiguous position Indian cinema holds. There is then a close investigation of the impact of the removal of the ban on dubbing films into Hindi in 1992. Hollywood was quick to attempt to move in on the Indian cinema market with film translations but has yet failed to capture the imagination of Indian audiences. She shows that there many internal contradictions within Bollywood, and calls it a wild card in the globalisation of the media. The final chapter of the section and indeed the whole book is another focus on film but the geographical setting is Russia. At a time of enormous economic and political change Kate Hudson considers a specific cultural form directly affected by such changes. In 'Mimicking Mammon? What Future for the Post-Communist Russian Film Industry?' Hudson demonstrates the ways in which culture and economic factors have been closely interconnected since the collapse of the Soviet Union. Film within Soviet and Russian society is described, in particular its role in the building of Soviet identity, its place in cinema history and its contemporary crisis as it tries to forge a new cultural role and struggle against the ravages of liberal market economics. The future of what has been hailed as some of the greatest experimental cinema looks very bleak and is a salutary reminder of the fragility of some cultural forms in the face of major global change.

'Culture' in its many manifestations is debated, analysed and interrogated within the following chapters. There are contradictions between chapters, readers are encouraged to question the concepts presented, and there is a recognition that we have all been writing about a fundamental yet complex and diverse concept. The significance and meaning of cultural issues for different people in different parts of the contemporary world have been discussed. The dynamism and fluidity, the messiness of culture have been illustrated through considerations of many global changes. We hope you enjoy making your own cultural journeys through this collection.

I

CULTURE AND DEVELOPMENT

1

CLASSIC CONCEPTIONS OF CULTURE

Peter Worsley

INTRODUCTION

Outside of the natural sciences, the word 'culture' is used in two main ways. The first, and oldest, uses the term to describe the 'fine' arts – not any kind of art, but only certain kinds of music ('classical'), painting, sculpture and literature created by an intellectual élite and consumed largely by the upper classes and the highly-educated middle class. Thus the book review sections in the 'quality' newspapers, or events like the Edinburgh Festival, are often described as 'cultural' phenomena.

The second usage is much wider, and often much less specific. This is the idea of 'culture' as a way of life. At the broadest level this may refer to almost everything that distinguishes human beings from animals. Here 'culture' is contrasted with 'nature' or biology. But culture in this sense is also used to refer to the way of life of a particular population. Thus we might refer to the culture of a community, or a nation, or a tribe, or a religious group or even a continent.

This second usage has been very important for some social science disciplines, while it has largely been avoided or ignored by others. It has been developed particularly by anthropology. In sociology and some parts of geography it has been rather less used, and in economics and political science it has, until very recently, rarely made an appearance.

Any discussion of the concept of culture in the social sciences, therefore, has to begin with approaches which have emerged within anthropology and to a lesser extent sociology. Later, though, I will briefly discuss 'cultural studies'. This evolved out of a quite separate tradition, that of literary criticism, and combines both usages of culture – not altogether satisfactorily. In Chapter 3 I turn to the implications of the traditional absence of culture in economics and political science for understanding 'Third World' development.

CULTURE IN ANTHROPOLOGY

For anthropologists in the USA, the concept of culture has always been at the centre of their kind of inquiry. As long ago as 1952, for instance, two leading American anthropologists, A.L. Kroeber and Clyde Kluckhohn, collected no fewer than 164 different definitions of the term in the anthropological literature. They concluded, however, that there was a common element to all of them. Culture consisted of 'patterns of and for behaviour ... acquired and transmitted by symbols ... the essential core [being] ideas and especially their attached values' (Archer 1988: 357).

British scholars have traditionally been less concerned with etymology, perhaps reflecting their notorious cultural trait of disliking philosophy. Even Margaret Archer, for instance, one of the very few sociologists who have taken the concept of culture seriously, gives little attention to defining it, and dismisses anthropological debate on the matter as mere 'definitional wrangling' (Archer 1988: 2). But Kroeber and Kluckhohn and other American anthropologists were not just debating about scholastic or fatuous things – about how many angels could stand on the point of a pin. They were asking what special kind of concepts were needed (if any) in order to think scientifically about society.

The founding father of modern American anthropology, Franz Boas, had been educated in Germany, and had therefore been exposed to an intellectual tradition which stemmed from the idealist philosophy of Immanuel Kant. When he emigrated to the USA, he took these ideas with him. He also influenced Carl Sauer, another immigrant German intellectual who founded the 'Berkeley School' of geography (Jackson 1989).

Boas had not begun his intellectual career as a social scientist at all: he wrote his Ph.D thesis on the colour of sea-water. Because of that training, he accepted that all scientists, whatever they were studying, needed to follow a rigorous scientific method – common procedures for constructing hypotheses, testing them by collecting relevant data, and so on. Yet Kant had also insisted that there was a fundamental difference in kind between those sciences which dealt with natural phenomena – rocks, gases, metals or animals, for instance – and those which dealt with human beings. The former were called *Naturwissenschaften* (the study/science of nature); the latter, the *Geisteswissenschaften* (the study/science of behaviour informed by complex mental processes). It is worth noting that the term *Geist* is often misunderstood, and used as if it applied only to 'high' art, such as Beethoven symphonies. In fact, as Rickman points out, 'a Whitehall farce is [just] as much a product of Geist' (Rickman 1979: 60).

Humans, though, did not fit into this dichotomy very neatly. As animals, they were the subject-matter of biology and of physical anthropology – we could study their brains or their physique. But they differed from other animals because they were members of cultural communities which cumulatively developed and elaborated ideas, and handed them on from generation to generation.

American anthropologists still train their students to think of anthropology as having four 'tracks': physical anthropology, linguistics, archaeology, and cultural anthropology. But since the last three are all concerned with culture, they necessarily entail different methods of investigation from physical anthropology. It was awareness of this that led Boas to formulate his concept of culture. Modern British anthropology, conversely, followed a different philosophical tradition – nineteenth-century 'positivism' – which took the natural sciences as the model for science in general. When Darwinism conquered biology, this seemed to be the final confirmation that using scientific method meant following the methods used in the natural sciences.

Hence in Britain, anthropology was defined by one of its major founders, Radcliffe-Brown, as 'social' anthropology, focused on the study of social structure. American anthropology, on the other hand, has always been cultural anthropology (Radcliffe-Brown 1952). Bronislaw Malinowski, the other major founding father of social anthropology in Britain, was something of an exception. He came closer than other British anthropologists to the approach of Americans in that he objected to Radcliffe-Brown's insistence that anthropology was about 'social structure', and maintained that culture was the crucial general concept. For him, each culture formed a 'coherent whole' (Malinowski 1946). However, in his view the institutions of society derived from underlying bio-social needs – something which few other anthropologists in North America, let alone Britain, were prepared to accept.

In addition to their agreement that culture was their central subject-matter, American anthropologists in the 1930s also maintained that there were cultures, plural. In this respect like Malinowski, they tended to think of each society as having a distinctive culture all of its own – one society, one culture. Each culture, Alfred Kroeber insisted, was a totality, and what gave each culture its distinctive character was that it was built around certain shared ideas and values. This distinctive, superorganic 'ethos' of a given society largely determined the content of its members' lives (Kroeber 1960 [1948]). Ruth Benedict, in a work which influenced an entire generation, described a number of tribal societies as each having its own 'pattern' (though she also thought that all such patterns could be classified as either 'Dionysian' or 'Apollonian') (Benedict 1935). Later, more idealist anthropologists went even further – values were not just 'deep', but 'ineffable' – beyond scrutiny or question; some described them as 'sacred'. They were, therefore, unchallengeable and, unproblematically, all-determining. And because ideas and values were held in common – shared – they provided society with what has been described as its 'overarching' intellectual framework – its way of looking at the world, and provided individuals with a code of moral behaviour, even etiquette.

Values and ideas are, of course, ideal phenomena (in the sense used by Maurice Godelier, the French anthropologist, who uses the term *idéel* instead of *Geist*, but means the same thing, i.e. that they are things of the mind). However, the classic American anthropological conception of culture was also idealist in a second sense. Ideas and values were treated as an impersonal, asocial force, independent of human agency and social interests which determined behaviour. The element of power – how ideas are imposed, internalised or resisted – was left unexamined.

CULTURE IN OTHER SOCIAL SCIENCES

Meanwhile, in the other social sciences, for the most part, mental phenomena were largely ignored. In what has been called 'functionalist Marxism', for instance, society was said to be founded on a 'material base'. So, dominant ideologies were merely devices through which ruling classes justified their exploitation of those they ruled, while the thinking of the lower orders was seen, simply, as conditioned by their exposure to the dominant 'ideology' – which they then internalised. Similarly, orthodox economics and political science, both of which had developed under the influence of nineteenth-century positivism, gave pride of place to the market and the State, but paid little attention to the concept of culture. Yet as Sahlins has reminded economists, and even anthropologists, the drive to boundless acquisition is not some innate human propensity, or, as theologians believe, a moral weakness, a sin. Rather, it is peculiar to certain kinds of market economy and class society, particularly modern capitalism. Western economic institutions and Western economic theory, therefore, are informed by cultural assumptions (Sahlins 1972).

In the enormous literature on modern capitalism, such issues are hardly discussed. Only writers on the law of property point out that capitalism in fact depends on two crucial, ideal elements – in Godelier's sense of *idéel* – two sets of normative propositions. One, the sanctity of the private ownership of wealth (including vast and vastly unequal wealth) and two, the correlative belief that it is justifiable to transmit that wealth to one's kin or to any other legatee whom one may designate. But in many cultures, these ideas would be as incomprehensible and unacceptable as the notion that a person is entitled to own air, or has a right to property in it, or to market it but that there are others who don't own air, and that their sufferings are justifiable according to some cosmic scheme of things.

When they did pay attention to the *idéel*, most theorists laid their main emphasis upon only one aspect of culture, the cognitive, upon thinking, which they therefore labelled 'ideology', not 'culture'. They left out of the picture the two other crucial ways in which behaviour is guided by values and norms – to use the terms introduced by Kant in the eighteenth century, the cognitive and the normative – fortunately translated into everyday language, by Hannah Arendt, as knowing, judging and acting (Worsley 1984).

However, people do not spend their whole time thinking about the world in some abstract way, or just in order to get their minds straight. Rather, they acquire and develop ideas they have come across, or had instilled into them, in order to do things – to cope with the demands of everyday life and to try and get other people to act in the ways they want them to.

In sociology, true, there was a place for culture in what was the dominant theoretical trend from the 1930s through the 1950s – 'functionalism'. For Talcott Parsons, the main theoretician of this school, the social system was only one of three dimensions of what he called, as a whole, 'social action'. The others were the personality system of the individual actors, and the cultural system 'built into their actions' (Parsons 1951: 6). Subsequently, when there was a (justifiable) reaction against functionalism, the central place which Parsons had given to culture was unfortunately thrown out with the bath-water.

Following Max Weber, whose thinking he had encountered when studying in Germany, Parsons broke down the social system into political institutions – through which power is exerted – and economic institutions – through which goods and services are produced and distributed. The production and distribution of culture, however, was confined, in his (as in Weber's) writings almost exclusively to the study of the ideas and values contained in the sacred books of the great world religions – Judaism, Christianity, Buddhism, Hinduism and Islam. More 'popular' kinds of thought were ignored.

Twenty years intervened between Parsons' *The Social System* and the next serious attempt to incorporate the concept of culture into general sociological theory – Zygmunt Bauman's *Culture as Praxis* (Bauman 1973). Bauman distinguishes three major uses of the concept: what he called the universalistic conception of culture ('Culture with a capital C') which marks off the boundaries between human society and the rest; the normative/aesthetic distinction between 'high' and 'low' cultures; and the conception of culture as the source of differences between societies or between communities within the same society.

Another two decades of resounding silence finally ended with the appearance of Margaret Archer's *Culture and Agency* in 1988. In sociology, she noted, culture was still a 'poor relation' of 'structure'. Firstly, the whole of thinking (and behaviour) was not reducible simply to 'ideology', dominant or otherwise. Secondly, those who did use the wider concept of culture used it in 'wildly vacillating' ways. Some treated culture as all-important – a supremely independent 'prime mover'; others as a mere epiphenomenon, in 'supine dependence on other social institutions' (Archer 1988: 1). For the first set of people, social integration comes about as a consequence of the 'downwards' pressure of ideology; for the second set, the solidarities entailed in everyday social life generate consciousness both of common identity and of differences from other groups – 'upwards', as it were.

For Archer, people were not just pulled this way and that by ideas alone, whether they came from on top, or 'upwards' out of their social interdependence. Rather, human agency – the ways in which people act in society – is crucial, for though people certainly behave in one way rather than another because of the ideas they hold, they are exposed to various sets of ideas. They also borrow bits and pieces from several sources, and re-interpret all of these in various ways. Culture, then, is not a rigid framework which imposes itself on people as an irresistible and unambiguous force – a kind of script they are doomed to act in accordance with. People do not just 'swallow' culture like some kind of pre-packaged medicine.

In taking this line of argument, Archer was rejecting a pervasive view among sociologists which, following the approach of the founder of modern sociology, Auguste Comte, totally rejected the idea that either individuals or groups collectively were able to influence history or evolution. In urban-industrial society, popular thinking had degenerated into a confused mess. Human history was in fact determined by forces most people knew nothing about, but which social scientists were now claiming to have worked out, using the methods of 'positive' science. What mattered was not culture, but social structure; morality, too, should be based on science (including social science); and human progress in the future would be determined primarily by advances in technology (Archer 1990: 117). The repercussions of this latter certainty in the context of development are examined by Gordon Wilson in Chapter 5 below.

It is important to emphasise that such views were prevalent not just amongst sociologists, but amongst other political analysts and policy-makers. Moreover, they have retained a currency, notwithstanding the extraordinary upheavals of the 1990s.

In the 1960s, for instance, Clark Kerr and his colleagues argued that as societies became industrialised, all of them, including both the USA and the then USSR, would necessarily 'converge' both in terms of social structure and in terms of their cultural values (Kerr *et al.* 1962). The Soviet Union would become more liberal; the capitalist world readier to accept the need for planning and state welfare. So debate and political controversy about which direction societies should move along, then, was just so much 'ideology' and a waste of time. Thus culture was a mere reflection of structural processes.

Also, along similar lines, Lipset argued that 'the fundamental problems of the industrial revolution had been solved' (Lipset 1969: 406). By the 1970s, such ideas had been become a new 'orthodoxy'. According to Daniel Bell, other kinds of thought were irrelevant and *dépassé*. Now, thankfully, we had come to 'The End of Ideology' (Bell 1962). By the 1980s, technological advances in computers and telecommunications were heralded as the beginning of a new era – a global 'information society' (Toffler 1981) – a modern version of nineteenth-century ideas about technology as the crucial determinant of progress which Archer has sarcastically described as 'positivism with peripherals'. With the disintegration of the major alternative to Western capitalism, the communist world, debate about alternatives seemed meaningless; as Fukuyama put it, history had definitively come to an 'end'. The future could only be one variant or another of the capitalist market economy and its associated social institutions, notably pluralist parliamentary democracy and a high degree of 'open' social mobility. These would henceforth become dominant in all societies (Fukuyama 1992).

SUB-CULTURE AND COUNTER-CULTURE

Meanwhile, the functionalist orthodoxy in anthropology had not gone unchallenged. Even in the 1930s, there had been several differences within American cultural anthropology: differences between those who saw cultures as integral value-systems, and those who emphasised the co-existence of plural sets of values within any culture; between those who emphasised culture as legacy and inheritance and those who argued that the essence of human existence was innovation and creativity; and differences over the weight to be given to social change rather than to continuity. There were also different ways of thinking about time. There were differences between those who taught their students to look at tribal societies as if they were taking 'photographs' at a single point in time (because they assumed cultures to be unchanging and stable) and people like archaeologists, who necessarily had to deal with the evidence of change. Archaeologists identified change in technologically simple societies such as the 'Anasazi' at Mesa Verde, Colorado, who had lived there for centuries and then suddenly disappeared in a couple of generations. They could show that even great, long-lived empires – the Mayan, Mohenjo Daro or Angkor Wat – had eventually passed away.

Time was considered similarly in sociology, where critics of functionalism denounced Parsons because values in his model seemed to be eternal and unchallengeable. If so, how could change come about? Parsons' answer put the emphasis not upon conflict between one set of ideas and another, but on inadequate socialisation. Ideas and values might be weakly held because people (the young especially) had been inadequately exposed to them – badly trained or educated by their elders, for instance; and because cultural elements were never completely integrated with one another and therefore did not form a consistent whole. Even though there might be 'pattern-consistency' within different parts of the social system (say, a single 'official', logical and consistent way of thinking about the world, or a generally used art-style), these might not fit coherently with other elements of culture or society. He thus recognised that populations with different value-systems might nevertheless co-exist, despite some degree of conflict between them, within the same society.

This approach was used in studies of ethnic minorities. Groups with customs very different from those of the majority or of other minorities might live together in 'ghettos' within a city and follow 'ways of life' with their own rules. Conceptualisations of ethnic minority cultures were taken a great deal further in the work of the Chicago School of urban sociology in the 1920s and 1930s, and of a whole succession of schools now generally lumped together as 'deviancy theory'. Social coherence, for these writers, now lay at the level of the sub-culture rather than at the societal level. This school drew upon a philosophical tradition that was different from both German idealism and British empiricist positivism – the American 'pragmatist' philosophy of George Herbert Mead.

The classic functionalist emphasis upon the coherence and persistence of culture is familiar enough today to anyone who visits stately homes and castles in England, Scotland or Wales, which are depicted as evidence of a common 'heritage'. Since this emphasises continuities in a sanitised past which we are all assumed to share in common (whatever the actual positions of our ancestors might have been), it is a basically conservative image of culture, of culture as something shared in common, learned, acquired and transmitted from generation to generation, an image of continuity rather than of change.

The shift away from functionalism occurred as a result of growing criticism within sociology, particularly by the Chicago School. Anthropologists tended be more resistant to abandoning the functionalist approach, and continued to produce ethnographies written in a timeless present about peoples who appeared to live in separated cultural units. However, Leach's study of Political Systems of Highland Burma cruelly exposed the inadequacies of the model (Leach 1954). Secondly, there was a growing influence from a new kind of literary theory which culminated in the emergence of 'cultural studies'.

Under these influences, the initial conception of sub-culture was extended to quite other kinds of sub-cultures than ethnic communities, even to 'deviant', and sometimes criminal populations – jazzmen in South Side ghettos, marihuana-smokers and street gangs each had their own sub-cultures.

To the bearers of WASP culture, of course, these would not have been accepted as being any kind of culture, 'sub-' or otherwise. Theirs had always been an élitist conception of culture. Even when culture was used to refer mainly to the 'fine' arts rather than forms of behaviour ('way of life'), there was no place in this model for 'low'-life – jazz and jitterbugging were simply not culture at all. In England, T.S. Eliot had developed a variant of this kind of model of culture, which, paradoxically (since he was a poet and not a social scientist), was not limited to the arts, but included ways of life and also had a place for social class and even for the culture of the lower orders – albeit an inferior one. Each social group has an élite, he wrote, instancing people 'concerned' with art, science and philosophy, while 'men (sic) of action', too, had their own élite. Together, they constituted the élite-in-general. But he did recognise that both the upper classes and ordinary people had their own characteristic cultural activities and interests – the upper classes had Henley Regatta, Cowes, the twelfth of August, and Wensleydale cheese; the *hoi polloi* more vulgar activities and tastes – the Cup Final, 'the dogs', pin-tables and darts, and boiled cabbage (Eliot 1948).

The new emphasis on sub-cultures offered a challenge to this kind of élitism, insofar as it helped undermine the traditional élitist conception of culture as the 'fine' arts. It was radicalising because it was relativising and pluralistic, discounting the idea that there could only be one kind of culture. Now, all kinds of groups, even so-called deviant ones, had been shown to have their kinds of culture, whether in terms of way of life or, more narrowly, of artistic self-expression. Popular music might still be distinguished from 'classical' music, even still regarded as inferior, but at least jazz and reggae have now come to be accepted as kinds of music (not just 'jungle noises'), alongside Bartok quartets, and are widely enjoyed even among the white upper and middle classes, despite their social distance from black culture. A much more radical critique of cultural élitism came, not surprisingly, from a very different theoretical source – Marxism – in the writings of the Italian Communist, Antonio Gramsci (1891–1937), which became widely available in English in the 1970s.

GRAMSCI, CULTURE, 'CLASS' AND HEGEMONY

In Italy (and elsewhere), the Catholic Church, though primarily a religious institution, played an extremely powerful role in political life and was the prime regulator of the everyday conduct of the faithful. It thus provided a 'hegemonic' ideology which contributed to the stability of the social order. Yet the importance of such cultural institutions and ideas was absent in most social theory, which – Marxism included – was, for the most part, merely political economy.

The Church, the main institution through which intellectuals were recruited and trained to produce and reproduce culture, was a very complex structure. The people who did the training and those whom they trained to carry the doctrines of the Church to the people were intellectuals. But though the Church might claim to be unitary and universal, in reality it had to compete with other religions and was also attached to different secular states, whilst the Papacy was itself also an Italian state as well as a church. The social functions of the clergy, moreover, varied according to their social backgrounds and in accordance with their differing roles in society. They were recruited from different social classes, and went through a very intense training in various kinds of institutions – schools, theological colleges and universities, seminaries and nunneries. They then filled different positions within a Church that was not only very hierarchical but also internally segmented into a plethora of rival Orders, each with its own institutions and distinctive interpretations of Christian ideas. Socially, therefore, the clergy ranged from 'hedge priests', dedicated to a life of poverty and to saving the souls of the poor, to 'proud prelates' and worldly cardinals. Some of these Orders, like the Jesuit or Franciscan Orders, were often effectively independent of Rome, especially in the colonial world.

The clergy did not, therefore, simply express the ideas of the ruling classes directly or in some unambiguous way, as Marx had argued. Rather, class struggles, Gramsci argued, often took place over cultural issues as well as over economic and political ones, and even in industrial disputes, ideas about what constituted, say, 'a fair day's pay for a fair day's work' involved arguments over definitions of 'fairness' which were essentially cultural. There was, therefore, a cultural dimension to economic and political class struggle. Nor did these struggles necessarily take place between the exploited and the bourgeoisie directly. They occurred also between rival intellectual groupings which competed for the hearts, minds and souls of the masses. Though the latter suffered inequality and oppression, how they responded was not determined simply by their material class position. They commonly accepted the prestigious, historically and cosmically dominant ideas of the Church, or, in modern times, voted Conservative. But even the politically orthodox and the faithful in the congregations were exposed at times to alternative, even radical interpretations of Christianity formulated by deviant religious thinkers, which spoke to their discontents. If successful, these counter-ideologies could give rise to organised movements and sometimes themselves become 'hegemonic'.

Intellectuals, then, mediate, in varying and complex ways, between the ruling class and those they rule and exploit. In modern times, it is the 'media', not the Church, which are the main producers and disseminators of ideas. Gramsci distinguished between two kinds of intellectuals. 'Organic' intellectuals were closely and consciously attached to the interests of a class (which might include a revolutionary class). Many of them were often little more than part of the 'cultural apparatus' of the State, such as civil servants who produce and disseminate 'information' and what we now call 'misinformation' – propaganda. But there was another type of intellectual – 'traditional' intellectuals – who had their own institutions, often very historic, notably within the Church and in the universities, and could therefore be intellectually and socially independent of the ruling class. So today, tabloid journalists, from a Gramscian point of view, are mainly mere 'hired pens' peddling conservative ideas; while others – the 'quality' Press – encourage their readers to think critically and take up popular causes. As with the clergy in medieval times, their social functions, then, depend on the connections they have with one class or another.

Being an Italian, Gramsci was also influenced by Machiavelli's distinction between 'force' and 'fraud'. The Prince, Machiavelli had argued, could rule by force alone and use it to crush opposition, even in modern industrial societies. But as Bismarck (never one who hesitated to use 'blood and iron') remarked, 'You can do anything with bayonets except sit on them'. So in order to undermine support for socialism, he introduced a Welfare State in Germany long before the Liberals did so in Britain. Gramsci argued that in order to guarantee the continuing dominance of the propertied and the 'political class' and to ensure continued confidence in the market, social stability and cultural domination – a 'hegemonic' ideology – were necessary. People have to be persuaded to give their 'spontaneous' consent to this ideology – to accept that their rulers have a right to rule, and/or that they can do nothing about changing who rules them. In Gramsci's Marxism, therefore, the internalisation of this kind of 'political culture' was a vital element in political life.

However, the exploited classes, he insists, also develop their own counter-cultural forms of resistance. Though they suffer inequality and oppression, how they respond to those conditions depends on how they interpret and explain them, or have them interpreted and explained to them. Their responses, that is, cannot simply be 'read off' from their position in the class structure.

FROM 'HEGEMONY' TO 'CULTURAL STUDIES'

Since the vast majority of the world's sociologists have always been American, the ideas of the Chicago School, including the concept of 'sub-culture', have had far more influence than Gramsci's theories. The concept of 'counter-culture', for example, has been used by very few writers apart from Theodore Roszak (Roszak 1970), writing at the time of student revolt in the USA, and, more generally, by Wertheim, in the form of what he called 'counterpoint' (Wertheim 1974). The concept of sub-culture, of course, is relevant primarily at a sub-societal level – when looking at the life-style of small communities, for instance. But the concept of counter-culture is far more useful when the focus shifts to relationships on a wider scale – to the relationship of sub-cultures to dominant cultures within the same society – which sub-cultural theory rarely tackles.

Not surprisingly, it was another Marxist, Raymond Williams, who took up some of Gramsci's ideas. He drew upon a similar 'humanist' tradition of Marxism, long neglected during the Stalinist period, and combined it with an older tradition of English 'literary criticism'. In the eighteenth and nineteenth centuries, images of human society had been much more developmental. The Enlightenment, for example, saw the history of humankind as the story of progress and the gradual extension of reason over human affairs; the nineteenth century, in terms of the all-conquering theory of evolution. The emphasis was not upon culture as binding legacy and tradition, but on culture as the history of human innovation, creativity, critical scientific thought and cumulative collective achievement. Williams combined these Marxist and Enlightenment traditions into a theory of what he came to call 'cultural materialism' (Williams 1958, 1961, 1962, 1974, 1981, 1983).

The term 'cultural materialism' had also been used in a quite different way, by the anthropologist, Marvin Harris. He argued that, in the final analysis, material needs determine the ways in which people think and behave – a 'reductionist' argument, since the production of even the most 'primitive' of material objects, such as stone axes, not just such obviously sophisticated instruments as CD-ROMs, involves the exercise of intelligence and skill. We cannot, indeed, understand what an object is, very often, unless we appreciate what it means to those who produce it and what they use it for – in other words, its cultural and social meanings. Furthermore, people devote a great deal of effort to producing immaterial culture, from music to literature. Either way, the *idéel* – ideas and values – is involved. At the end of the day, culture is still ideal, not material. But, Williams reminds us, books and violins are also material objects, and have to be produced. Though he began by studying the relationship between accepted 'great' literature and society, he later shifted to the study of mass culture.

This is an important point. However, Williams' ideas about cultural studies, though often enlightening, are also sometimes confusing. Thus in some of his writings he focuses on how cultural production takes place in modern society, and who controls it. In the modern world, ownership and control of the 'media' take two main forms – private ownership of newspapers and of more modern and even more powerful means of mass communication, notably television, on the one hand, and collective, usually State or Party, ownership and/or control, on the other. Ownership and control, Williams shows, are sometimes used primarily because they constitute what one media mogul called a 'licence to print money'. But they can also be used because they make possible the ideological indoctrination of the masses (or both may be the case) (Williams 1962).

At other times, Williams uses the term 'cultural materialism' to describe the production of culture as distinct from the mere ownership of the means by which it is produced. In particular he was interested in the study of 'mediation': the place of producers of culture (writers, editors, film-makers); their relationship to those who employ them; and the degree of 'relative autonomy' they may possess; the content of what they produce including the ideas they communicate about everything in social life, from sex-roles to dress-styles, not just about specifically political issues.

The interplay between culture and power which stood at the heart of Gramsci's thinking thus became somewhat obscured in Raymond Williams' writings, and more so by subsequent writers who built upon his ideas. Some, like Stuart Hall and the historians at the Centre for Contemporary Cultural Studies at Birmingham in the 1970s, did study social class, the distribution of political power, etc. (Hall and Jefferson 1976; Willis 1977). Others effectively returned to a separation between culture and society. Where Williams had struggled to link culture and power, 'post-modernists' influenced by the work of Lyotard (1986) from the 1980s treated cultures merely as 'discourse', and tended to leave politics, history and economics out of their analysis (see many of the chapters in During 1993). It seems that they seriously limited the capacities of many students of cultural studies to interpret social life.

CONCLUSION

Discourse theory has spread from cultural studies into other disciplines, notably anthropology where it has become, to use Gramsci's term, 'hegemonic'. For those of us who still think that inequality and oppression are important, this has led to a disturbing retreat from reality, and it is fundamentally anti-political. In my own writings (Worsley 1964, 1984, 1997), I have tried to preserve and build upon Gramsci's insights into the relationship between culture and power, and to avoid treating culture as a category 'in itself'.

For me, it is not that some aspects of human existence are more 'cultural' than others, or that some human beings are 'cultural' and others are not (a view which informs media reporting of 'exotic' or 'barbaric' behaviour in the Third World; see Atkinson in Chapter 8 below). All behaviour and relationships are informed by ideas and beliefs, in any society, anywhere, at any time.

In the simple-minded 'segmental' conception of culture, on the other hand, it is assumed that culture is either a discrete sphere or a residual category. In this kind of thinking, behaviour is thought of as governed by economic or political principles, or by cultural 'factors' (for example, people's religious beliefs or their conceptions of their own ethnicity and that of others). This is a trap – we should not fall back on this concept of culture as a separate sphere only to be resorted to when we find that other kinds of explanation, economic, political etc., prove inadequate. Rather, we need to see culture as a dimension of all social action, including economic and political. As I put it some years ago:

> We need to avoid ... [both] the assumption that the 'cultural' is a separate sphere, [and] that it is causally secondary (merely 'superstructural'). It is, in fact, the realm of those crucial institutions in which the ideas we live by are produced and through which they are communicated – and penetrate even the economy.
>
> (Worsley 1984: 60)

2

GLOBALISED CULTURE: THE TRIUMPH OF THE WEST?

John Tomlinson

GLOBAL AND GLOBALISED CULTURE

The idea of a single, unified culture encompassing the whole world has a long and, so far as I know, relatively undocumented history. An inventory of the various historical dreams, visions and speculations about a global culture would, I suppose, have to include at least those of: the imperial projects of the ancient 'world empires' such as China or Rome; the great proselytising 'world religions' and the communities of faith established around them – Christendom, the Ummah Islam etc.; the utopian global visions of early socialists such as Saint-Simon; the various movements dedicated to establishing world peace; the ideas, beginning in the nineteenth century, of enthusiasts for artificial 'international' languages such as Esperanto; and many more.[1] These ideas clearly differ from each other in all sorts of ways. For example, some (probably most) were simply naïve, unproblematised, often dogmatic, projections of a particular cultural outlook onto a 'universal' screen, while others were driven by the desire to reconcile cultural differences and to usher in a new, pacified ideal home for humankind. But two things unite all these visions. First, that they all approached the idea of a single global culture with enthusiasm, and second that none of them came anywhere near to seeing it achieved.

The ideas of a global culture in the air today – in the intellectual and critical discourses of the 1990s – are different. They are not, in the main, visionary or utopian ideas.[2] Rather they are speculations that arise in response to processes that we can actually see occurring around us. These processes, which are generally referred to collectively as 'globalisation', seem, whether we like it or not, to be tying us all – nations, communities, individuals – closer together. It is in the context of globalisation, then, that current discussions of an emergent global culture assume a different significance from earlier speculations. It is not only that the current social, economic and technological context makes a global culture in some senses more plausibly attainable – a concrete possibility rather than a mere dream. It is also that this very sense of imminence brings with it anxieties, uncertainties and suspicions.

Talk of a global culture today is just as likely, probably more likely, to focus on its dystopian aspect, to construct it as a threat rather than a promise. This, at any rate, is the sort of talk I want to consider here. To grasp its close relation to the processes of globalisation and to distinguish it from earlier traditions of thought, I shall refer in what follows to the idea of a *globalised* rather than global culture. A globalised culture refers here specifically to the way in which people, integrating the general signs of an increasing interdependence that characterises the globalisation process with other critical positions and assumptions, have constructed a pessimistic 'master scenario' (Hannerz 1991) of cultural domination. This is the speculative discourse that I want to criticise.

In order to develop this discussion in a relatively short piece, I shall have to leave on one side some pretty big and thorny related and contextual issues. Most of these relate to the way in which the notion of globalisation has been theorised. Though I shall offer later a brief description of what globalisation broadly means, it must be recognised that there are all sorts of unresolved controversies in globalisation theory which this discussion will necessarily rub up against from time to time without explicitly recognising. If readers recognise these, and develop the argument themselves, so much the better!

GLOBALISED CULTURE AS WESTERNISED CULTURE

The argument I want to focus on can be set out quite briefly in outline, though we shall see that it contains some crucial assumptions that will require unpacking presently. It goes like this. The globalised culture that is currently emerging is not a global culture in any utopian sense. It is not a culture that has arisen out of the mutual experiences and needs of all of humanity. It does not draw equally on the world's diverse cultural traditions. It is neither inclusive, balanced, nor, in the best sense, synthesising. Rather, globalised culture is the installation, world-wide, of one particular culture born out of one particular, privileged historical experience. It is, in short, simply the global extension of *Western* culture. The broad implications – and the causes of critical concern – are that: (a) this process is homogenising, that it threatens the obliteration of the world's rich cultural diversity; (b) that it visits the various cultural ills of the West on other cultures; (c) that this is a particular threat to the fragile and vulnerable cultures of peripheral, 'Third World' nations; and (d) that it is part and parcel of wider forms of domination – those involved in the ever-widening grip of transnational capitalism and those involved in the maintenance of post-colonial relations of (economic and cultural) dependency.

This is, of course, to view the globalisation process through a now familiar critical prism – that of the critique of Western 'cultural imperialism' (Friedman 1994; Hannerz 1991; and McQuail 1994). As I and others have argued elsewhere (Boyd-Barrett 1982; Schlesinger 1991; Sinclair 1992; Tomlinson 1991, 1995, 1997), this is a peculiarly vexed and often confused critical discourse which rolls a number of complex questions up together. In the space available here I shall have to take for granted most of this criticism. But before I come to my central argument it will be useful just to mention a couple of the most common objections to the general idea of Westernisation, so as to distinguish them from the specific, rather different line of criticism I want to follow later.

What do people mean when they talk about 'Westernisation'? A whole range of things: the consumer culture of Western capitalism with its now all-too-familiar icons (McDonald's, Coca-Cola, Levi Jeans), the spread of European languages (particularly English), styles of dress, eating habits, architecture and music, the adoption of an urban lifestyle based around industrial production, a pattern of cultural experience dominated by the mass media, a range of cultural values and attitudes – about personal liberty, gender and sexuality, human rights, the political process, religion, scientific and technological rationality and so on. Now, although all of these aspects of 'the West' can be found throughout the world today, they clearly do not exist as an indivisible package. To take but one example, an acceptance of the technological culture of the West and of aspects of its consumerism may well co-exist with a vigorous rejection of its sexual permissiveness and its generally secular outlook – as is common in many Islamic societies. A prime instance of this contradiction is the current attempt to regulate or even ban the use of satellite television receivers in countries like Iran, since they are seen by the authorities to be the source of various images of Western decadence. This sort of cultural-protectionist legislation is almost impossible to implement, partly because of the huge numbers of dishes involved (estimated at more than 500,000 in Tehran alone) but also because use of this technology is vital for education and scientific research. Restriction of access is thus resisted by these constituencies within the intelligentsia who might otherwise hold quite 'conventional' Muslim attitudes towards, for example, images of sexuality or nudity in Western television programmes (Haeri 1994).

So there is obviously a need to *discriminate* between various aspects of what is totalised as 'Westernisation', and such discrimination will reveal a much more complex picture: some cultural goods will be broader in their appeal than others, some values and attitudes easily adopted while others are actively resisted or found simply odd or irrelevant. All this will vary from society to society and between different groupings and divisions – class, age, gender, urban/rural, etc. – within societies. The first objection to the idea of Westernisation, then, is that it is too broad a generalisation. Its rhetorical force is bought at the price of glossing over a multitude of complexities, exceptional cases and contradictions. This criticism also connects with another one: that the Westernisation/homogenisation/cultural imperialism thesis itself, ironically, displays a sort of Western ethnocentrism (Hannerz 1991; Tomlinson 1991, 1995). Ulf Hannerz puts this point nicely:

> The global homogenisation scenario focuses on things that we, as observers and commentators from the centre, are very familiar with: our fast foods, our soft drinks, our sitcoms. The idea that they are or will be everywhere, and enduringly powerful everywhere, makes our culture even more important and worth arguing about, and relieves us of the real strains of having to engage with other living, complicated, puzzling cultures.
>
> (Hannerz 1991: 109)

A second set of objections concerns the way in which Westernisation suggests a rather crude model of the one-way flow of cultural influence. This criticism has – rightly – been the one most consistently applied to the whole cultural imperialism idea. Culture, it is argued, simply does not transfer in this way. Movement between cultural/geographical areas always involves translation, mutation and adaptation as the 'receiving culture' brings its own cultural resources to bear, in dialectical fashion, upon 'cultural imports' (Appadurai 1990; Garcia Canclini 1995; Lull 1995; Robins 1991; Tomlinson 1991). So, as Jesus Martin-Barbero describes the process of cultural influence in Latin America: 'The steady, predictable tempo of homogenising development [is] upset by the counter-tempo of profound differences and cultural discontinuities' (1993: 149). What follows from this argument is not simply the point that the Westernisation thesis underestimates the cultural resilience and dynamism of non-Western cultures, their capacity to 'indigenise' Western imports. It is also draws attention to the *counter-flow* of cultural influence – for instance in 'world music' (Abu-Lughod 1991) – from the periphery to the centre. Indeed this dialectical conception of culture can be further developed so as to undermine the sense of the West as a stable homogeneous cultural entity. As Pieterse puts it: 'It ... implies an argument with Westernisation: the West itself may be viewed as a mixture and Western culture as Creole culture' (1995: 54). Of course, the ultimate implication is that whatever globalised culture is emerging, it will not bear the stamp of any particular cultural-geographical or national identity, but will be *essentially* a hybrid, *mestizaje*, 'cut-and-mix' culture(Pieterse 1995; Garcia Canclini 1995).

These sorts of criticism quickly take some of the wind out of the sails of the Westernisation argument, at least in its most dramatic, polemical formulations. However, they do not entirely resolve the issue of the contemporary cultural power of the West. For it could very reasonably be argued that, when all is said and done and all these criticisms met, Western cultural practices and institutions still remain firmly in the driving seat of global cultural development. No amount of attention to the processes of cultural reception and translation, no anthropological scruples about the complexities of particular local contexts, no dialectical theorising, can argue away the massive and everywhere manifest power of Western capitalism, both as a general cultural configuration (the commodification of everyday experience, consumer culture) and as a specific set of global cultural industries (CNN, Times-Warner, News International). What, it might be asked, is this, if not evidence of some sort of Western cultural hegemony? What ensues from this is a sort of critical stand-off. Both positions are convincing within their own terms, but seem somehow not precisely to engage.

To try to take the argument a little further I want, in what follows, to focus on one particular, largely implicit assumption that seems to be embedded in the idea of globalised culture as Westernised culture. This is the assumption that the process of globalisation is *continuous* with the long, steady, historical rise of Western cultural dominance. By this I mean that the sort of cultural power generally attributed to the West today is seen as of the same *order* of power that was manifest in, say, the great imperial expansion of European powers from the seventeenth century onwards. So, for example, this implicit understanding of globalised culture would see the massive and undeniable spread of Western cultural goods – 'Coca-colonisation' – as, at least broadly, part of the same process of domination as that which characterised the *actual* colonisation of much of the rest of the world by the West. I do not mean that no distinction is made between the obviously coercive and often bloody history of Western colonial expansion and the 'soft' cultural imperialism of McDonald's hamburgers, Michael Jackson videos or Hollywood movies. But I think it is often the case that these and many other instances of Western cultural power get lumped together – 'totalised' – by the term 'Westernisation' and that the result is an impression of the inexorable advance – even the 'triumph' of the West.

It is this particular totalising assumption that I think could benefit from being unpacked and critically examined. This is for two reasons: first because it mistakes the nature of the globalisation process and secondly because by conflating a number of different issues it overstates the general cultural power of the West. I do not want to deny that the West is in a certain sense 'culturally powerful', but I do want to suggest that this power, which is closely aligned with its technological, industrial and economic power, is not the whole story. It does not amount to the implicit claim that 'the way of life' of the West is now becoming installed, via globalisation, as the unchallengeable cultural model for all of humanity.

Indeed, as I shall now go on to argue, there are ways in which the globalisation process, properly conceived, can be shown to be actively problematic for the continuation of Western cultural dominance: to signal not the triumph of the West, but its imminent decline.

GLOBALISATION AS THE 'DECLINE OF THE WEST'

In this section I want to examine some observations by two important and influential British social theorists, Anthony Giddens and Zygmunt Bauman, which connect globalisation with the decline, rather than the triumph of the West. Giddens in particular has been highly influential in theorising the globalisation process and in relating this to the wider debate about the nature of social modernity. Neither of these particular arguments, however, is developed at any great length and there will not be space here to develop them much further. I present them simply as suggestive ways of thinking against the grain of the arguments we have so far reviewed. First, however, it will be useful to say a little more about the nature of the globalisation process itself.

Probably the most important thing to be clear about is that globalisation is not *itself* the emergence of a globalised culture. Rather, it refers to the complex pattern of interconnections and interdependencies that have arisen in the late-modern world. Globalisation is heavy with implications for all spheres of social existence – the economic, the political, the environmental, as well as the cultural. In all these spheres it has the effect of tying 'local' life to 'global' structures, processes and events. So, for example, the economic fate of local communities – levels of economic activity, employment prospects, standard of living – is increasingly tied to a capitalist production system and market that is global in scope and operation – to global 'market forces'. Similarly, our local environment (and consequently our health and physical quality of life) is subject to risks arising at a global level – global warming, ozone depletion, eco-disasters with global 'fall-out' such as Chernobyl.

What these aspects of globalisation represent, then, is a rapidly growing context of global interdependence which already 'unites' us all, if only in the sense of making us all subject to certain common global influences, processes, opportunities and risks. But clearly this sort of 'structural unity' does not of itself imply the emergence either of a common 'global culture' (in the utopian sense) or of the globalised (Westernised) culture we have been discussing. Neither, it should be added, does this interdependence imply a levelling out of advantages and disadvantages globally. Globalisation is generally agreed to be an *uneven* process in which neither the opportunities nor the risks are evenly distributed (McGrew 1992; Massey 1994). But, again, this does not mean that it necessarily reproduces – or will reproduce – the precise historical patterns of inequality supposed in the dualism of the 'West versus the Rest'. More complex, contradictory patterns of winners and losers in the globalisation process may be emerging.

Another important aspect of globalisation is, of course, the increasing level of social-cultural *awareness* of global interdependence. As Robertson puts it, globalisation 'refers both to the compression of the world and the intensification of consciousness of the world as a whole' (1992: 8). Our sense of the rest of the world and of our connections with it are 'brought home to us' routinely via globalising media and communications technology. Now, of course, it can be argued that the *contents* of these images of a wider world are often highly selective and restricted ones. For instance, it has long been observed that the picture of developing countries portrayed on Western televisions tends to be restricted to 'the narrow agenda of conflicts and catastrophes' (Cleasby 1995: iii). Thus as Peter Adamson of UNICEF has argued, with 'no equivalent sense of the norms in poor countries to set against this constant reporting of the exceptional ... the cumulative effect of the way in which the developing world is portrayed by the media is grossly misleading' (quoted, Cleasby 1995: iii).

However, the point I want to stress is the routine *availability* of distanciated imagery (however accurate) that globalising media technology provides. In the affluent Western world we take the experience provided by such technology pretty much for granted. We *expect* to have instant images of events happening in every remote corner of the world on our television screens. It is with no sense of wonder that we pick up the phone and speak to people on other continents. We just as quickly – some of us – become used to communicating globally on the Internet and the 'World Wide Web'. So globalisation seems also to involve, as Giddens puts it (1990: 187), the extension of our 'phenomenal worlds' from the local to the global. Of course, access to this technology is obviously not evenly distributed and so we must avoid the mistake of extrapolating from the Western experience. But, on the other hand, it would be equally misleading to treat such communications technology, and the experience it affords as, somehow, the exclusive property of the West. Again, we have to recognise the possibility that globalisation may be producing shifting global and local patterns in what has been called the 'information rich and the information poor'.

Perhaps the most widely recognised property of globalisation amongst those who have theorised it is its *ambiguous* nature – its mixture of risk and opportunity, its 'dialectical' counterposing of generalising and particularising tendencies, its confusing capacity both to enable and to disempower.

It is within this broad conceptualisation of globalisation that Anthony Giddens writes of '[t]he gradual decline in European or Western global hegemony, the other side of which is the increasing expansion of modern influences world-wide', of 'the declining grip of the West over the rest of the world' or of 'the evaporating of the privileged position of the West' (1990: 51–3). What can he mean by this?

Well, Giddens has written a good deal about the globalisation process and at a fairly high level of abstraction and really these comments need to be read in the context of his overall theorisation of the globalising nature of modernity (Giddens 1990, 1991, 1994a, 1994b). But to put it at its simplest, his argument is that, though the process of 'globalising modernity' may have *begun* in the extension of Western institutions, the very fact of the current global ubiquity of these institutions (capitalism, industrialism, the nation-state system and so on) – in a sense the West's 'success' in disseminating its institutional forms – also represents a decline in the differentials between it and the rest of the world, thus a loss of the West's (once unique) social/cultural 'edge'. As he puts the point in a more recent discussion:

> The first phase of globalisation was plainly governed, primarily, by the expansion of the West, and institutions which originated in the West. No other civilisation made anything like as pervasive an impact on the world, or shaped it so much in its own image ... Although still dominated by Western power, globalisation today can no longer be spoken of only as a matter of one-way imperialism ... increasingly there is no obvious 'direction' to globalisation at all and its ramifications are more or less ever present. The current phase of globalisation, then, should not be confused with the preceding one, whose structures it acts increasingly to subvert.
>
> (Giddens 1994b: 96)

There are various ways in which this 'loss of privilege' and even the 'subversion' of Western power may be understood. For example, it might be pointed out that certain parts of what we were used to calling the 'Third World' are now actually more advanced – technologically, industrially, economically – than some parts of the West. The comparison here might be, for example, between the so-called 'Asian Tiger' economies and some of the economically depressed heavy-industrial regions of Europe or the US. And there might be a complex causal relationship between the rise and decline of such regions connected by a globalised capitalist market (Giddens 1990: 65, 1994a: 65). Or, to put this slightly differently, it might be argued that capitalism has no 'loyalty' to its birthplace, and so provides no guarantees that the geographical patterns of dominance established in early modernity – the elective affinity between the interests of capitalism and of the West – will continue (Tomlinson 1997). There are signs of this, for example, in the increasingly uneasy relation between the capitalist money markets and the governments of Western nation-states – the periodic currency crises besetting the Western industrial nations. A rather spectacular instance of the capitalist system deserting the West could be seen in the débâcle of Britain's oldest merchant bank, Baring Brothers, in February 1995 as a result of its high risk globalising speculations carried out, appropriately enough, on the Singapore market.

On a more directly cultural level, the loss of privilege of the West can be seen in the shifting orientation and self-understanding of the discipline of anthropology, the academic discipline which, perhaps more than any other, displays the cultural assumptions on which the West has presumed to organise a discourse about other cultures. As Giddens points out (1994b: 97), anthropology in its formative stage was a prime example of the West's self-assured assumption of cultural superiority. Because of its 'evolutionary' assumptions, early taxonomising anthropology established itself as a practice to which the West had exclusive rights – the 'interrogation' of all other cultures. Other cultures were there, like the flora and fauna of the natural world, to be catalogued and observed, but there was no thought that they could ever *themselves* engage in the practice – they were categorised as 'if not inert, no more than a "subject" of enquiry' (1994b: 97).

Early anthropology was part of the cultural armoury of an imperialist West during 'early globalisation' precisely because it had not developed its inner logic. As this emerges, with the recognition of the integrity of traditions, the knowledgeability of all cultural agents and the growing sense of 'cultural relativism', so anthropology becomes simultaneously both a more modest and humble undertaking and, significantly, a globalised practice. Present-day anthropologists have to approach their study in the role of the *ingenu* – the innocent abroad – rather than as the confident explorer and taxonomist. Without the assurance of a taken-for-granted superior cultural 'home-base', anthropological study becomes a matter of 'learning how to go on' rather than of detached, *de haut en bas* interrogation. Not only this, it becomes clear that in this later phase of globalisation, *all* cultures have a thoroughly reflexive anthropological sensibility: 'In British Columbia the present day Kwakiutl are busy reconstructing their traditional culture using [Franz] Boas' monographs as their guide, while Australian Aboriginals and other groups across the world are contesting land-rights on the basis of parallel anthropological studies' (Giddens 1994b: 100).

So, the trajectory of the development of anthropology, as Giddens puts it, 'leads to its effective dissolution today' (1994b: 97). This could also stand, more broadly, for the way in which current globalisation subverts and undermines the cultural power of the West from which it first emerged.

To conclude this section I want to comment briefly on an interesting distinction that Zygmunt Bauman makes between the 'global' and the 'universal':

> Modernity once deemed itself *universal*. Now it thinks of itself instead as *global*. Behind the change of term hides a watershed in the history of modern self-awareness and self-confidence. Universal was to be the rule of reason – the order of things that would replace the slavery to passions with the autonomy of rational beings, superstition and ignorance with truth, tribulations of the drifting plankton with self-made and thoroughly monitored history-by-design. 'Globality', in contrast, means merely that everyone everywhere may feed on McDonald's hamburgers and watch the latest made-for-TV docudrama. Universality was a proud project, a Herculean mission to perform. Globality in contrast, is a meek acquiescence to what is happening 'out there' ...
>
> (Bauman 1995: 24)

Mapping this onto Giddens' distinction between early and late globalisation, the key difference becomes that between a Western culture with pretensions to universalism and one without.[3] The globalisation of the West's cultural practices is simply occurring without any real sense either that this is part of its collective project or 'mission', or that these practices are, indeed, the tokens of an ideal human civilisation. Early globalisation involves the self-conscious cultural project of universality, whilst late globalisation – globality – is mere ubiquity.

Now, whilst it may be argued that Bauman erects a rather contrived dualism here between the 'high cultural' project of enlightenment rationalism and some rather specific 'popular cultural' practices, I do think his stress on the cultural self-image and self-confidence of the West is an important one. The specific doubts he detects that now 'sap the ethical confidence and self-righteousness of the West' tend to be those of the intellectual. These are doubts about the capacity of the Enlightenment project ever to deliver full emancipation for all human beings, about 'whether the wedlock between the growth of rational control and the growth of social and personal autonomy, that crux of modern strategy, was not ill-conceived from the start ... ' (1995: 29)

However we can read the idea of the loss of Western self-confidence in more mundane ways. Bauman's description of globality as a meek acquiescence to what is happening 'out there' may be a little overstated, but it does grasp something of the spirit in which ordinary people in the West probably experience the global spread of their 'own' cultural practices. Indeed a lot probably hangs on the extent to which Westerners actually feel 'ownership' of the sorts of cultural practices that, typically, get globalised. Although this is an immensely complicated issue, my guess is that there is only a very low level of correspondence between people's routine interaction with the contemporary 'culture industry' and their sense of having a distinctive *Western* cultural identity, let alone feeling proud of it. It seems more likely to me that things like McDonald's restaurants are experienced as simply 'there' in our cultural environments: things we use and have become familiar and perhaps comfortable with, but which we do not – either literally or culturally – 'own'. In this sense the decline of Western cultural self-confidence may align with the structural properties of globalising modernity – the 'disembedding' (to use Giddens' term) of institutions from contexts of local to global control. In a world in which increasingly our mundane 'local' experience is governed by events and processes at a distance, it may become difficult to maintain a sharp sense of (at least 'mass') culture as distinctively 'the way we do things' in the West – to understand these practices as having any particular connection with our specific histories and traditions. Thus, far from grasping globalised culture in the complacent, proprietorial way that may have been associated with, say, the *Pax Britannica*, late-modern Westerners may experience it as a largely undifferentiated, 'placeless' modernity to which they relate effortlessly, but without much sense of either personal involvement or of control.

CONCLUSION

The arguments reviewed above are obviously not conclusive and leave many issues unresolved. In particular the complex phenomenology of cultural identity in a globalised world requires far more extensive and nuanced treatment than has been possible here. What I have tried to offer is simply a glimpse of alternative ways of thinking about the complex cultural issues forced upon us by the globalisation process. Nothing in this is meant to deny the continuing *economic* power of the West, nor even that particular, limited, sense of 'cultural' power that proceeds from this – the power of Western transnational capitalism to distribute its goods around the world. Nor, to be clear, do these arguments entail the idea of a simple 'turning of the tables'; the 'decline of the West' does not mean the inevitable 'rise' of any other particular hegemonic power (no matter how tempting it is to speculate about the 'Asian Tigers' and so forth). In the short term at least, much of the 'Third World' will probably continue to be marginalised by globalising technologies (Massey 1994). But, to look beyond this, these reflections do suggest that what is happening in globalisation is not a process firmly in the cultural grip of the West and that, therefore, the global future is much more radically open than the discourses of homogenisation and cultural imperialism suggest. We surely need to find new critical models to engage with the emerging 'power geometry' of globalisation, but we will not find these by rummaging through the theoretical box of tricks labelled 'Westernisation'.

NOTES

1. On some of the historical – particularly nineteenth-century – utopian visions of global unity see Armytage (1968). More generally, see Robertson (1992) on the broad history of ideas about globalisation.
2. Although some are. The stance of some of the writing on 'cyberspace' – for example *Wired* magazine – might be described as awe-struck techno-utopianist, making extravagant claims for the power of globalising communications technology. This applies even more so to some of the material actually transmitted on the Internet and the World Wide Web. I shall not deal with this very specific discourse here. For a critical discussion, see Stallabrass (1995).
3. In the interests of brevity I ignore here the ideological issues surrounding the discourse of universalism. Bauman is, of course, aware of the discursive position of power – including the power of patriarchy – from which the Enlightenment notion of universalism arose and his comparison is not meant to be celebratory of this earlier cultural self-confidence. But this is not to deny, of course, that globalisation might *eventually* issue in some 'universally' shared cosmopolitan human values recognising commonality of interests and a respect for cultural difference (see Giddens 1994a: 253). These would, of course, have to draw on non-Western as well as Western traditions and perspectives (Therborn 1995: 137).

3

CULTURE AND DEVELOPMENT THEORY

Peter Worsley

INTRODUCTION

Development theory emerged as a separate body of ideas following the Second World War. From its inception, it necessarily had to deal with peoples having a wide variety of ways of life and outlooks. However, few writers put culture at the heart of their analyses, and even anthropologists tended to see their subject-matter as something that was disappearing before their own eyes. It was assumed that, with the end of colonialism and the adoption of the correct policies, 'traditional' cultures would disappear and the world would become rapidly 'modernised', a view reflected in the title of Lerner's book of 1958, *The Passing of Traditional Society*.

As a consequence, development theories tended to emphasise the state, planning, the market, labour-flows, money-supply or commoditisation, etc. as if these things were not themselves the cultural constructs of a particular kind of civilisation rather than concepts and institutions of universal validity. So development analysts and practitioners commonly ignored religion, kinship, ethnicity or the arts, and thought of their economic and political models as acultural. To some extent, this was a healthy reaction to racist ideas prevalent in colonial times, but it meant that the complexity and diversity of human social life was lost. As we shall see, this has continued to be a weakness of what has come to be called 'development studies'.

Scientists and technical specialists are particularly inclined to the view that economic development has little to do with social systems or cultural values, but depends predominantly on introducing the right technical inputs (machinery, seed-strains, fertilisers, tractors, pesticides), since these are seen as having an intrinsic pure logic of their own. In the design of their projects, therefore, 'questions of cultural incompatibility, contradiction or resistance', Archer remarks, 'do not arise' (Archer 1990: 99), and motivations other than economic ones, such as pricing policy or wages and incomes policy, are rarely discussed. Reform of systems of land-ownership, for instance, or of the ways in which capital is channelled, from money-lenders to banks, which only lend to those with collateral, are treated as 'non-economic' and irrelevant, while attempts of the poor to organise themselves are not supported, even when, as in the *sertão* of Brazil, armed repression is used to keep the peasants and landless labourers under control.

MODERNISATION THEORY AND THE PREDOMINANCE OF ECONOMICS

For a decade or two after the Second World War a single theory was dominant – what was later called 'modernisation theory'. Its best-known version was W.W. Rostow's 'Non-Communist Manifesto', an evolutionist model which postulated that any society aiming at 'development' would have to go through several 'stages of growth'. Most commentators treat him, wrongly, as simply an economistic/technological determinist, though in fact he went out of his way to insist that economic forces and motives are neither unique nor necessarily the most important determinants of the course of history. He also argued that:

> economic change has political and social consequences, economic change in itself ... [can result from] political and social as well as economic forces. And in terms of human motivation, many of the most profound economic changes [are often] consequences of non-economic human motives.
>
> (Rostow 1960: 2)

For Rostow, therefore, economic development required not only appropriate economic, technological and demographic conditions, but also appropriate social institutions and value-systems. Only when all of these were present would development be possible. Similar ideas were to be found in the writings of the West Indian and Nobel Prize-winning economist, Arthur Lewis (Lewis 1954).

Though Rostow and Lewis recognised the importance of ideas and values, most development practitioners and academics who read their work did not pay much attention to this aspect of it. Rather, they were impressed with what they saw as their hard-nosed economic approach. The subtleties of their arguments were ignored; progress meant economic growth, and went through stages. In crude forms, and most of them were crude, this simple formula was a modern version of nineteenth-century positivism, imbued with the same kind of facile optimism. The path pioneered by Western Europe and the United States was the path to follow; if not, things would go wrong. The worst scenario would be to follow the Soviet model. For those who stuck to the correct path, however, underdevelopment would be eradicated and be replaced by affluence.

Modernisation theory dominated thinking about development not only in academia but became the ruling orthodoxy in the major think-tanks, international aid agencies, and transnational institutions in Washington and the UN where world development policy was shaped. To the extent that attention was paid to culture, the basic assumption was that what was needed was some equivalent (not necessarily Christian) of the Protestant ethic, which had provided ideas and values crucial to modernisation in the West.

The obverse was the assumption that failure to develop could result not just from going down the communist path, but also because of the influence of negative cultural factors, i.e. not having a Protestant ethic. For Daniel Lerner, 'traditional' cultures were barriers to development – the entrepreneurial spirit and social mobility could not develop, for instance, in societies based upon what Maquet called 'the premise of inequality' (Maquet 1961), or where irrational religious thinking is dominant. Development, others argued, is frustrated because 'traditional' political authorities are corrupt – they favour their kin and thereby inhibit the development both of the free market and of political democracy. Political parties are simply 'clientelistic' networks or tribes writ large. Foreign aid will therefore simply be funnelled to relatives of the members of the ruling élites, or to their own ethnic group only.

Those development specialists who did think about cultural factors assumed that what was needed was the whole package of modern Western values and social institutions. Adoption of the 'Western way of life' was assumed to be the way forward, though they were usually careful not to say so too publicly and presented their strategies in 'neutral', language. There was though resistance to such 'encouragements'.

With the emergence of new states in Africa and Asia in the 1950s, new political élites came to power, most of their leaders trained in the West. Their experience of, and struggle against, colonialism, however, caused them to distance themselves from the former colonial powers. India, for example, drew inspiration from Soviet-style state planning. In Africa, the new élites built all-encompassing one-party states modelled on the Soviet Communist Party in order to mobilise and control the population.

Development policy, now, could also be

managed by nationals who controlled the state apparatus, not by foreign development experts. These indigenous planners, however, did not take culture any more seriously. Indeed, they were ferocious but contradictory 'modernisers'. In their private lives (e.g. within the extended family), they were often as traditionalistic as anyone, and in public, too, traditional values were often celebrated. Ministers wore traditional clothes and were greeted by ululating troops of women and schoolchildren. Nevertheless, when it came to planning development, traditional ideas and values were largely ignored, even where the secular and socialist ideas which officially shaped policy were manifestly superficial.

CRACKS IN THE DEVELOPMENT IDEALS

But by the late 1960s, the contradictions and limitations of development planning, whether under the aegis of international organisation or by national governments and planning institutions, were becoming difficult to ignore. Although there was some significant improvement in social welfare provisioning during the boom years after the Second World War, it was evident that global poverty was not going to disappear; social stratification seemed to be worsening, and state planning often went hand-in-hand with political oppression. These problems became particularly acute in Latin America where most countries had been independent for over a century.

Latin Americans had long experience of brutal dictatorships which had used their power to impose 'order'. Altruistic civilian planners had been discovering that modernisation policies had had consequences they never intended and which the theory could not explain. The goods that modernisation theory had promised, even when produced, had failed to reach those most in need. Planners, though normally nationalists in politics, and anti-communist, and who followed normally Western Fordist and Keynesian models, had nevertheless been sufficiently impressed by the Soviet planning model, which had made the USSR a superpower in only two generations. Their readiness to use the State as the crucial instrument for bringing about economic growth was shared by those who became their masters, the new post-colonial élites. They included, for example, Latin American military dictators who, though usually simply seen as brutally repressive (which they were), had also been trained in the USA, not just as soldiers, but as managers of the large-scale, high-tech organisations which modern armed forces are. Once 'internal war' had eliminated political opposition – the major obstacles to forced-march economic growth – what has been called 'violent modernisation' could be unleashed. Military juntas created a new nightmare state in which political repression was combined with modern state planning.

Revolutionaries like Che Guevara had long argued that underdevelopment was not a natural condition – due, say, to lack of natural resources. Rather, poor countries had been 'underdeveloped' during hundreds of years of imperialism. By the 1980s, even those who had been major advocates of Western-style modernisation, like Daniel Bell, now began lamenting that one hundred years of industrialism had spawned 'an industrial monstrosity ... Junk culture [was] the spoilt brat of affluence'. 'Information society', equally, seemed to promise only a social order that nobody wanted. Their solutions were not very impressive. Bell, for instance, suggested the idea that religion might provide a way out (Archer 1988: 102).

Those who had been suspicious of planners and of modernisation theory from the beginning were not at all surprised at the effects of this kind of modernisation on the lives of ordinary people and their consequences for the stability of society. A new generation of Green critics pointed to the rape of natural resources; to appalling levels of pollution in cities like Mexico City and Rio de Janeiro; and not just to the persisting poverty of the millions in the rural backlands of Brazil, for example, but also to the increasing polarisation in the distribution of wealth in the cities and the concentration of political power in the hands of a few.

ALTERNATIVES: WORLD-SYSTEM AND DEPENDENCY THEORIES

The time was ripe, then, for a successor to modernisation theory, which finally emerged in the shape of world-system theory and 'dependency' theory (Frank 1969; Wallerstein 1979). Through these theories the world was no longer seen, as in classical economics, as a collection of national economies, or, politically, as a set of autonomous states, but as a single global market dominated by a limited number of giant transnational corporations, far more powerful than most formally independent states. Only a few, in fact, had sufficient power to operate effectively on the new global market, at the 'centre' of the new world order. The rest were only of 'peripheral' importance. Many, myself included (Worsley 1980), have discussed the writings of Frank and Wallerstein as if they had been produced by a single author (whom we may call 'Frankenstein'). But Wallerstein had always insisted that the dichotomy of 'centre' and 'periphery' was inadequate to capture the complexities of development. A further category, that of 'semi-periphery', was needed. Nevertheless world-systems/dependency theory paid no more attention to culture than the modernisation models it was critiquing.

The world-system model, especially in Frank's version, discounted programmes of short-term or limited reform. Total transformation of the entire world-order was what was needed, though how this might be achieved was not spelled out. The communist states, the obvious challengers to the West's dominance, were equally dismissed. These weaknesses, combined with the lack of attention to culture, meant that a new approach to development, less structural and radical, began to emerge in the late 1970s, which did have a place for reforms based on meeting 'basic needs'. The emphasis was on the improvement of educational provision, housing and health facilities at grassroots level, rather than political and economic total transformation. Such an approach even began to influence policy in world development organisations such as the OECD and the ILO. Others took up issues like genocide, 'Green' ecological issues and human rights, which went beyond political economy.

However, even on the world-system theorists' own chosen battlefield, political economy, critics attacked dependency theory for failing to recognise that 'non-dependent' economic growth was taking place in some formerly 'Third World' countries. There was little place in Frank's Manichaean model of 'underdevelopment or revolution' for the increasing number of formerly 'underdeveloped' countries that were either 'newly-industrialising', or had already become industrialised, nor for the process by which such change had come about. The 'Four Little Tigers' in East Asia, plus Brazil and Mexico, some thought, were the beginning of an evolutionary process which, eventually, would result in the 'End of the Third World', to quote the title of one influential book (Harris 1986).

END OF THE THIRD WORLD?

What Harris was primarily attacking was the 'ideology' of 'Third Worldism' – the illusion that small countries could develop a 'third way', independently of either communism or capitalism. The reality, he argued, following earlier world-system theory, was that they were being increasingly sucked into the new global economy as mere components in a world system of production. What kind of 'alternative' economic development could be developed after a successful revolution, in poor countries, locked, as he himself constantly insisted, into a world-system in which they would still be weak primary producers (and political lepers)? Or what kinds of relationships, political or economic, with other revolutionary or radical states, might new, post-revolutionary regimes develop in a world under the domination of the major industrial powers? These questions he argued were never answered by 'Third Worldism'.

In the post-World War II epoch, the emergence of large numbers of newly independent states which had formerly been colonial possessions of the Western powers culminated in the formation of the group of Non-Aligned Powers. Several successful armed revolutions in the Third World – notably in the most populated country in the world, China, and in

Vietnam and Cuba – seemed to indicate that a new era was beginning in which the political domination of the West over the rest of the world was over. The power of the new OPEC oil-producing bloc, too, suggested that it was possible for countries of the former Third World to challenge the West's economic domination. But after the Cuban revolution, the capitalist powers in concert succeeded in defeating any further revolutionary challenges, the crushing of Che Guevara's action in Bolivia being the turning-point. Radical Third World governments were 'destabilised', as in Nicaragua, while most of Black Africa has degenerated into one-party rule or chaos. The collapse of the Soviet Union removed the major external support and the major alternative model for revolutionary and radical alternatives to world capitalism in the Third World. Today, revolution is no longer on the agenda in any but a handful of small countries. The former Soviet Empire has been replaced by a collection of states aiming at capitalist transformation and embracing liberal market economics, although, as current events in Russia show, often at great social, economic and political costs.

Theorists with impeccable radical credentials like Harris also pointed out the converse – the growth not just of capitalism, but of industrial capitalism in countries like Mexico and Brazil in Latin America and the 'Four Little Tigers' in East Asia (South Korea, Hong Kong, Singapore and Taiwan), together with the transformation of desert sheikhdoms into some of the richest countries on earth. These successes seemed to prove the validity of modernisation theory, and gave rise to a mood of 'triumphalism', an atmosphere in which Western modernisation was exported across the world through successive programmes like Point Four, the Alliance For Progress, and the Peace Corps. Yet Harris pointed out, long before the recent financial crises of the world economy, that countries like Mexico and Brazil were saddled with enormous foreign debt, and were therefore extremely vulnerable to shifts in world-prices or to the periodic flight of foreign capital whenever there was panic in the West about the health of the economy or the political stability of the country. His model, though, seemed to envisage that the process would go on, until the 'end of the Third World'. It had little to say, though, about the vast majority of poor, agrarian countries which were not industrialising, but remaining what they had always been – suppliers of cheap labour and primary goods.

COULD CULTURE BE A FACTOR? TAKING CULTURE SERIOUSLY

More recent radical 'globalist' theory has started to pay attention to those countries which could no longer be adequately described as 'newly-industrialising' because they had already become industrialised platforms for the world-wide operations of transnational corporations in which new regional divisions of labour have occurred. Some ex-Third World countries, such as Hong Kong, have not only become 'branch-plants' of their parent corporations but have also acquired advanced technology; a few have even become research and development 'cores' of their regions (Henderson 1989).

Why, it was now asked, had these countries, and not others, 'taken off', and had cultural factors anything to do with it? Their achievements, even some economists began to argue, might have something to do with cultural factors. Culture, at long last, was being taken seriously, even by economists. Where an earlier generation had blamed Japan's (and China's) backwardness on Confucianism, now it was said to be the key to an understanding of the Japanese 'miracle', providing an appropriate work-ethic – deference and loyalty towards superiors; identification with the corporation; the absence, even, of expectations of holidays; and placing a high value upon education (see Chapter 11 of this volume, by Roger Goodman). High-tech electronic innovation and the rationalisation of the organisation of production were contributory, but not sufficient explanations.

A key theoretical work, Leslie Sklair's *Sociology of the Global System*, explicitly allocated as much conceptual space to culture (what he calls 'culture-ideology') as to economic and political factors. Even so, culture is still treated in the 'arts' sense, though given an economistic twist, since it is treated primarily as consumerism and as the means by which First World tastes and wants are exported, profitably, to the Third. The Third World is merely a consumer-market for things Western, for example music. Yet it is a particular kind of music, growing out of Black culture, that has generated most of the popular musics of the twentieth century, from jazz to reggae and beyond, while today 'world music', a new kind of synthesis of musics from both North and South, has spread, via global technology, perhaps for the first time in human history, to a mass audience of young people in both hemispheres. Its main market, inevitably, is the wealthy North, but its main cultural content comes from the South.

A more enlightening conception of culture, which sheds light on the interplay between imposed Western forms and indigenous ones, had, in fact, long been available in Robert Redfield's distinction between 'Great' and 'Little' Traditions; on the one hand, the great world-religions – Islam, Christianity, Buddhism, Hinduism; on the other, the local cults of the ancestors and the land (Redfield 1960). In the face of the global power of modern communications and capital, 'Little Traditions' are, of course, very fragile, their religions, for instance, vulnerable to the onslaught of high-tech Christian fundamentalism funded by millions of dollars. But as Osman has shown for Malaysia, older and smaller cultures – pre-Islamic 'little' local traditions, particularly cults of the earth and of the ancestors – have co-existed with and interpenetrated for centuries, often for millennia, the more scholastic Great Traditions (Osman 1989). And the Great Traditions – Hinduism, Buddhism, Islam – certainly do not simply disappear with the spread of global capitalism. Rather, they undergo revival, transformation and synthesis.

It is impossible to understand such processes simply in terms of political economy, for the concept which is needed is missing – culture. To say that is not for one moment to underestimate the importance of economic and political power. But the impact, and the ultimate failure of modern imperialism and colonialism, for instance, were never solely economic, nor were political outcomes determined simply by the military-political fact that we had the Gatling gun and they did not. Some peoples, for instance, realising that they could not hope to resist in material terms turned to cultural resistance. The forms they adopted were never uniform, always a dialectic between the culture of the conqueror and their own cultural traditions. But there were many different elements in indigenous cultures, even in 'simple' societies, which they could turn to.

In more complex cultures such as India, a country usually thought of as quintessentially 'other-worldly', Chattopadhyay has argued that atheism is, in fact, an ancient philosophical tradition (1973). Wolf demonstrates that there has always been intense dispute over rights to land and about the relative pre-eminence of the Brahmin priests and the Kshatriyas who disposed of earthly power (1982: 46–7). Through what Srinivas has called 'Sanskritisation', and through hypergamy (marrying 'up'), some individual and collective upward mobility through the caste-system was possible (Srinivas 1952). And though foreigners often still think of Hinduism or Islam as unchanging, in India it was often Western scholars and administrators who codified and institutionalised 'frozen' readings of caste and of religion conveyed to them by a conservative priesthood and by ruling castes, which reflected their social position and their vested interests.

In reality, non-Brahmans in the South had struggled throughout the entire period of British rule. They struggled not just against the colonisers, but also against the claims of the priesthood to hold office in the temples by hereditary right; over rates of remuneration and the services provided by the priests and temple servants and other personnel, and over control of religious endowments. In the twentieth century, these socio-cultural struggles continued: Harijans stepped up their campaign for the right to be admitted to the great temple of Minakshi in Madurai, South India (Fuller 1984). That temple, like the great Temple of the Tooth at Kandy, Sri Lanka, which most people in the West think of solely as a religious institution, is also a complex organisation which disposes of great wealth generated by large flows of income from pilgrims. Temples are therefore important sources of political and social power, and control over these economic resources and the management of the temples is the subject of much competition between different sets of religious professionals. They are sources of political power, too, for party politicians involve themselves in temple affairs, notably by sponsoring rituals, in order to bolster their standing in the world outside the temple.

This, then, was neither unchanging tradition, nor a case of 'religion-in-itself'. It involved cultural struggles between the privileged and the underprivileged over access to cultural rather than material goods. Likewise, in twentieth-century Bali, reformers fought indigenous aristocrats and religious leaders, backed by the Dutch, who worked together to preserve both the traditional privileges of the high castes as well as many newly invented ones, struggles which produced complex divisions rather than a simple anti-colonialism. During World War II, for instance, some nationalists appealed to traditional culture as a source of solidarity in the struggle against Japan; others saw the entire social order as archaic and compromised. After independence, for instance, some modernisers attacked both the caste-system and Hinduism; others tried to convert the pantheon of Hindu gods ('ten thousand', it has been said) into a monotheistic world-religion, identified with a major power, India, and therefore fit to stand alongside Islam or Christianity (Vickers 1989: 150–5, 164–7).

Hence, just as there was never a unitary coloniser – they were Portuguese, British, French or Spanish; some had been soldiers, others missionaries, some men of wealth and ancient lineage, others adventurers from impoverished areas, some mine-owners, others industrial entrepreneurs or plantation-owners – so, too, there was never simply a unitary 'colonial subject'. Indigenous cultural traditions were equally plural, and therefore diverse, and different interest-groups made different analyses of how to cope with colonial rule and adopted different strategies – from accommodation to resistance – whichever they thought would benefit them most. Resistance, from the attempt to restore Moghul rule in India to the rise of Islamic fundamentalism today, often draws upon traditional sources.

In post-Shah Iran, Zubaida has shown, what is usually seen simply as backwards-looking 'fundamentalism' is, in fact, a modern continuation of debates which have taken place for centuries over the status of the *ulama*, not just about their status as custodians of a purely religious orthodoxy, but in terms of their role *vis-à-vis* the state. Khomeini's reading of Islamic doctrine is therefore only one latter-day modern variation on an ancient theme. His conception of the 'nation' as a political force, and of the right of the mullahs to rule and legislate, or the more radical-populist conception of Ali Shari'ati of the people as masses, victims of oppression and exploitation and therefore the main agent for bringing about social change, are both part of an historic debate, and distinctly novel interpretations of Islam – as novel as Khomeini's use of audio tapes to spread his message from exile in Paris (Zubaida 1989).

All societies are open to foreign ideas, whether these are borrowed or imposed on them. But these always have to be adapted to existing, local cultures. The result is a dialectic; not imposition or the blind acceptance of ideas imported from abroad, but a synthesis of cultures, a hybridity. Internationalist 'proletarian' communism, for instance, when introduced into colonial peasant societies, generated new kinds of Russian, Chinese, Cuban or Vietnamese 'communism', which Marx would neither have recognised nor approved of. His conception of the international solidarity of all workers was thus replaced by militantly nationalist communist states; in the extreme, China and Vietnam went to war against each other, not against capitalism. For a long time, too, the major enemy of the USSR was China.

It would be impossible, equally, to explain major social changes in modern Western society simply in terms of political economy. The emergence of the women's movement, or of student protest in the late 1960s, for instance, were outcomes of profound cultural changes when, for the first time in history, young people, and particularly women, had access to mass higher education in societies that permitted them considerable mobility –

though never enough to threaten the position of tight-knit élites from the 'Establishment' in British society or newer dynasties such as the Kennedys or the Nehrus.

All this is not to argue that everything is determined by culture – a notion that has been called 'culturalism'. It is to say that social science is wider than political economy; wider, too, than a purely structural sociology devoid of culture.

In explaining such horrific happenings as the recent war in Chechnya, then, we should not fall into the 'culturalist' trap. There are very important cultural/ethnic dimensions to that war. In Russian cities, there is a hatred of southern 'Blacks', and the Caucasian 'Mafia' is commonly blamed for the rise of corruption, crime and the drug-trade. At the level of the State, the Russian government fears that 'fundamentalism' might spread right across its southern borders. But all this cannot be laid at the door of Islam or of intransigent Chechen nationalists. It has a lengthy political background, from the Tsarist conquest of the Caucasus to Stalin's deportation of the population in World War II, while today, the Russian government has an economic motive – to preserve its control over the crucial oil-pipeline from the Caspian to the Black Sea.

Recourse to culturalist explanations has been common when people, notably media commentators, try to explain genocide in terms that have been called 'primordialism'. Former Yugoslavia, for instance, was for centuries, indeed, a 'shatter-zone' between the Catholic cultures of Croatia and Slovenia coming from the north (Austria-Hungary and Germany); Orthodox Serbia in the West; and the former Ottoman, Islamic Empire in the South. These were, and are, real cultural differences. But they were converted into murderous enmities by individuals, at home and abroad, who wanted to maintain and expand their political power by exploiting those differences.

The other major genocide commonly explained in 'primordialist' terms is Rwanda/Burundi. In the 'interlacustrine' societies of East Africa, cattle-owning ruling castes dominated their agriculturist subjects of different ethnic origins (and poorer pastoralists, too) for centuries. In Ankole, adjoining Rwanda, for example, the Bairu agriculturists (equivalent to the Hutu in Rwanda) had to pay tribute to their Hima lords and masters (the equivalent of the Tutsi). But there was also a cultural dimension even to the economy, for the Bairu were forbidden to own productive cows. Socially, they were forbidden to marry Hima. So this was a caste, not just a class society (Oberg 1940). For centuries the Tutsi steadily extended their control; under Belgian colonial rule their position was further strengthened. After independence, the society had become so polarised that racist Hutu, who had gained power in elections (in which they were the majority), were able to use the power of the modern state and army to launch a campaign of genocidal slaughter against their former masters (Prunier 1995).

These cases have led to an uncritical extension of 'primordialism' – the assumption that hostility, including violent conflict, is inevitable not just whenever one ethnic community exploits another, but even when they simply live side-by-side. But Bosnia and Rwanda/Burundi are the exceptions rather than the rule. In south-central Europe, indeed in most parts of the globe, innumerable ethnic communities live side-by-side, within multinational states, with consciousness of and pride in their own cultures, occupying their own niches in the economy, but without violent conflict between them, let alone genocide. In East Africa, Rwanda/Burundi is a very exceptional, extreme case of 'totalitarian', overlapping, historic inequality in every sphere of life – the political, the economic, and the cultural. Most states are multi-ethnic and inter-ethnic relationships are common.

Culture, then, is not so much a sector of social life, marked off from other sectors – notably the political and the economic – but a dimension of all social action, including economic and political life. So while we can all recognise that the Bank of England is primarily an 'economic' institution, Parliament a political one, and the BBC a 'cultural' institution, they are parts of a wider society, and therefore necessarily affect one another (as when governments control what is presented as 'the news' for political reasons). Each institution, too – a university, a business enterprise, a religious community – develops its own culture (or sub-culture), within the wider culture, which, in modern times, is the culture of the state. Development as a process has its own culture and this is an important recognition. As much development theory, policy and practice is generated within the West, it is inevitable that it comes loaded with cultural attributes found in the West. Hence, as subsequent chapters show, there are cultural clashes inevitably written into contemporary development and this has to be, and indeed is being, challenged.

NEW WAYS OF THINKING ABOUT DEVELOPMENT?[1]

The unintended consequences of an unregulated world economy and of a culturally and politically diverse and uneven world order go far beyond the economic – financial crises and crises of over- and under-production. Disasters such as El Niño in the Pacific, the desertification of large parts of Africa, or the destruction of the world's rainforests, can no longer be attributed to natural causes or to 'primitive' agricultural practices or inadequate technology. Such disasters derived, often, from very high-tech practices on the part of multinational corporations based in 'advanced' states in collusion with their political allies and economic partners, the kleptocratic élites who control some of the states in the 'Third World', an alliance in which pursuit of profit is unconstrained by the formal political boundaries of the nation-state. Nor is the continuing poverty of tens of millions of people confined to (still-sizeable) rural backwaters. Now the majority of the world's population lives in towns and cities.

Even with the advent of mass consumerism, non-economists, such as international charity organisations, have long questioned whether GDP or consumption per capita constitutes development in a meaningful, human sense of the term. 'Structural adjustment', adopted at the insistence of the World Bank or the IMF, was certainly a success for a new, wealthy class. But the privatisation of state enterprises and the removal of subsidies for basic commodities, notably food, resulted in immiseration for millions world-wide and huge increases in unemployment. From the mid-1980s this could not be ignored by anyone involved with development. At the World Bank the responses to the publication of *Adjustment with a Human Face* in 1987 (Cornia, Jolly and Stewart) led to a rethinking of structural adjustment and in the 1990s there have been renewed concerns about poverty alleviation and 'institutional capacity'. Institutions are here being understood in terms of rules of behaviour affecting 'transaction costs' in market exchange (for example, that individuals do not necessarily respond to price incentives alone and even when they do will not necessarily respond in the ways neo-classical economic theory assumes). These ideas are clearly influenced by the so-called new institution economics, which in certain respects can be seen as a way of trying to incorporate culture into orthodox economic theory. Douglass North for example has stated:

> We [economists] need to know much more about culturally derived norms of behaviour and how they interact with formal rules to get better answers to such issues. We are just beginning the serious study of institutions
> (North, quoted by Putnam 1993: 181)

Increasingly, at least outside of the World Bank and the IMF, it seems clear that true development should be measured not just in terms of economic criteria – the size of the national product, or of increasing production, or growing consumption of imported First World 'goods' – but in terms of who gets what, and of whether what they get via the market is really what they need to enrich their quality of life. A better measure would take into account, for instance, access to basic health services or to education. The political power of the state or, rather, the predatory élites who control it, is not a meaningful measure of political development. Political development, rather, should be measured in terms of the empowerment of the ordinary citizenry and their right and freedom to develop their own institutions. An important example of how such ideas have been incorporated into mainstream development thinking has been the introduction of the Human Development Report published annually by the

United Nations Development Programme since the 1990s. This report includes an index which combines a range of factors other than purely economic indicators which are then used to indicate levels of development.

Throughout the 1990s the United Nations has become much more involved than hitherto in situations of on-going conflict, particularly conflicts occurring within member states of the UN system. The 1994 Human Development Report drew attention to the ways in which 'inter-ethnic' strife was crippling the development prospects in many parts of the world. Indeed it stated that half the members of the UN had recently undergone some form of inter-ethnic strife.

There is indeed a growing pessimism that has replaced the optimism following the end of the Cold War. There is a scepticism that traditional economistic strategies can provide an adequate basis for social progress. The débâcle in Somalia, in 1993, proved to be an important turning point in that it seemed to confirm that 'international' interventions that took no account of local culture were doomed to total failure. Sadly the lesson learned, at least so far, was not to find out about local ways of life and then act intelligently, but effectively to abrogate responsibility – a fact that was brought home shockingly during the Rwanda genocide of 1994.

Plate 3.1 Uganda/Sudanese repatriation. These are new arrivals from Nimule, Sudan, en route between Bibia and Arinyapia, where a large new reception centre has been built.

Source: Photographer, J.M. Goudstikker. Reproduced by kind permission of UNHCR, March 1989.

Not surprisingly, in the Non-Governmental Organisation sector and among academics, critiques of orthodox approaches to development have gone much further. Some of the most scathing in recent years have been produced by a group of thinkers sometimes referred to as the 'post-development school' (Crush 1995a; Escobar 1995; Ferguson 1990; Rahnema and Bawtree 1997; Sachs 1992; Tucker 1997). Some of these scholars go so far as to argue that the whole notion of development is counter-productive and imbued with culturally imperialist assumptions. All place great emphasis on the varieties of ways of being human and the dangers of any form of cultural homogenisation. Not surprisingly, a new mode of criticism has emerged and there are several contributors. Development theory, they tend argue, grew out of the subjugation of the rest of the world by European colonialism. What was called 'the Enlightenment' in Europe – an epoch of scientific advance – was in fact also a period of bloodshed, intolerance and religious wars at home, and of genocide and slavery abroad. In the process, entirely new categories of humankind – such as 'Indians' in Latin America – were invented and imposed on the great variety of societies and cultures the Europeans encountered. This relationship later became reduced to a simple binary formula: Europe (and, later, the 'West') and European rationality were seen as uniquely progressive; all other societies and cultures were lumped together as 'traditional', their knowledges and cultures unworthy of attention, and denigrated as the antithesis of knowledge. Civilising these 'People without History' (Wolf 1982); by force was necessary, it was the White man's world-historic mission.

Modern development theory, it is argued, is a continuation of this binary opposition.

In conventional development theory, growth is perceived as a linear sequence of stages, and development can only come about by following the path pioneered in the West: the adoption of Western political forms – the management of society by a state based on parliamentary democracy and bureaucracy – and, in the economic sphere, the market as the key framework. The ultimate emphasis, then, is upon economic development 'in itself', with subsidiary attention to political institutions, and no serious place for culture. The post-development school argue, like world-system and dependency theorists before them, that true social progress would require no less than the radical transformation of world society and not simply of the underdeveloped countries. Like almost everyone else they recognise that the alternative paths developed by communist states have proved to be largely dead-end and that radical transformation through planned revolution looks far less likely than further economic crisis and continued wars. In the longer run, when pushed to discuss their views of a future world, they suggest the possibility of either a global apartheid in which rich countries increasingly disengage and barricade themselves away from marginal parts of the world, or that an integrated global middle-class will eventually facilitate the development of adequate welfare services (Sachs 1992).

However, the thrust of the work of post-development thinkers is to discuss the specifities of cultural change and resistance in particular contexts. Although grouped together, for example in books like *The Post-Development Reader* (Rahnema and Bawtree 1997), their approaches are not without important differences. The school is by no means integrated or dogmatic in its approach and this is reflected in the ways in which culture as an analytical construct is used. James Ferguson (1990), for example, uses discourse theory not in the sterile and apolitical manner that I discussed at the end of Chapter 1, but as a way of exploring how the World Bank employs a particular 'gaze' as a mechanism of asserting power. He demonstrates that World Bank Reports on development in Lesotho can easily be shown to be factually nonsense but that this has little or no effect on development planning. In fact the anti-politics machine depoliticises all that it comes into contact with, pushes political realities out of sight and at the same time surreptitiously expands its own power and ability to keep political control (Apthorpe 1996: 167).

Perhaps the most influential of all the post-development thinkers, Arturo Escobar (1995), on the other hand, has attempted to develop a new grand theory, which is seen as a form of culture-based political economy. A danger with Escobar's approach is that he appears to want to replace one 'meta-narrative' with another. As Apthorpe (1996) has pointed out, this may both be over-ambition and misplaced ambition; social science cannot be 'usurpatory of political power' (Apthorpe 1996: 171). Clearly there are significant limitations with the post-development school, notably their mix of ecological mysticism with ethno-romanticism. However, the way in which they have placed culture and power at the centre of their analysis seems to be a positive and significant contribution. This link between culture and power is particularly important because it has enabled most of the post-development thinkers to avoid for the most part the trap of culturalism. This trap, however, is not avoided by some of the other writers who have contributed to recent discussions of culture and development, a prominent example being the recent work of Samuel Huntington (1996).

Huntington predicts that culture will become a fundamental source of conflict in the post-Cold War world. He then provides an extremely crude and highly problematic representation (both in writing and in maps) of world cultures and civilisations: Western, Confucian, Japanese, Islamic, Hindu, Orthodox, Latin American, African and Buddhist. Whole continents and countries are placed under these banners. His blurring of categories and definitions, cultural history (Western), geography (Latin American) and religion (Hindu) is obvious and one has to question whether these world divisions actually mean anything. There are also more subtle problems with such an approach, not least the lack of recognition of diversity within countries, regions and continents – India, for example, is far from being completely Hindu, and what does Latin American actually mean in cultural terms? Nevertheless Huntington's thesis has been very important and was described by Kissinger as 'one of the most important books ... since the end of the Cold War' (*The Economist*, 1996: 25–30). This is a worrying conclusion about a book which fails to relate adequately and thoughtfully culture to real-life, lived conditions. Culture does have an important influence but it is unlikely to have become *more* important than government or economic forces in the years since 1989, and in any case it is far from a unitary entity fixed within territorial boundaries.

Another influential political commentator Francis Fukuyama (1995) has also now turned to cultural explanations of human action and social change, suggesting that the integration of the global economy has heightened the significance of cultural differences. He agrees with Huntington that these will loom larger from now on 'as all societies will have to ... deal not only with internal problems but with the outside world' (1995: 5). However, Fukyama differs from Huntington in that he states that cultural differences will not necessarily be the source of conflict. On the contrary, he suggests that the rivalry arising from the interactions of different cultures can bring about creative change, and rich cross-cultural stimulation (1995: 6). As we have seen, this is hardly an original suggestion; nevertheless it makes Fukyama's analysis slightly less culturally deterministic. It is worth noting that, like Huntington, he falls into the trap of treating cultures as 'distinctive and functional' (1995: 6). However, almost on the same page (1995:5), he effectively, but perhaps usefully, contradicts himself, because he acknowledges that the growing significance of culture moves beyond national boundaries and 'extends into the realms of the global economy and international order'.

The influences of 'new institutional' approaches to economics, of the post-development school, and of Huntington and Fukuyama's ideas all suggest that 'culture' is moving into the mainstream of debates about development. This is at least partly to do with a struggle to find modes of explanation for social action following the end of the Cold War and the erosion of post-Cold War euphoria. The world seems to have become more difficult to understand, or at least more difficult to fit into simple paradigms. Turning to culture is probably less an enthusiastic recognition of an overlooked way of seeing things than it is an aspect of a crisis of confidence. Nevertheless, for those of us who long argued that culture must be taken seriously, it offers a space to be heard, and some grounds for hoping that the reductive and misleading simplification of so much development thinking may be in the process of, at last, being abandoned.

NOTE

1 This final section has been written in collaboration with the editors.

II

QUESTIONING CULTURAL ASSUMPTIONS

4

MODERNISATION VERSUS THE ENVIRONMENT?

Shifting objectives of progress

Alan Thomas

INTRODUCTION

There is a strong argument that development historically has depended on a change to a global culture of modernisation. This, the argument continues, is what underpins the dominance of Western-style capitalist industrialisation and to date there is no alternative route to development.

Since the downfall of the Soviet Union the main challenge to this view may appear to come from global environmentalism. In the industrialised north, anti-road, anti-nuclear and anti-pollution protests proliferate alongside lifestyle-change movements such as that towards green consumerism, while in the 'developing' south, indigenous peoples and others struggle for control over their own environments (for more on the view of the 'north–south divide' used in this chapter, see Thomas 1992: 6, or The South Commission 1990: 1–4). Solidarity between the northern and southern parts of this global movement is illustrated by examples such as the reported message of support sent by fax from the Ogoni people in Nigeria to the protesters attempting to prevent the building of the Newbury by-pass in the UK (Goodwin and Guest 1996).

This challenge is expressed at a different level by the increasing influence of non-governmental organisations (NGOs), from both north and south, on environmental policies at local, national and international levels. Local examples of large-scale water engineering projects being halted can be cited along with international agreements such as that on targets for reducing carbon gas emissions. Such policies are in effective opposition to the assumptions underlying global industrial society and modernisation and if adopted consistently would imply rethinking the basis of development.

The global environmental movement can be taken to include an international cadre of expert scientists and officials of international agencies as well as the staff and activists of environmental NGOs and the people's movements, north and south, mentioned above. This movement can be seen as promoting a new way of thinking in which environmental and development concerns are no longer in opposition but need to be considered together. For many, the 1992 United Nations Conference on Environment and Development (UNCED) or 'Earth Summit' at Rio de Janeiro (Brazil) epitomised this new way of thinking. Despite the disappointing lack of many clearly agreed practical outcomes, it has been hailed as a watershed in terms of how human progress is conceptualised.

This chapter casts doubt on the notion of global environmentalism as a single movement. It points out that the cultural basis of dissent is very different for southern as opposed to northern environmentalists. There may often be a suspicion on the southern side that environmentalism has been invented in the north as an excuse for 'pulling up the ladder' and preventing those in the south from attaining northern levels of affluence, thus preserving the lion's share of the world's resources for the north. On the other hand, northern environmentalists may fear the aspirations of their southern counterparts in that the energy implications, say, of meeting those aspirations appear to entail a global catastrophe. Partly because of this difference, it is all too easy for ill-judged attempts to link northern and southern NGOs actually to have negative effects.

Thus environmentalism as a global political movement contains fissures which weaken its potential as a threat to the Western culture of industrialisation and modernisation. These fissures could even be seen at UNCED. Nonetheless, the need to combine considerations of environment and development into a new conception of progress remains imperative. To be politically effective in the long run, a global movement based on such a new concept would have to be built on people's aspirations in both north and south, not just on the views of a cadre of experts, and would have to communicate across the gap between those for whom the environment means global habitat and wilderness and those for whom it is their everyday source of livelihood.

The rest of this chapter is in five sections. The first two explain the cultural content of the idea of 'modernisation' and the view of the environment which is entailed. The third and fourth sections give the basis of northern and then southern environmentalist opposition to modernisation and show how different they are from each other. The final section explores the possibilities and dangers of north-south environmentalist alliances, with some examples, and discusses the imperative for a new conception of progress and where it might be found.

THE IDEA OF 'MODERNISATION'

It has been pointed out how modernisation, industrialisation and development often appear effectively to be synonyms (e.g. Allen 1992a: 381). The concept 'development' embodies considerable ambiguity. In particular there is a basic tension between the normative idea that development means human progress and the view of development as an historical process which can be studied empirically. In the first case the definition of development depends on the views of those proposing it as to what constitutes progress, something never divorced from questions of power and interests. The second view leads to observations such as that development has inevitably involved industrialisation and that the process has always been 'awful' for many at the same time as being necessary for progress. Kitching (1982, 1989), for example, argues that the only way to raise living standards for the mass of the population is to increase productivity to an extent that can only be achieved through industrialisation.

> ... industrialisation cannot be avoided or run away from, either in theory or in practice. Those who try to do so, in the name of loyalty to the peasantry and the poor, are likely to end up offering no effective help to 'the people', and seeing the process of industrialisation occur in any case, under the anarchic sway of international capital.
>
> (Kitching 1989: 179)

Of course, those in positions of power in an industrialised world tend to promote a view of progress which elevates the value of the products of industrialisation. The combination of this normative view with the proposition that industrial development is not only necessary in order to achieve this form of progress but also historically inevitable creates a dominant ideology of development which certainly corresponds closely to the interests of those in such powerful positions.

The idea of being or becoming 'modern' finds a place in both the normative and the analytical approaches to development. In the first place, it implies a whole bundle of cultural values, artefacts and attributes likely to appear desirable particularly to those living in less modern or less 'developed' societies or communities. Obvious examples range from the ubiquitous Coca-Cola to private motor cars, piped domestic water and modern curative health care. In this sense, to become modern epitomises a particular, very widespread and powerful, normative view of development, which goes deeper than simply a desire for the products of industrialisation to include a rather uncritical acceptance of the role of science, the market, urbanisation and so on. 'In a world dominated by advanced capitalist economies, all aspects of Western society are elevated to represent the ideal of what development is trying to achieve' (Potter and Thomas 1992: 119).

To say that this view is 'powerful' leads to the place of the idea of modernity in the other approach. Here it is argued that modern aspirations and the adoption of a modern approach, for example in turning an informal economic activity into a 'modern' business enterprise, are driving forces which form crucial parts of the historical process of development. Thus one of the central arguments of the US 'modernisation theorists' of the 1950s and 1960s was that development in the south required the adoption of more 'achievement motivation' by its people (McClelland 1963).

Modernisation theory as such is now rather unfashionable, perhaps because of its tendency towards racist stereotyping. However, just as the proposition that industrialisation is necessary for development is still to be refuted, it remains the case that the trappings and the real benefits of modernisation are visible signs of 'development' that tend to define people's aspirations. This means that alternative views of what constitutes progress are hard to maintain.

THE 'MODERN' VIEW OF THE ENVIRONMENT

What view of the environment is embodied in the ideas of modernisation and the necessity of industrialisation?

It is difficult to answer this question in terms of a single, coherent view. Partly this is because, until fairly recently, modernisers and developmentalists of all political persuasions tended to ignore environmental issues for various reasons. Many were adherents of what Colby calls 'frontier economics', the most extremely anthropocentric of five 'paradigms of environmental management in development' (Colby 1990: 8), seeing the environment as something out there and beyond what has already been utilised, with unknown and unlimited resources waiting to be exploited. One could argue over whether value derives from use and exchange or only from human labour, but either position effectively denied the possibility of intrinsic value attaching to resources or the environment generally. However, such unreconstructed views are now rare: 'For a long time, the natural environment was treated as a boundless resource. Since the 1960s, however, the bottom of the well has gradually been coming into view' (Glasbergen and Cörvers 1995: 2).

Although the environment can no longer be taken for granted, there is as yet no worked-out alternative to modernisation. Since the 1960s, the second of Colby's five paradigms, 'environmental protection', has been dominant, in which remedial actions and damage limitation measures are added on so that industrial development and economic growth can continue. Those whose ideas of progress still correspond more or less to those of modernity tend to use a rather ad-hoc combination of notions about the environment, rather than having a single coherent view.

First, the idea of *man's mastery over nature* is deep-seated, often traced back to the first book of Genesis:

> ... and God said unto them, ... replenish the earth and subdue it: and have dominion over the fish of the sea, and over the fowl of the air, and over every living thing that moveth upon the earth.
>
> (Gen. 1, 28)

Although this quote refers to 'them' (male and female), this idea is often posed deliberately in terms of the *mastery* of *man* – perhaps with the implication that women have a different relationship to nature. Indeed according to eco-feminists such as Plumwood (1993) women have been conceptualised as part of nature so that they could be subject to male 'reason'. However, the main point is that in this view 'the resources of nature are to be harnessed to satisfy human wants and needs, however defined' (van de Laar 1996: 4). In fact, this was originally a generally positive view in which 'human transformation of the environment was considered a necessary and creative activity' (Woodhouse 1992b: 97).

An extension of the above idea sees *the environment as a resource bank and a pollution sink*. If human activities are typified by industrialised processes in which inputs are transformed into outputs, it is easy to conceive of relations between human activities and the environment in these two parts. Here the human-environment relationship is not quite so one-sided as in the previous notion of *mastery over nature*. The effects of both resource use and pollution on the environment may need to be weighed against the benefits of a particular production process to humanity.

A third notion is *the alienation of the environment from the direct experience of individuals*. Modernisation emphasises the individual both as an entrepreneur and as a customer or source of demand for modern products. One result is that individual rather than collective consciousness determines how the environment is experienced. Modern individuals, via purchasing decisions or jobs or through where they live, certainly relate to institutions (such as shops, employers and local councils) which use natural resources and pollute. However, neither resource use nor pollution immediately relates to individuals' own everyday sources of livelihood, so they do not identify closely with these aspects of environment.

There is also *the aesthetic or amenity view of the environment*. Both the built and the 'natural' environment should ideally be tidy, peaceful, beautiful, awe-inspiring or evoke one or other of a long list of positive emotions. In fact in a modern society people tend to relate most directly and personally to the environment in terms of how pleasant the places are where they live or go on holiday.

Finally, the modern approach to environmental issues is generally one of *the application of science and technology to problem-solving*. The relative depletion of a particular resource or a specific local pollution threat to health or recreation are examples of how particular breakdowns in the relationship between human activities and the environment can be treated as bounded problems. Finding solutions is the province of experts and technicians, rather than ordinary people. (In these cases, solutions could be technological innovations to allow a less concentrated source of a mineral to be exploited or to substitute a new material for a certain use; or regulation to penalise polluting industries and encourage new mechanisms for waste treatment).

One might suppose that the increasing evidence of interrelated factors causing damage to the environment on a world scale would lead to the adoption of more of a 'relationship-maintaining' approach (Vickers 1970). However, 'the *force majeure* of a wounded environment [will not] compel change, because catastrophe will be cumulative and strike in little bits' (Middleton, O'Keefe and Moyo 1993: 10). Hence each oil tanker disaster or each call for a new dam or motorway extension can remain a separate problem for the experts to deal with, and the cumulative issue need never be considered in an holistic fashion.

To a large extent the picture I have painted looks like a 'northern' view of the environment. However, I prefer to call it a 'modernisation' view. It also largely corresponds to what Middleton, O'Keefe and Moyo call 'some curious twentieth-century consumerist account of nature' (1993: 6).

Whatever the label used, one should note that versions or aspects of this type of view are also found in the south. Southern governments and élites are likely to view their environment in terms of natural resource endowment to be exploited in order to modernise and industrialise. Some, including several Southeast and East Asian countries as well as Botswana, have been successfully developmentalist in these terms, although as always the élites have benefited more than others. Thus the Malaysian government's well-publicised intention to become a fully industrialised country by the year 2020 depends on its asserting its sovereignty over its forests and hence its right to export timber, stressing that 'economic growth is needed first, to generate the ability to clean up the environment later' (Eccleston 1996a: 134).[1]

Middle-class urbanised populations or skilled wage workers in many parts of the south may already have 'modernised' lifestyles with both similar material benefits and similar types of environmental concerns to their counterparts in the north. However, many people in the south, as well as not insignificant numbers in the north, still have more direct relationships to their environments than the typical modernised person whose relationship to their environment is described above as alienated. These direct relationships with the environment come from participation in primary commodity production, whether as wage labourers in mining, forestry or agriculture or as petty commodity or even 'subsistence' producers. However, it is likely that the aspirations of such people are towards some more modern type of livelihood.

Malawian poet Frank Chipasula captures this well when, after describing a contemporary who has grown fat and now wears three-piece suits and 'entertain(s) company with imported spirits /bought with the people's tax money', he complains: 'I am still where you left me, strapped ever to my hoe /in the dust ... ' (Chipasula 1990).

Only a small minority can attain a 'modernised' lifestyle through corruption, and petty commodity producers and wage workers may, like Chipasula, be aware of some of the negative features and contradictions of such a lifestyle. Nevertheless, modernisation, with its attendant 'consumerist' account of the environment, provides very visible and desirable symbols of the material benefits that the related style of development could bring, as well as providing the actual cultural basis of those societies which have been fully industrialised.

NORTHERN ENVIRONMENTALIST OPPOSITION TO MODERNISATION

The most extreme form of environmentalist opposition to modernisation and industrialisation comes from 'deep ecology', another of the five paradigms set out by Colby (1990: 8). Van de Laar characterises deep ecology as a social movement (not to be confused with the science of ecology) as follows:

> Deep ecologists approach the relationship between man (sic) and nature from a non-anthropocentric perspective. Instead, it sees man as just one of many species living in complex eco-systems. Their 'bio-centric' or 'harmonious' view of the relationship between man and nature often means putting *Man* (sic) *under nature*, the reverse of the frontier economics hierarchy.
>
> (van de Laar 1996: 6)

Deep ecology includes both what van de Laar calls 'wilderness preservationism' and various quasi-religious and holistic notions about the Earth as a single eco-system, including James Lovelock's well-known 'Gaia hypothesis'. Less extreme northern environmentalist opposition to modernisation still generally encompasses these two ideas in viewing the environment.

The first notion, of *the environment as wilderness*, is closely related to ideas of conservation or preservation, of species and habitats. It has given rise to conservation-oriented NGOs, to the movement to set aside specific areas for the preservation of 'nature', as national parks or otherwise designated protected areas, and to growing numbers of protest and lobbying groups and informal networks such as those among anti-road campaigners.

An immediate motivation for supporting such groups is a reaction against loss or possible loss of amenity. To some extent it is as a result of the work of conservation groups themselves that people go beyond this view to think in terms of threats to 'the wild'. As Middleton, O'Keefe and Moyo (1993: 209) put it, some of the larger northern NGOs have 'fostered the popular conservatory perception of "the wild" ... they have sold the wild to a world anxious to believe that it exists and which spends substantial sums of money to look at it through a camera lens'. Although people may feel angry when pollution or development threatens wilderness areas, this view of the environment is one from which they are alienated just as much as from the source-sink view. In fact wilderness is by definition alien – 'the wild' does not relate to people's everyday experience or to how livelihoods should ideally be made in the modern world.

Thus, apart from a few professionals, the work of northern conservationist NGOs does not relate to the immediate individual economic interests of their members. If someone works for a modern industrial firm, the resources used most likely come from far away, and the environmental impact of any pollution is not likely to affect that individual or threaten his or her own livelihood directly. Even if it does, the effect in terms of loss of amenity may be set against the benefits in terms of employment and hence access to the modern lifestyle. From this one can see that conservation groups are far from representing local economic interests or even communities, and are likely to be opposed by both business and working-class interests in localities which depend on particular industries.

Recent increased awareness of *the environment as an issue of global forces* has perhaps more than anything caused it to enter the modern consciousness as an area of major concern. This tends to be combined with a rather vague notion of the interlinked, holistic nature of environmental issues. The most obvious examples are the threat of climate change through global warming resulting from the greenhouse effect and the depletion of the ozone layer. Loss of bio-diversity and the pollution of international waters may also be portrayed in terms of global forces, and tropical deforestation has become a popular issue in northern countries more because the idea of forests as the 'lungs of the world' has caught the imagination than because of sympathy for the displacement or loss of livelihood of local people in previously forested areas.

Global environmental issues conceived in terms of global forces provide a relatively new motivating factor for northern activists. Friends of the Earth and Greenpeace are the two best-known examples of environmentalist NGOs which have grown in size and prominence as a result of the recent growth in this concern. To a large extent questions like global warming and ozone depletion have come to the fore as concerns for modernised populations because there is some scientific agreement on these forces and their likely effects – a scientific basis for concern about the future of a society whose own development has come about largely as a result of applying science (Yearley 1994: 162–3).

As with the conservationist concerns, these issues of global forces are also far outside the direct experience of individuals. As Chris Rose of Greenpeace put it, 'Greenpeace and environmental issues were "out-there", wherever that luminous world was, somewhere down the cathode ray tube [i.e. on the television]' (Rose 1993: 292). Clearly, in this respect as well, the NGOs and the northern environmental 'movement' of which they are part are not built on the direct economic interests of their members and supporters.

However, in another sense both aspects of northern environmentalism do relate to the interests of many of those living in modernised societies. Such people often find that their immediate material needs are quite well met through networks of exchange that leave them not only alienated from the resource bank and pollution sink aspects of the environment but also prone to suffer from alienation at community and personal psychological levels. The amenity, aesthetic and even spiritual value of the wilderness aspect of the environment, which is likely to mean very little to someone whose immediate material needs are not satisfied, may be of particular importance as a counter to such alienation, and could thus be said to relate to the modern individual's interests. This idea is certainly a modern one, both in the sense of relating to modernisation and in being relatively new. It can be traced back only about a century, to writers such as John Stuart Mill:

> Solitude ... in the presence of natural beauty or grandeur, is the cradle of thoughts or aspirations which are not only good for the individual, but which society could ill do without. Nor is there much satisfaction in contemplating the world with nothing left to the spontaneous activity of nature.
>
> (cited in Barbier 1989: 13)

Even if one takes a purely economic view of interests, when needs are met through a network of exchange relationships which are global in extent, then it is much too limited to say that interests are only direct and local. It is the most industrialised, most modern economies which rely to the greatest extent on the exploitation of resources at a global level and on the capacity of the globe as a whole to act as a pollution sink. Thus it is those whose livelihoods are embedded in those modern economies who stand to lose if the global economic system is threatened by global environmental catastrophe. Hence in this sense also one can say that the modern individual's interests are involved when she or he joins or decides to support an environmentalist NGO.

A final aspect of the northern form of environmentalist opposition to modernisation relates to the fact that, not surprisingly, most people in industrialised countries react to environmental issues with feelings of powerlessness, perhaps heightened by the way that it is individual rather than collective action which is given precedence in the modern way of thinking. According to Chris Rose of Greenpeace:

> There is now a widely-observed loss of faith and confidence in western countries in institutions and processes (political parties, trades unions, local authorities) which formerly enabled people to feel they had social agency (influence). ... [T]his has led to pervasive public feelings of anxiety and helplessness.
> (Rose 1993: 293)

Rose also argues that many of the scientific arguments about the importance of issues such as ozone depletion have been won and that 'providing more evidence of environmental problems can actually set up a cycle of despair which drives people out of NGOs ... ' (1993: 293). Hence Greenpeace and other NGOs now need to 'enforce solutions' through 'interventions' (p. 295) and do 'whatever it takes to deliver real *change*' (p. 297; emphasis in original). In other words, while for many individuals the NGOs they support are as distant from their everyday lives as the environment itself, these NGOs have to be seen to make a difference in order to maintain that support.

SOUTHERN ENVIRONMENTALISTS AND MODERNISATION

I explained above how the modernised or consumerist accounts of the environment are found also in the south, among members of government and élites, middle-class urbanites, wage workers and rural petty commodity producers. There are also some examples of indigenous but modern environmentalist NGOs in the south, with similar concerns about amenity, conservation and global environmental issues as their northern counterparts. In Botswana, for example, littering from drinks cans is a major concern that has allowed a new urban NGO, Environment Watch Botswana, to gain a lot of support with its call for separation and recycling of household wastes. In Zimbabwe, Environment 2000 was set up following a 'Save the Rhino' campaign and later led a campaign against a proposed hydroelectric dam on the Zambezi river largely on grounds of loss of habitat for certain endangered species.

On the whole, however, where conservationist NGOs are found in the south they are either branches of northern NGOs or else formed by expatriate European or settler groups. Some of these NGOs appear to represent partial interests such as those of tourism companies or private hunting operators. Thus, in another example from Zimbabwe, the success of the Zambezi Society in its campaign to force the Mobil Oil Company to undertake its oil prospecting activities in the Zambezi valley in a more environmentally sensitive manner can be explained by reference to the Zambezi Society's power base in white commercial interests (Thomas 1995: 6).

Such conservationist concerns are not typical in the south. Generally, for many of the very different types of southern people for whom modernisation represents an ideal, the environment is not a problem, unless it is a northern or 'white man's problem'.

As noted already, environmental 'problems' in this sense are identified on the one hand with conservation and on the other with a global-level concern over global issues. To many southerners the former can imply that animals, rivers, and so on are apparently to be treated as more important than people. In many southern countries conservation regulations can be traced back to authoritarian colonial policies under which agricultural and pastoral practices were 'policed'. In what is now known as Zimbabwe, for instance, enforced contour ploughing and the outlawing of riverbank cultivation both symbolised white power, and these regulations were deliberately flouted as part of the struggle for independence. In many parts of the world, state ownership of all wildlife and designation of protected areas such as national parks, so that it becomes a criminal act to kill wild animals or to utilise the products of forests, is similarly resented.[2]

Northern-style conservationist NGOs may well be seen as the descendants of those who implemented such hated practices in the past, and what appears to conservationists as necessary protective measures may be construed as devaluing local people's knowledge, immediate needs and aspirations.[3]

The question of global environmental issues also leaves many southerners cold. They appear far distant from local concerns of poverty or livelihood, or even national economic development. The formulation of such issues in terms of global forces effectively defines them in terms of the need to defend the world system as it is, and defending the status quo means entrenching the privileges and power of the north. It is hardly surprising, then, that southern reactions at the Earth Summit amounted to the response: 'What, in return for help in these matters, is the north prepared to do for the south?' (Middleton, O'Keefe and Moyo 1993: 9).[4] At Rio and subsequently, southern governments and NGOs had difficulty persuading northern institutions to include desertification as a global issue, since it does not involve physical forces or ecological feedback on a global scale. However, in the end a convention on desertification may have been one thing the north was 'prepared to do'; even then, in the Convention to Combat Desertification (CCD), desertification and drought are not defined as global issues but as 'problems of global dimension in that they affect all areas of the world and that joint action of the international community is needed ... ' (UN General Assembly 1994).

In the south, the environment is likely to figure as a political issue in a very different way from the global and conservationist issue it is in the north. The environment is the whole milieu in which people have to make a living. Questions of amenity, wildlife, and the impact of big engineering projects, which might bear a specific 'environmental issue' label in the north, are not separated in people's minds from everyday matters of importance like water, agricultural production or other means of livelihood, and health. As noted above, many southerners have a direct relationship with their local environment as a source of resources, through their involvement in primary commodity production. This means, as van de Laar points out, that 'resource depletion is often felt more severely than pollution effects, and it is the poor who are affected most' (1996: 8). Desertification, mentioned above, is perhaps the archetypal 'southern' environmental issue, since it relates to the degradation of one of the most basic resources necessary for subsistence, namely land.

Not surprisingly, then, in some southern countries such as India, 'ecology movements' have risen from the lower classes (Bandyopadhyay and Shiva 1988). In most countries of the south environmental problems in rural areas are perceived largely in terms of land tenure and competition for land. However, although the effects of land degradation and other forms of resource depletion may be local, and local questions such as forms of tenure and management may make things worse, the issue may well be global in terms of underlying socio-economic causes. One of the successes of the NGOs in the negotiations leading to the CCD was to gain recognition of factors such as global trading patterns and international debt as contributing to land degradation. The need for bigger yields to offset declining terms of trade and for more export earnings to service debts can lead to over-grazing or over-exploitation of poor arable soils. Similarly in the case of tropical forests, while it may often be forest dwellers who are the immediate agents of deforestation, broader socio-economic forces are at work, which end up leaving them no choice if they are to survive. Hence the 'ecology movements', like community-based movements or NGOs elsewhere in the south, oppose the exploitation of people as much as excessive resource exploitation and have a direct interest in the need for locally sustainable livelihoods.

One should not forget how the environment is viewed by poor wage workers as well as the under- or un-employed in the south. Particularly in cities and other urban areas, but also in the countryside, such people have poor access to resources, poor health, poor housing – for them the environment is a question of *dis*amenity. Although, as I have noted, resource depletion is the most important form of environmental issue in the south, pollution, where it occurs, is much more likely than in the north to have immediate and serious local effects, with pollutants often damaging the health of those who labour in the very industrial processes that produce them.

Thus a southern environmentalist critique of modernisation focuses on immediate effects on people and livelihoods rather than on global forces and is based on direct experience rather than scientific evidence. One form of this critique rejects development altogether (Alvares 1994); while another advocates 'resource management' (the third of Colby's paradigms).

From India, Alvares argues that, while:

> [t]he idea of 'development' ... has been closely identified with those of modernity, progress and emancipation [and] has successfully maintained an aura of indisputable inexorability normally associated with the law of falling objects, nevertheless 'development' is a label for plunder and violence, a mechanism of triage.
>
> (Alvares 1994: 1)

He quotes a similar argument from a Mexican writer who concludes that 'development stinks' (Esteva 1985: 135).

Rather like Kitching (see above), Alvares argues that industrialisation has always been based on 'mass pain' (1994: 6). For the majority there are no benefits to outweigh the costs. In this argument, the negative aspects of modern industrial development, including mass displacement of populations (for example by huge dam projects), mass unemployment alongside increases in labour productivity of modern manufacturing, the many cases of pollution damaging both human health and the environment, and more general global environmental concerns such as deforestation or global warming, are all explained as inevitable costs of progress by those in powerful positions. Alvares illustrates this argument by citing Adolf Jann, president of Hoffman La Roche, who, when asked about the 1976 accident in which poisonous dioxin gas escaped from his company plant at Seveso, Italy, commented that ' ... progress can lead sometimes to some inconvenience' (quoted in Adams 1984: 85).[5]

By contrast, the resource management perspective looks for ways in which local people's livelihoods can be guaranteed through sustainable utilisation of resources. NGOs working from this perspective aim both to lobby for the internalisation of environmental costs in large-scale projects and for people's participation in natural resource management. Both the 'Joint Forest Management Programmes' (JFMP) in various Indian states and the Zimbabwean 'CAMPFIRE' programme of local community-based resource management schemes, in which communities benefit financially from participating in the management of wildlife and other resources, are examples where southern NGOs have become involved from such a perspective. Northern development NGOs (as opposed to those with a conservationist or global environmentalist perspective) also tend to align with the idea of resource management and often support southern partner NGOs in projects based on this notion.

THE POSSIBILITIES AND DANGERS OF NORTH–SOUTH ENVIRONMENTALIST ALLIANCES

Much has been written about the potential for alliances between environmentalists and particularly environmentalist NGOs of north and south. Bramble and Porter (1992: 348) described a number of case studies of policy influence attributable to the activities of such alliances, notably that between Brazilian rubber-tappers and US environmentalists, which campaigned to slow the rate of deforestation in Brazil by targeting Multilateral Development Banks and the United States Congress.

Eccleston (1996b) points out how, for southern NGOs, collaboration with northern NGOs can mean both access to additional resources and an opening up of political space. He notes that environmentalists generally accept the idea of such collaboration without question. One might go further and say that it represents a move towards an ideal of globalised environmentalism which for some environmentalists is almost the only hope of a social movement powerful enough to counter the forces of industrial capital together with the culture of modernisation. Eccleston also warns, however, that such collaboration can be a two-edged weapon, possibly galvanising the policy-making system into more effective action, possibly even leading to suppression of the southern environmentalist organisations concerned.

Here I want to argue that the dangers of north-south environmentalist collaboration go beyond tactical questions. As I have shown, the economic and cultural basis of environmentalist opposition to modernisation is very different in the north and in the south. Hence co-operation on a particular issue or campaign may be a matter of a one-off coincidence of objectives and there can be no assumption of the general commonalty of values and interests that would be necessary for a global movement to maintain coherence over a long period. In fact, collaboration between one set of northern and southern NGOs, based on, say, the resource management perspective, may only serve to show up sharply their differences with another set of NGOs, northern or southern, who may be deep ecologists or anti-development. Several examples illustrate this argument.

First, the CAMPFIRE projects, mentioned above, are promoted by a group of six institutions including two Zimbabwean government departments, a University department, two Zimbabwean NGOs including the CAMPFIRE Association itself and the World Wide Fund for Nature (WWF). However, the model of wildlife and other resource management which they put forward is at odds with that espoused by leading US conservationist NGOs and encapsulated in the international Convention on International Trade in Endangered Species (CITES), which now bans trade in elephant ivory and rhino horn. Members of the CAMPFIRE Association have attended CITES meetings to present their experience of community utilisation of wildlife resources and the idea that such utilisation could be extended to community participation in elephant management based on communities benefiting from regulated trade in ivory. However, certain northern NGOs which lobby to keep the complete ban on trading in ivory, notably the Environmental Investigation Agency (EIA), have become very hostile to CAMPFIRE and the NGOs which promote the CAMPFIRE projects.

Second, JFMP, also mentioned above, in the state of Karnataka, is opposed by Pandurang Hegde, an Indian activist with the *Appiko Forest Movement*, inspired by the famous *Chipko* forest-protection movement of the lower Himalayas. Hegde, who can perhaps be seen as an anti-developmentalist, believes that the participation offered under JFMP is a sham and that people should be left to manage local forests themselves with no idea of increasing cash benefits from forest products that they could have part of. He finds it easy to get support from northern conservationist NGOs, while other local NGOs attempt to work within the JFMP framework, along with northern development NGOs such as Oxfam and aid agencies like Britain's Department for International Development (the Labour Government's expansion of the former Overseas Development Administration).

A third example is in the NGO campaign against the Southern Okavango Integrated Water Development Project (SOIWDP) in north-western Botswana. The Okavango Delta is one of the world's greatest wetlands, world-famous for its wildlife, of increasing importance for tourism, and the provider of livelihoods for over 100,000 people of a variety of ethnic groups. The SOIWDP was to be a multi-sectoral project providing water for commercial irrigation, village-based flood recession agriculture, and communities and livestock throughout the area, but particularly for the town of Maun and surrounding areas, and for the De Beers diamond mine at Orapa.

When earth-moving equipment began to arrive in Maun in late 1990, a new local community-based environmental action group was formed, the Tshomorelo Okavango Conservation Trust (TOCT), which mounted a very effective local campaign. In parallel, local and southern African environmentalists allied to TOCT contacted Greenpeace International and other international NGOs. Greenpeace in turn wrote to the Botswana government and on its invitation undertook a study tour and recommended shelving the project pending further scientific enquiry. The same environmentalist activists appear to have used the media in southern Africa and Botswana to spread the rumours that Greenpeace might add a 'Diamonds are for Death' campaign to its anti-furs work or lobby within the European Community for a reduction in Botswana's beef quota. Although there is no evidence that Greenpeace made any specific threats, Botswana government spokespersons to this day seem happy to give credence to this rumour and to paint Greenpeace as an unrepresentative, interfering, northern organisation.

Whether as a result of the campaigns by TOCT and Greenpeace or simply in response to general local opposition, the Botswana Government suspended the SOIWDP and commissioned an independent study through the World Conservation Union (WCN), and when its report was negative the project was 'terminated'.

One interpretation of these events is that the anti-SOIWDP campaign was led by expatriates in Maun with interest specifically in wildlife conservation and connections to safari companies. In this view the local NGOs such as TOCT that became involved were not really representative of local opinion and the international NGOs such as Greenpeace took up the issue without really understanding the local situation. Nevertheless, the threat from international lobbying was sufficient to make the Botswana government drop the project, at least for the time being. In this interpretation, confrontation between NGOs and government was more significant than scientific consideration of the issue. Even where NGOs might have apparently been collaborating with the government, as with the WCN assessment of the SOIWDP, the result was pointed disagreement, with the leader of the WCN team accusing the Botswana government of ignoring their alternative recommendations in order to keep open the option of putting forward a similar project at a future date.

The way that additional political space can be created by moving to an international level can be seen from both the campaign for water from the Zambezi and the campaign against the SOIWDP. In the latter case, the issue may appear at a national level to be a rather small local debate about the best way of utilising resources in a region. It becomes internationally a debate about the conservation of an ecosystem of global significance.

However, Eccleston (1996b) also notes how internationalising an issue can galvanise the policy-making system (in this case the various institutions and forces in favour of large-scale development of the water resources of the Okavango) into becoming more effective proponents of their cause. There are certainly signs in Botswana of government spokesmen talking up the impact of the suggested threat from Greenpeace International in terms of northern conservationist interests wanting to prevent development. To the extent that local conservationists (especially expatriates and other whites) can be portrayed as in league with Greenpeace, then their alliance with local community interests in the area around the Okavango may be shaken. A new proposal, such as that by Namibia to extract water from the Okavango upstream of the delta, may not necessarily meet with such united opposition.

A NEW CONCEPTION OF PROGRESS?

Having reached this point in the chapter I am faced with the problem of how to conclude. There seem to be two choices. I could follow what might seem to be the logic of the previous sections and conclude that the global environmental movement is not and cannot be a single movement because of the differences in interests and hence cultural basis of the environmentalist oppositions to modernisation in different parts of the world. Or, if I want a more positive conclusion, I could try to make some suggestions for how conflicts and splits between parts of the movement, northern and southern, might be avoided, mended or transcended.

The second approach might seem to call for a proposal of a new environmentalist paradigm which could unite the disparate groups and communities who oppose aspects of modernisation. One candidate would be the fifth of the paradigms in Colby's framework, viz.: 'eco-development', which stands somewhere between resource management and the deep ecology movement. Unfortunately this paradigm is not very well developed and according to Colby '[the] magnitude of changes require[s] new consciousness'. This leaves one wondering where the underlying values of that new consciousness will come from and how they could become attractive enough to unite those so different as the anti-road activists alienated from mainstream life in a northern country like the UK, professional conservationist staff in large NGOs, northern and southern, and forest dwellers and other rural inhabitants in various parts of the world whose access to resources and way of life is threatened by commercial or other large-scale developments.

If this chapter has led some readers to appreciate better the differences between such groups in terms of the different basis of their 'environmentalism', that is in itself a good enough outcome and I need not worry too much about this concluding section. So I can add some tentative thoughts in the hope that they will contribute to the debate and leave it at that.

My main thought in this context is that, since the various varieties of environmentalists are all opposed to modernisation, and since modernisation is a particular normative view of development and thus a particular idea of what constitutes human progress, then the basis of a new consciousness which would unite them would have to be a new conception of progress. In other words, there is a need for positive thinking on how individuals, and humanity in general, can move forward, not just dissatisfaction with or opposition to the currently dominant paradigm of development. Such a new conception of progress would have to 'speak to' the conditions of all the various environmentalists and those they represent with their different relationships to the environment in different parts of the world. It could not be seen as an imposition of one set of values over another.

Thus with reference, say, to the local inhabitants of the Okavango Delta region, a type of progress may be sought which builds on existing uses of resources and the local knowledge behind these uses. This would contrast both with the notion that human communities, like wildlife, should be conserved in their present state, which is completely anti-development, and with having modern development projects forced on the area from outside in which some might take part and from which material benefits might ensue, but from which others would be likely to be excluded.

Again, with reference to the supporters of a large northern NGO like Greenpeace, or to the staff and activists of such NGOs, there may be a need to move beyond a critique of modern industrialisation and its attendant projects like the SOIWDP, to promoting new positive values that might provide some common ground between these supporters and, say, the local inhabitants of the Okavango.

What would be the values underlying such a new conception of progress? They would have to be capable of competing, in cultural terms, with the consumerist and materialist aspirations set up by the ideal of modernity. The new conception would have to offer some kind of mechanism relating individuals' striving for progress for themselves and their families and communities with progress for humanity in general, in at least as powerful a way as the market mechanism does in respect of modernisation as a development process. It would have to address the problem of how an individual can feel they are making a difference, to their own situation and more generally.

Values such as diversity, tolerance, solidarity, participation, dispersed public action and so on can be put forward. How to build a powerful mechanism for progress on such values is extremely unclear, however. Perhaps increasing awareness of the importance of these questions is at least a step in the right direction.

NOTES

1. Of course the recent economic 'changes' in Malaysia have created serious concerns about the stability of its economic growth and the possibility of reaching its 2020 target. It is likely that in the attempt to recapture elements of economic prosperity the environment of the country will be placed under increasing threat.
2. Some commentators saw the Gulf War as a resource war in which the USA acted as global policeman defending the north's access to the world's resources and as directly in line with colonial police defending the colonialists' access to the resources of 'their' colony (see, for example, Middleton, O'Keefe and Moyo 1993: 8).
3. It is ironic to note, as a comparison, that, in the north, national parks such as Snowdonia in the UK often

depend on the maintenance of certain 'traditional' forms of livelihood in order to maintain their now-expected 'natural', 'wild' appearance – and may find it difficult to keep those 'traditional' practices going since they are often uneconomic in a modern society.

4 More recently at the global meeting, agreements were reached only at the eleventh hour as the countries of the south tried to safeguard some commitment from the north for greater help to reach international targets on things like CO_2 emissions.

5 Such incidents are, if anything, more common in southern countries, with well-known examples like the 1984 gas disaster at Union Carbide's Bhopal plant being merely the tip of the iceberg.

5

LOCAL KNOWLEDGES AND CHANGING TECHNOLOGIES

Gordon Wilson

> Technologies are seen to be embedded in, and to carry, social values, institutional forms and culture.
>
> (Anderson 1985: 57)

> Socio-economic progress means emancipation from the whims of mother nature and also from the domination of the already established. And, it seems that only human-made technology resources can provide the foundation for such progress ... Our (Asian) culture prizes respect for the elders/learned and upright morality. Now, we should add technology orientation, self-discipline, and hard work to this culture, and get on with the challenge of managing technological change for improving the quality of life of our people.
>
> (Sharif 1994: 114)

Technology is not usually thought of as a cultural product. Music, poetry, dance, drama and film, yes, but technology? Nevertheless, as the first quotation above suggests, like these conventional forms, technology influences and is influenced by cultural norms and values.

That this is so becomes obvious after a few moments' reflection. Technology is the application of knowledge to practical, 'problem-solving' tasks that involve organisations of people and machines (or other hardware). Moreover, it requires people in many different ways.

Take the Green revolution in India, which enabled the country to move towards self-sufficiency in grain, and which has been hailed by many as a technological achievement. Yet it needed a huge and co-ordinated effort on the part of politicians, scientists, technical extension workers and local farmers to devise and implement it. Local farmers were particularly important in adapting the new high-yielding seeds from the research stations to local conditions by multiplying the seeds themselves and selecting those plants that did well (Pacey 1990: 192–3).

It is useful to think of the people-hardware interaction as a technological system. People also belong to other systems: political systems, economic systems and cultural systems of shared meanings. The commonalty of people to these systems means that they interact dynamically and it should not surprise anybody that technology carries with it culture, and often a global hegemonic culture which, as Sharif (1994) puts it in the quotation above, includes among its shared meanings, self-discipline, hard work and 'emancipation from the whims of mother nature'!

Technology is also about the application of knowledge and, given that knowledge is experienced through our shared meanings about the world, this represents another way of establishing the link between technology and culture. For, what may be classified as knowledge by me, sitting in front of a word processor in an urban English environment, may be inadequate for people living in, say, a Ugandan village. This has important implications when we consider technology.

In other words, knowledge itself is a

cultural construction and what one culture considers valuable, another may dismiss. Consider the villagers of Laropi, Uganda. They are engaged in a daily struggle *against* affliction and *for* health (Allen 1992b: 217–48). They do take on board, by and large, Western-promoted[1] and Western-supplied (often as international aid) biomedical technology such as drugs when confronted by disease and, in doing so, they take on board 'knowledge' embodied in the technology, which is underpinned by beliefs about the ability of science to understand and transform nature. This pretty well approximates to my own beliefs about knowledge. Also, being a liberal-minded person, I can further accept that there is much local knowledge embodied in the use of local plants and herbs for healing, upon which the villagers again rely. This is experiential knowledge of what works in a practical sense. Although, unlike biomedical knowledge, it does not derive from generalised principles of science, it is essentially technical in nature.

In other words, I can easily accept as rational any knowledge-base that states there are physical causes of disease and physical remedies. But the villagers of Laropi go further, invoking spiritual causes, which include ancestral invocation, wild spirits and sorcery. Behind these spiritual beliefs lies what they believe to be the ultimate causes of affliction, which are anti-social or amoral behaviour by a person or people. Thus, ancestral spirits may be invoked because of problems with family relationships; maltreatment can result in seizure by wild spirits from the bush; and sorcery is a manifestation of general amoral behaviour.

Allen tells the story of a boy in Laropi who fell ill and was medically diagnosed as having sleeping sickness. But the grandmother of the child was convinced that the real problem was that the dead ancestors of the family had not been 'fed' at the ancestral shrine since the family returned from Sudan. Others suspected that sorcery was at work, as many people were jealous of the boy's father, who held an influential administrative position. Pressure was placed on him to consult various kinds of local healers and to use herbal remedies. He was also persuaded to arrange the sacrifice of a sheep at the ancestral shrine to placate his aged mother. Since he did not have a sheep, a chicken was killed, and a sheep or goat promised to the dead ancestors at a later date. Eventually the child was taken to hospital and his condition subsequently improved. No one doubted that the drugs administered there helped cure him, but many nevertheless believed that this alone would not have worked (Allen 1992b: 231–2).

From the perspective of the people of Laropi, therefore, spiritual beliefs and technical knowledge about physical causes and remedies are part of the same continuum for making sense of, and coming to terms with, the world. They both influence action and thus both are knowledge. That makes me think hard about my own beliefs, which derive from my culture – in other words, just how culture-bound my own knowledge is.

Knowledge is embodied in technology: in the decisions about what constitutes the problems that technology might solve; in the choice of, design of, and actual hardware that is developed; and in the eventual use of that hardware. And because knowledge is experienced through our culture, this too becomes embodied in technology, which explains the acceptability or otherwise of technological change in a society. It also explains the dominance of some technologies which reflect global cultural hegemony, another feature of which is to describe developing countries as 'technologically laggard' (Deniozos 1994) or as characterised by 'backwardness ... a reflection of their technological poverty' (Sharif 1994).

These points are illustrated when we consider the gender consequences that arise from the adoption of new technologies. A study for the United Nations on the impact of scientific and technological progress on employment and work conditions in various trades reported that:

> In every case where machinery was introduced in activities traditionally done by women, men either completely replaced women or the activity became sub-divided and men took over the tasks that used the technology and required greater skill while women were relegated to less skilled, menial tasks. These shifts were accompanied by loss

> of income-earning opportunities or marginalisation and lower income for women ...
>
> In Java, when rice mills were introduced, women who had earned their only monetary income from hand milling were displaced as men assumed the positions in the factories. In Korea, when the government installed rice mills, men in the mills did jobs previously done by women.
>
> (Anderson 1985: 61)

Underlying these processes are assumed beliefs that men work with technology and women do not, or that men do 'modern' work while women only work in the subsistence sectors. But the transferred technologies are also 'blind' to values of the recipient societies, values which may, for example, determine whether or not women may work outside of their homes.

Anderson also reports that these consequences of technological transfer are not necessarily negative, precisely because they do disrupt cultural norms:

> Technologies which gather women in certain areas, such as grain mills ... can facilitate social activity and opportunities for education. For example, women in Asia have received literacy training while they wait for their rice to be ground at mills and women in Africa have received nutrition training while waiting in line at clinics.
>
> (Anderson 1985: 66)

Clearly, local knowledge, experienced through culture, has important implications for technological change. The rest of this chapter discusses these implications.

SOCIAL ORGANISATION, KNOWLEDGE AND LIVELIHOODS

The application of knowledge, whether it be to the conception, design, manufacture or use of machinery, requires organisation, itself influenced by social and moral beliefs. Allen (1992b: 246) makes the point that many Third World societies maintain values which contrast with the individualism of Western society. Local farming practices in many parts of Africa, for example, depend on communal (free) labour at key points in the cycle. Changing livelihood patterns in rural areas, which usually involve commoditisation of labour, are leading to individualisation, however. If communal labour becomes unavailable because of these changes, the most likely result is the degradation of the farming practice. Sowing and weeding may not get done at the right time, or clearing fallows may not be undertaken (Woodhouse 1992a: 182–6).

The point here is that structural processes taking place in society lead to changing shared meanings and loss of cohesion among old forms of social organisation upon which local technological, in this case agricultural, practices depend. These old forms are often based on household or wider kinship ties, themselves changing. Yet, in their place, new forms do emerge, based no longer perhaps on kin, but on mutual social and economic interests, and pursuing specific goals. Woodhouse refers, in this context, to the *foyers* in Senegal and women's self-help groups in Machakos District, Kenya. The *foyers* are youth organisations which transformed themselves from cultural groups into lobbying agencies to gain independent access to irrigated farming for young people. The women's self-help groups have been instrumental in reasserting women's property rights via, for example, savings clubs. These groups have become well-known for their water conservation terraces, hand-dug voluntarily on steeply-sloping farmland (Woodhouse 1992a: 191).

People who set up small-scale, informal-sector, manufacturing enterprises frequently rely on older social and family networks for intra-firm organisation and for inter-firm and firm–customer trading. These networks prove crucial for initial survival and coping in that they form a basis for *trust* beyond, and often instead of, material, impersonal contracts between owner–worker and between traders. Indeed, studies of clusters of small-scale enterprises in urban areas stress their potential for 'collective efficiency', one of the essential ingredients being the ease of trust that is critical to their functioning in terms of sharing information and tools, providing each other with services and goods and other inter-firm linkages:

> The general argument is that an effective market economy requires a strong code of behaviour, including social sanctions (laws

> are not enough). Such regulatory mechanisms function particularly well in communities with a common identity, be it based on kin, caste, ethnicity, religion, professional affiliation or an experience of collective suffering.
> (Schmitz 1995: 559)

A study of silk-reeling enterprises in five villages of southern India found that an unusually high number of Scheduled Caste[2] ex-labourers had been able to set up in business themselves and become successful entrepreneurs. The study cites the example of the man who never attended school and started his reeling career at the age of seven. In 1986 he set up his own business. Most of his labourers were relatives or fellow reelers who worked for him because of friendship and kinship. All were proud because they had become independent of the Upper Caste reeler for whom they had previously worked (Mayoux 1993).

Networks based on *ascribed* trust (i.e. those based on kin, caste, ethnicity or religion) can, however, act as impediments to further economic growth of the enterprise because of the mutual obligations involved. A manufacturing enterprise may find it difficult to acquire and use new machinery that will improve productivity because it may displace workers. This will be particularly difficult if there are kinship ties between owners and workers, but the issue of whether work is primarily about productivity or people and their well-being tends to surface wherever there are co-operative forms of work organisation, as illustrated in Box 1.

The cultural/family networks, so essential for initial trading of an enterprise, eventually limit it in the sense that they can be exclusive, thus making trading outside of the network difficult (Holmström 1994). One study of garment manufacturing districts in Nairobi found divisions based on ethnicity (Asians predominating in mass-producing firms; Africans in mini-manufacturing), kinship and gender. African women, many of them rural migrants, owned nearly all the mini-manufacturing firms and the study noted:

> The typical mini-manufacturer is using her networks well. The problem is that most of these entrepreneurs are poorly educated African women whose networks have limited power to uplift a business. Some lead back to rural markets where competition with second-hand clothes keeps profit margins extremely low.
> (McCormick 1994: 27)

As a result of these limiting factors, more impersonal, professional networks may appear, although they are often still based on socio-cultural ties. Thus in a study of industrial development in Bangalore, India, it was suggested that bases for trust that go beyond 'kinship and 'birth community' are:

> links between business associates; colleagues who worked together in big firms, or in public agencies, before starting their own business; or members of social-and-philanthropic clubs like Lions. These people are open to new ideas and anxious to hear about new methods (e.g. Japanese management).
> (Holmström 1994)

Schmitz goes further, however, in a study of shoe manufacturing in the Sinos Valley (south Brazil), where he charts a changing foundation from trustworthiness being ascribed to being *earned*. Thus, economic necessity, rather than socio-cultural ties, becomes the basis of trust and Schmitz concludes:

> It seems that operating in the world market has both eroded and created trust. It has undermined the socio-cultural ties and now demands new ties – from those who want to succeed in it. Those new ties are based on conscious investment in inter-firm relationships. The business partners do not necessarily have to change – but the foundation of trust does. Could it be that making this transition is what distinguishes growing from declining clusters? Only further research will tell.
> (Schmitz 1995: 559)

Schmitz writes within what is essentially a global modernisation paradigm. If firms wish to succeed in the 'world market' they have to climb on board and, in doing so, jettison old and incorporate new relationships. But, even if they want to, climbing on board is usually not possible for the poor and less powerful who can remain locked into kinship and other ascribed networks that are themselves poor. They are thus unable to progress beyond immediate coping and survival, such as the

BOX 1: THE CHEZIYA BAKERY, MASVINGO, ZIMBABWE

Plate 5.1 Cheziya Bakery, in Masvingo, Zimbabwe. Labour-intensive, Fordist production styles are subtly adapted, resulting in a highly flexible and effective work-force.

Source: Photographer, Gordon Wilson.

The sixteen-member Cheziya Co-operative Bakery was started after a worker buy-out from its private owner in 1983. Although largely unmechanised (only the initial mixing of the ingredients that form the bread dough is done mechanically), production is highly organised on the shop floor in order to produce at least 720 loaves over a four-hour shift. Production is characterised by some division of labour between tasks, but also an over-riding and voluntary flexibility among the workers to help overcome certain bottlenecks, such as the final moulding of the dough and placing it in bread tins to rise. There is also a hierarchical division of labour with shift supervisors and day-to-day decision-making delegated to a production manager. Even the production manager will participate in shop-floor tasks, however, such as unloading the baked loaves from their tins onto trolleys, sometimes enlisting the help of customers who have called in.

In 1991 a bread-moulding machine was bought but has rarely been used. It mechanises an operation that involves all the workers on the shift and, indeed, when it was used, one or two of the workers would be sitting around doing nothing.

The bakery decided not to use it, therefore, so that everyone could be gainfully employed. The only exception to this decision would be when business is exceptionally brisk and its use becomes essential. Such conditions have been extremely rare since its purchase.

The point is that on purely economic grounds the bread-moulding machine would be used all the time with under-employed workers laid off, as not only does it increase the productivity of the enterprise, it represents capital investment not being utilised. In fact, in this instance, the bakery has made a welfare rather than a technical decision, based on the co-operative's aim of providing for its members.

(Wilson 1995b, 1996)

woman involved in garment mini-manufacturing in Nairobi, 'with only a primary education whose chief linkages are to a peasant-farming family in a distant rural area' (McCormick 1994:10).

TECHNOLOGY, KNOWLEDGE AND THE POOR

Knowledge is critical for technology, but its cultural construction raises questions like:

> What value is placed on different forms of knowledge (and by whom)?
> Who has access to, or power to exploit, knowledge?

For poor people, the knowledge that they do possess is often undervalued, and they neither have access to, nor the power to exploit, valuable new knowledge. This is particularly true of the knowledge possessed by poor women. Introducing a book of case studies on women and technical innovation, Appleton argues:

> Compared to men, women in all the studies were disadvantaged in the amount of control that they had over their lives: they had less time available and less disposable income, had received less education, participated less in political decision-making, were less able to move about, interact with other technology users and obtain information. All these factors helped to constrain women's access to technological information and to new technologies. They also contributed to a lower self-image for women, which meant that neither women nor men regarded women as producers and users of technology or as having valuable technical skills and knowledge.
>
> (Appleton 1995: 8)

Appleton also identifies one of the factors that contribute to the relative invisibility of women's technical knowledge as 'the cultural perception of what is "technical" and what is "domestic"'. Thus, women may carry out a large number of technically-related activities, often within a single day – farming, processing crops and food, weaving, sewing, collecting wood and water, cooking, fishing, tending animals, and caring for the sick and children. They tend to be classed as domestic, rather than as technical, activities, however, and domestic work is rarely regarded by outsiders, or by women themselves, as requiring technical skills and knowledge (Appleton 1995: 7).

Nevertheless, poor people do react creatively, using the knowledge that they possess, to changing opportunities and threats to their livelihoods. Three case studies of farming in sub-Saharan Africa showed how it has changed as economic options for rural people have changed, and in particular how small-scale farmers have responded to market opportunities:

> They have adopted new cash crops (cocoa, groundnuts, maize, vegetables) and new technologies (ploughs, irrigation, hybrid maize and fertiliser), migrated to new areas and even switched to new foods (maize, cassava and rice) in pursuit of these opportunities.
>
> (Woodhouse 1992a: 181)

Woodhouse makes the further point that what may appear 'traditional' in farming may be the product of conscious 'preservation', such as the use by colonial states of customary land tenure to suppress the emergence of land markets, or of deterioration, such as the development of subsistence production as a result of withdrawal of farm labour or collapse of agricultural markets.

For poor people generally, however, and particularly for poor women, creativity is likely to be tempered by risk considerations. As a consequence, technical innovation is often incremental and may be rejected. Appleton makes this point strongly with respect to women:

> Women's perceptions, use and adaptation of technology are shaped by their evaluation of risk. Decisions to reject particular technologies are often dismissed as signs of women being 'resistant to change' or 'conservative', but such decisions are based on women's knowledge of their environments, available resources, priorities and the *risks that they can afford to take*. Women have a wider portfolio of responsibilities and activities, including care of family and children, and they balance risks and priorities for technical innovation across all those activities.
>
> (Appleton 1995: 10; emphasis in original)

Thus, what can seem to outsiders to be 'irrational' or 'culturally determined', such as 'resistance to change' and 'conservatism', make for sound sense once the material environments in which poor people operate and the interplay of these environments with their shared meanings of daily life are fully understood.

Appleton also draws the distinction between technical changes that are 'defensive' responses to deteriorating livelihood situations (and which are likely to be risk-averse) and those that are 'more proactive – creating or taking advantages of opportunities to improve livelihoods' (Appleton 1995: 301). But, whatever the response, people use and adapt their prior knowledge and assimilate new knowledge, the latter then becoming part of their body of knowledge. Thus, the distinction frequently made between 'indigenous' and 'imported' knowledge is blurred. Local knowledge accumulates over time as a result of interaction with one's close environment and with the rest of the world. The interaction may, at certain times, be greater than others, such as times of colonisation or of the interjection of a development project.

There is, however, a distinction to be made, which is especially important in the context of technology transfer, between 'prior' and 'new' knowledge. It has been argued that for new knowledge to be assimilated easily and usefully it should build upon, rather than reject, prior knowledge; in other words, build upon what is known, rather than what is not known (Wilson 1995a, 1995b, 1996). *Technology blending* and *lynchpin projects* are examples of intervention project-types that consciously start from this premise. Technology blending has been defined as 'the constructive integration of emerging technologies with traditional, low-income sectors, or for the production of basic-needs goods and services regardless of scale of operation' (Bhalla and James 1991: 479). Lynchpin projects seek to mechanise 'traditional' methods without abandoning their fundamental principles, and an example would be the motorisation of 'traditional' agricultural implements (instead of importing tractors) (Ogbini 1990).

The danger in this discussion lies in assuming that knowledge (and the shared meanings that go with it) flows only one way – from the 'knowledge-rich' in the West to the largely 'ignorant' poor people in the Third World. But this assumption is changing, especially in Third World agriculture, where:

> [Since the middle 1980s] farmers have increasingly been recognised as themselves innovators and experimenters ... and perhaps most decisively, farmers have again and again been found to be rational and right in behaviour which at first seemed irrational and wrong to outside professional observers.
>
> (Chambers, Pacey and Thrupp 1989: xix)

The most well-known example of knowledge-reversal – i.e. it flows from poor to powerful – is the exploitation of the biodiversity of the rainforests. This exploitation is critically dependent on local knowledge, where the debate is now focusing on how it is valued by the transnational corporations who wish to exploit it (Pearce 1992). Another example concerns the nkejji fish of Uganda which is widely known among women in the Lake Victoria area as a preventative medicine against measles and for its high protein content. For years this knowledge was ignored by scientists and policy makers, to the extent that the Ugandan Government, acting on the advice of the United Nations Food and Agriculture Organisation, introduced Nile perch into Lake Victoria which almost drove the nkejji to extinction (Wekiya 1991). Belatedly, professional researchers have now confirmed the local women's knowledge of the importance of the fish, but:

> Who owns the knowledge? ... The women who knew that it could cure but didn't know why, or the researcher who heard from the women, did the research and discovered the reason why? The researcher, through the Ministry of Health nutrition clinics, is saying 'Feed your babies with nkejji, nkejji heals kwashiorkor and marasmus', so he has taken the credit and the women's knowledge is not recognised.
>
> (Wekiya 1991: 4–5)

Poor people engaging in new enterprises usually have little but their previous knowledge upon which to draw. They typically enter businesses that require little or no capital, and which require no new knowledge to start (although a willingness to learn is crucial if their businesses are to survive). Petty trading is perhaps the extreme example of this. One danger, however, of relying almost exclusively on prior knowledge at entry is that enterprises may be formed on the basis of what people can do, rather than whether or not there is any demand for their products. It also leads to widespread copying and to gendered enterprises. In the communal areas of Zimbabwe, for example, it has been reported that women predominate in beer brewing, basket making, tailoring, knitting, crochet and pottery; men in grain milling, leather tanning and products, brick making, housing construction, carpentry, tinsmithing and metal products (Helmsing 1991: 262–3). Thus, in Zimbabwe, women line the roadside near tourist attractions with an extensive array of crochet blankets and clay pots. Men may be found with wooden chairs and metal products outside the richer shopping centres in Harare. In both situations, production driven by what people can do (rather than by what they can sell) and widespread copying of one another's products leads to a mismatch between supply and demand. Very little is sold.

The introduction of industrial, Fordist production methods does require new knowledge, the technical aspects of which are relatively easy to acquire because of their repetitive nature, each worker typically performing a single task on a machine within

Plate 5.2 Metal-working in Mbare suburb of Harare, Zimbabwe. The 'informal sector' workshops of Mbara cover a large area with significant clusters of metal-working, wood-working and other trades – learning and copying from, sharing and competing with, one another.

Source: Photographer, Gordon Wilson.

a strict division of labour. The gender implications of adopting industrial techniques have been discussed earlier, but even in terms solely of providing technical additions to knowledge, such techniques are limited. There are few opportunities to develop and acquire skills that go beyond the specific piece of equipment on which the worker has been trained (Bhagavan 1990).

Nevertheless, Fordist techniques with their rationalist approach to technology management, remain part of Western cultural hegemony, and their introduction into Third World countries is rarely questioned by major international development players (the World Bank and International Monetary Fund) or by many First and Third World governments. The introduction of these techniques can undoubtedly have negative consequences, but also, what often appears on the surface as a typical Fordist method, has been subtly modified, as in the Cheziya bakery featured in Box 1. Here the scale of production requires industrial production methods, including a broad division of labour. What can best be described as a co-operative spirit pervades, however, so that cheerful flexibility also becomes a key component of the labour process.

Entrepreneurial activity among small-scale, Third World manufacturers often involves copying technologies. There is general consensus, however, that knowledge acquired in this way soon hits a ceiling, and that a crucial point is where an enterprise is able to move from 'learning' by mastering acquired technology to 'understanding' by engaging in its adaptation (Quingrui and Xiaobo 1991), which enables 'knowledge accumulation' (Bhalla 1991: 79–84). Such adaptation aims to improve and develop the technology to its new circumstances, for example, to a smaller scale of production, or to the available raw materials (Dahlman 1989).

Although much of the prior and existing knowledge that such adaptations rely upon may be technical, it also invariably includes a broader knowledge of the socio-cultural milieu in which the enterprise operates, as in the example of the Takura milling company (Box 2).

Technological development in enterprises is ultimately predicated on the Western cultural assumption that improving productive efficiency is the prime function and there is at least notional acceptance of this, even among enterprises in the least-industrialised countries of sub-Saharan Africa. Productive efficiency is the amount of output per unit input, but what if we think of efficiency in completely different terms – such as amount of human fulfilment per unit input? Then the concept becomes problematic, with fulfilment being a balancing act between material and non-material factors. Thus, in sub-Saharan Africa, it is common to find an important social function for enterprises, as well as their obvious role in making a living, in that they are meeting places for people engaging in widespread and varied social interactions. (This debate about the social function of enterprises, has recently resurfaced in Western countries where infor-

BOX 2: THE TAKURA MILLING COMPANY, ZIMBABWE

The managing director of the Takura Milling Company is also a livestock farmer with over 600 pigs and runs a local bus company.

The milling company was started in February 1992 when, because of continuing drought in Zimbabwe, he became constantly worried about how he was going to feed his pigs. He felt he was 'heading for disaster'.

He hit upon the idea of buying a machine that would process maize (by dehulling it), knowing that the by-product from the operation would yield him valuable animal feed. His intention was that local maize farmers would bring their surplus grain (i.e. that designated for personal use and not for the state-owned Grain Marketing Board) to him for processing and he would benefit from the by-product they left behind.

He dipped into his savings and sold some of his pigs to raise money to buy the machinery and, thus, Takura Milling was born. The business expanded rapidly and the emphasis shifted away from the by-product to the actual processing. He obtained a maize allocation from the Grain Marketing Board and, by the end of 1992, production was in full swing with the company employing 84 workers who operated seven small dehulling machines.

He needed seven because the machines are normally manufactured and sold with personal, rather than commercial, use in mind. He therefore had to make some modifications so that the maize falls directly into bags from the machines without any intermediate handling. This improved both productivity and quality without changing the high labour-intensity of the operation. The managing director said he didn't mind this as he wanted to provide jobs for local people, some of whom had never worked for a wage before.

> Shortly before the end of 1992, the managing director progressed to sunflower oil processing, again using a small mill to which he planned modifications.
>
> (Wilson 1993)

mation technologies are opening up possibilities for dispensing with a central workplace and working from home instead.)

The Garabwe tailoring co-operative in Zimbabwe is a case in point. This worker co-operative has twelve women members and grew out of a social and savings club which was formed on national independence in 1980. Its main business is producing school uniforms which it sells at local schools. Teams of three make a complete garment and within a team there is division of labour. Thus, a more-competent member will do the pattern-cutting and there is labour division within the different aspects of tacking and sewing.

Production often borders on the chaotic, however, with unfinished garments being literally tossed from one member of the team to another and over the head of a third. It is further disrupted as tools are invariably lost among the debris on the work tables, some of the older workers are unable to thread needles and this has to be done for them by whoever is to hand, and sewing machines break down. Workers also halt production to go to the nearby shop, prepare lunch (eaten communally and preceded by Grace), tend the vegetable garden outside or have an extended chat with relatives and customers who call or simply with one another. There seems to be no resentment, however, among those who continue working during disruptions or who are technically more competent, and there is an implicit acceptance of the validity of social as well as productive needs at the co-operative.

The members periodically divide equally between them any excess income over expenditure irrespective of who has done what, and nothing is re-invested (either equipment renewal occurs as aid, or it does not occur at all). Although this division of proceeds provides their only cash income, the women who form the co-operative have multiple work roles, including heavy domestic responsibilities and their own fields to farm. The women work long hours, therefore, with the time at the co-operative sandwiched between time spent on their other roles. The co-operative is nevertheless strict about time-keeping. It fines members who arrive late for work (Wilson 1995a, 1995b, 1996).

From the Western rationalist viewpoint, production at Garabwe is a mess. On another level, however, the enterprise clearly fulfils several, interconnected needs, some of which are non-material. Actions are also contradictory as when workers are fined for turning up late but not for talking on the job. In general, however, the women seem to exercise considerable autonomy and power within this area of their working lives, which may be contrasted with the working conditions of their numerous counterparts engaged in factory garment manufacture throughout the world, working conditions which are geared entirely to achieving maximum productivity. For example, in a study of the Mactan Export Processing Zone in the Philippines, where garments and electronics predominate and the workforces are heavily feminised, Chant and McIlwaine write of the 'strict nature of daily work routines and the setting of challenging production quotas to achieve maximum productivity' (Chant and McIlwaine 1995: 163). Unlike Garabwe, there is little possibility of such enterprises entertaining other roles and the authors add: 'In the light of daily drudgery on the assembly line, women's only thoughts may be focused on the receipt of a salary at the end of the week' (1995: 166). Although women engaged in garment manufacturing in export processing zones may have slightly greater power in the home as a result of their independent remuneration (a general observation of women in remunerated employment, which is also echoed by the Garabwe women), they face ongoing subordination in the workplace.

Returning to the Garabwe co-operative, it has to be acknowledged that it has taken on board many of the trappings of Western production methods that we witness in garment factories in export processing zones in Third World countries, but often only notionally, and the physical, social and cultural conditions within the enterprise continually undermine its productive function. What we are witnessing at Garabwe, in fact, is a process of change in shared meanings where the nature and function of work is being redefined around its productive and social functions, which are themselves not fixed in relation to each other.

The women are active agents in the process of change – accepting, rejecting, modifying – but, like all processes, it is messy and sticky. It is in this context that apparent contradictions, like the one mentioned above concerning timekeeping and unproductive practices, must be viewed. Moreover, the process does not lend itself to the simple exhortations of the second of the opening quotations to this chapter. Within their context, the Garabwe women do have a 'technology orientation' (they work with machines and tools throughout their lives), they do have 'self-discipline' (they wouldn't survive if they didn't) and try suggesting to them that their lives are not 'hard work'!

The second of the opening quotations puts these exhortations in the context of 'managing technological change for improving the quality of life of our people'. It combines a crude *cultural determinism* with a *technical fix* view of development, where technology is assumed to be culturally neutral and can be made to do whatever we want it to do – as long as we buckle down. The view of the first of the opening quotations to this chapter (and that of the chapter author), however, is that technology, along with its twin supports, knowledge and organisation, is culturally constructed. Decisions on the use to which technology is put depend on social and cultural power, therefore, and its functioning depends, at least in part, on its embodied cultural assumptions. This means that, for some, 'quality of life' may be improved by technological change, but for others it will worsen.

There is a further consideration, however. Implicit in the view of the second opening quotation is that 'quality of life' is exclusively about economic well-being. Sharif indeed makes this explicit at the very start of his article: 'This paper begins with the well-accepted premise that technological progress provides the foundation for economic prosperity' (1994: 103). Quality of life is undoubtedly connected with economic well-being, but it has other dimensions, such as independence, self-respect and dignity as human beings. It is these other dimensions that the Garabwe women are striving to maintain at the same time as trying to improve their economic well-being. This creates tensions, as undoubtedly it also does for those engaged in the Cheziya bakery, the Takura milling company, the petty traders, the small enterprise clusters and the farmers far too briefly described in this chapter.

If we can generalise at all from the Garabwe experience, it is not a matter of *adding* technology orientation, self-discipline and hard work to what 'culture prizes', but more to do with *understanding* how these attributes exist in the present and are *experienced* through culture. It is through such an understanding that technology can make its valid contribution to development.

CONCLUSION

Hence technology can be regarded as a cultural product, embodying shared meanings about knowledge. This has implications for the acceptance and assimilation of technology into societies. The application of technologies requires organisation and this too depends on changing shared meanings. In agriculture, communal work practices based on extended kinship are breaking down, but new forms based on shared interest are appearing. In urban contexts, small enterprises may be dependent on kinship and friendship networks, but these can inhibit further development of an enterprise. There are cases where local knowledge is sometimes of technological value in a wider arena and there are unresolved issues concerning the ownership of this knowledge.

Contrary to the preconceptions of many 'development experts' poor people react creatively to changing livelihood opportunities. They assimilate new technologies most easily, however, when these build on prior or existing knowledge. The introduction of Fordist production methods rarely does this and these are further limited in that they only provide experience on specific, repetitive tasks. Indeed, they stifle local innovation, so causing frustration, and they may foster local levels of resistance. Small-scale enterprises often notionally accept industrial production methods, but then subtly modify them to reflect shared meanings about the nature and function of work. Productive efficiency may be lost because of the need to accommodate non-productive functions, or it may sometimes be enhanced because the changes introduce co-operative flexibility into labour processes.

NOTES

1 'Western' (and 'Third World'): by 'Western' or the 'West' I mean the economic, political, social and cultural structures that first came to prominence in Western Europe and North America, and which have, over a period of time, come to dominate the world order. The 'West' is contrasted in this chapter with the 'Third World', by which I mean those countries that are on the periphery of the system produced by the world order. These are 'those nations which, during the process of formation of the existing world order, did not become rich or industrialised' (Abdalla 1980: 6). Although there are problems with continuing to use the term, particularly since the demise of the Soviet-dominated 'Second World', 'Third World' still represents for me the possibility of an 'alternative (or at least a strong modification) to capitalist industrialisation' that can launch a concerted attack on poverty (Thomas 1992: 5), and thus has positive connotations as well as the more obvious negative ones.

2 Caste system: hereditary, hierarchically ranked social categories, based on the *dharma* – the sacred order of things – in Hinduism. This system structures the whole of social life (Bujra 1992: 164, 165). Scheduled Caste above refers to a grouping low down in the hierarchy.

6

UNDERSTANDING HEALTH

Murray Last

INTRODUCTION

No one would dismiss out of hand the WHO 1948 ideal (health as 'a state of complete physical, mental and social well-being') or the Alma Ata call in 1978 ('health for all by the year 2000'), let alone the 1993 World Bank Report's advice on extending public health – but everyone knows they simply aren't realisable. Indeed matters are getting worse, if anything. So what is to be done? First, we need to re-state what is actually happening (and not just what should happen) – and then, as an integral part of a new realism, to take into account what the people whose health we worry about are themselves doing and thinking. We may not agree with it. Yet more may be achieved, in terms of their happiness if not always of physical health, by helping people use their own style of public health more effectively – *in addition to* the public health work as it is commonly understood. This is not mere populism, but an acceptance of how modest are both our abilities to fulfil our ideals and our willingness to spend the kind of sums required for them. Nor is this to dismiss the value of rhetoric in psyching up a team (or in getting good policies accepted by the otherwise reluctant), but rather to argue that a hard look at what the other players are really doing is just as necessary.

It is impossible to wipe the slate clean of 'traditional public health' even were anyone to wish to. First, it is etched into society, and is not some superficial, marginal gloss; second, it encodes and structures the way people understand misfortune and inequality, and replaces the nagging fear of risk with a sense of hope and hidden protection. It makes social well-being a possibility. Instead we need to look closely at the nation's medical culture, with all its various healers as well as the Ministry of Health, as a single whole, and recognise how that medical culture works and the contributions made to well-being by each of its constituent parts. If the values and beliefs of a culture emphasise the group over the individual and the critical importance of the individual's place as a member of the group, then the culturally appropriate way is to secure the well-being of that group and to ensure the group's ability to keep up its links with the patient. This will, for example, involve investing in such public items as transport or communications, or providing longer-term security for children and spouse.

HEALTH FOR ANY BY THE YEAR 2000?

While the implications of globalisation may prove materially beneficial for a substantial segment of the world's population, and while material well-being undoubtedly is a factor in health, there are good reasons to believe that well-being overall – the satisfactions people have in being alive – may well not improve world-wide. In this context public health is not just important in promoting physical health but is a valuable pointer to social malaise in a community: for example, a resurgence of an epidemic disease such as measles implies the breakdown of the social controls that usually prevent it – hygiene, nutrition, resistance to infection, immunisation programmes, housing – and the fact that such a breakdown was possible points in turn to fault-lines emerging within the community.

The threat globalisation poses to health has at least two aspects. First, there is the rapid dissemination of new diseases or strains of diseases, and the re-activation of old diseases once thought to be under control, such as tuberculosis. Though biomedicine may eventually discover treatments for the new diseases and resistant strains, the numbers of those who fall ill in the meantime remain a significant public health problem. Viruses always spread faster (and start off earlier) than government controls. Secondly the stresses liable to be generated by incorporation into the global economy – the loss of jobs, reduction in wages and safety standards, migration in search of work and the sense of alienation it can cause – will take its toll on the health of many who are already, through poverty, less resistant to disease. Mental health has long been treated as a Cinderella when budgeting the scarce resources for health; yet the impact of mental illness, if only in terms of productivity, is too serious to ignore.[1] Admittedly mental illness is very hard to quantify, but statistics on health generally tend to underestimate actual experience. They reflect, for example, the more single-focus epidemiological, and not the more open-ended diagnostic, criteria used in labelling and reporting disease. Furthermore, the inevitable reliance on mortality statistics not only is apt to sideline morbidity but also eliminates sub-clinical conditions that are distressful but not necessarily dangerous to life. The 1993 World Bank Report recognised this and sought to correct it somewhat with its more sensitive Disability-Adjusted Life Year, but the Report's purpose was to offer only broad indicators for policy. It cannot be an index of distress, a 'social-strain gauge'.

Globalisation may seem a new, modern phenomenon – but from the perspective of its victims it can look no different from their forefathers' experience of previous empires. Historically the spread of a world religion such as Islam or Christianity and the subsequent imposition of European colonialism offered to local people membership of a (self-proclaimed) global system. It offered a single market complete with rules and a legal system to enforce them; there was a new system of communication (literacy and books, in various languages and scripts), a transport network of previously unparalleled speed, on both land and sea; new technologies of medicine and health, of irrigation, manufacturing and warfare. These empires also saw epidemics on a world scale – the Black Death, for example, in the fourteenth century, cholera in the nineteenth and influenza in the early twentieth centuries. Globalisation spelt death then; it does so now, with AIDS.[2]

Admittedly today's global system is even quicker-moving than earlier global systems and more complete in its coverage, and therefore less easy to escape. But it is seen by some as no more unequivocally good than its predecessors. Although no state today can opt out (as some did in the past), peoples within states have responded to globalisation in different ways: many have, of course, adopted it eagerly, but while some have rejected it outright, others have chosen the third way of negotiating a compromise. They maintain a modified version of their culture while becoming part of the wider system. Still others find themselves unwillingly marginalised, either as nations losing out in the world competition for investment and jobs, or as groups within a nation. The perspectives of all these out-groups are crucially important for public health. Not only must any programme designed to integrate them into the global system take account of their attitudes, but as potential foci of infection they pose a threat to everyone else in the system.

As we approach the year 2000 and move into the next century, the combined effects of recession and globalisation will both increase the numbers of those marginalised and deepen the deprivation of those already marginalised. At the national level, the ability of some governments to provide effective services to the marginalised is decreasing, as is people's willingness to heed government health campaigns.

In many countries there has been a slowing down of health improvements (such as infant mortality) because of repeated economic crises. A further, dire consequence of this failure of government and national bureaucracy is the local wars and civil strife that fuel high civilian casualties, first the war-wounded, then the steady flow of rural people injured by mines and other devices left over from war. The uncontrolled flow of automatic weapons into communities that were previously free of them means that such casualties will become increasingly commonplace.[3]

Marginalisation is not simply a matter of being on the fringes of the global economy. It includes those at the heart of the system too, in the huge conurbations, who live in quarters so densely populated and so fluid, that individuals can make themselves effectively invisible. Paradoxically, globalisation makes it easier to hide, with individuals switching identities and crossing old boundaries and thus defying surveillance – or at least surveillance compatible with the requirements of human rights and democratic politics. Yet the essence of global public health is to be able to measure and monitor trends, to immunise and preempt the spread of a disease, to provide the material context for health – housing, water, drainage, adequate child nutrition – and to do so comprehensively. The very fact that such a high percentage will miss, for whatever reasons, the public health net threatens the effectiveness of global programmes. In this, health is more vulnerable than other aspects of the global economy. The inherent ability of disease to spread makes globalisation essential in matters of health. In turn this requires levels of preventive, global surveillance that would be, if not impracticable, largely unacceptable in that 'preventive surveillance' requires mandatory checking on those that are symptom-free.

The concept of global health, and the acceptance of what is required in order to maintain it, is made initially possible by the almost universal acceptance of biomedicine as a culture-free, almost apolitical technology. The efficacy of antibiotics and analgesics, the power of surgical procedures to repair, say, crash victims under anaesthesia, the ability to transfuse blood – none are culture-dependent, given the right equipment and training. As a consequence, few refuse the possibilities medicine offers; no nations reject biomedicine outright. Politically, the value of lowered mortality rates for children and mothers is not a matter for debate, either at the national level or usually at the local level. People want medical services – the marginal no less than the rich and powerful – and want them to be affordable.

Though biomedicine can be presented as a technology which is largely culture-free, the really effective use of biomedical services requires a distinct style of life, a way of thinking and of doing things; it is related to wealth and education and the social status especially of women, but it is also more than these. Health requires more than the purchase of this drug, or attendance at that clinic. It means separating off health matters from other aspects of one's culture (such as religious beliefs and practices), as well as having the means to do so – accessible services and the money to pay for them. Health, in short, becomes an almost alien department of daily life: unlike aspects of one's own culture for which there are in-built means of sorting out the valuable from the fraudulent, the means of checking on these ultimately alien items are lacking.

In the global economy, though the prices of medical services might fall, so too may the quality of medicine on offer. The relaxation of trade and market restrictions should make it even simpler to dump expired or unsafe medicines upon an unsuspecting community; sub-standard or even fake medicines, with labelling of dosage and side effects in languages unreadable locally, are already a major problem. Similarly, without close scrutiny of certificates licensing practice, the quality of service is in danger. The patient is rarely in a position to check on either drug or doctor. In these circumstances, the quest for quality by patients and their kin becomes an over-riding concern; since high prices are taken to indicate high quality, these are the prices people believe they must pay. Hence the costs of health care rise sharply precisely because people cannot afford to make a mistake. The market in health is notoriously a distorted one: a purchaser cannot normally take the risk of buying the cheapest therapy – there may be no second chance. Thus unless the state (or some other, more effective agency) intervenes to control the medical marketplace, good medicine is soon priced beyond the reach of those who most need it.

Urban communities are no less at risk than rural ones. The very size and complexity of the urban medical market offers opportunities for practitioners and therapeutic systems that the rural communities cannot support. With self-medication an easy option, shop-bought remedies may be used to relieve the pain or stop a child crying, and so delay appropriate treatment; and recession can make such options the only ones available. The apparent prosperity of cities, be they in Asia or Latin America, also conceals stresses that are unsupportable without some outlet, for example in healing churches or possession cults. Therapies in such contexts turn commercial; with profits high relative to entry costs and overheads, they are easy to start up since clients, in the diverse culture of the city, do not necessarily expect to understand in detail the 'system' of healing on offer. Furthermore, given that biomedicine can often only contain but not cure the chronic 'diseases of civilisation' that afflict many who work in today's cities, here are new markets for alternative systems of treatment.

In short, the prospect of health for all by the year 2000 is not necessarily enhanced by the gradual unification of the world's economies. This is not to say that the WHO league tables may not show some marked improvements by aggregating epidemiological statistics nation by nation. But as the 1993 World Bank Report showed by disaggregating those statistics, there is a considerable range as well as inequality in the experience of ill health – and this is only ill health as defined biomedically. If we want to understand the levels of distress world-wide, should we not also seek to define ill health as it is experienced by the sick, rather than by the doctors who treat it (let alone by the epidemiologists who merely count it)? Patients' own definitions of health are indeed relevant here, not least since compliance is needed if good results are to be achieved and costs are to stay low.

HEALTH OR 'WELL BEING'?

It is important to remember that, just as most of us fear falling ill in remote places, so too for the marginalised the milieu where the global economy holds sway can seem a dangerously septic place, with a plethora of infections as well as pressures – and with treatments both alien and expensive. Integrating the marginal, so long as draconian measures in the name of public good are ruled out, requires offering them more to come in than what they can gain by staying out. It is not self-evident that the global economy is the better place for them to be. They may not necessarily measure 'gain' as we insiders do – in terms of wealth or personal freedom. Indeed the gain in health as we might see it can be at the expense of 'happiness' as they see it – and it is by no means certain whether the ultimate goal of most of us would not be happiness rather than merely health. Health, as we tend to define it, is a personal attribute limited to each of us as individuals whereas happiness can be understood as involving (among other things) other people as well as ourselves. A definition of health that included the social (and by extension, implicitly included such aspects of well-being as happiness) is basic to a broader, more widely applicable notion of public health.

The World Health Organisation in its founding mission statement of 1948 seemed to be saying just that: health, it said, 'is a state of complete physical, mental and social well-being and not merely the absence of disease or infirmity'. Such a normative prescription, unattainable as an ideal, is clearly not intended to replace more modest, working definitions of health such as René Dubos' remark that health is 'not the elimination of disease but a *modus vivendi* enabling imperfect men [sic] to achieve a rewarding and not too painful existence while they cope with an imperfect world' (1968: 67). Here his emphasis is not on perfect health – which otherwise puts old age, for example, or impairment automatically into the category of ill health, and makes the maintenance of youthfulness synonymous with proper health-seeking behaviour – but rather on the quality of life. But it is a quality of life that varies with circumstances. Similarly sensitive to context is the biomedical

BOX 1: IMAGES OF HEALTH

One of the most common images of health that people have is that we are like containers filled with the liquid of life: health is when we are full – of energy as well as happiness, well nourished and sleek; ill health is being less than full, drained by tiredness, sorrow, anxiety, looking ragged and emaciated. The work of a society's culture no less than its medicine is keeping people 'full' by preventing or stopping uncontrolled losses. The liquids of life are not for storage only, but for daily use; the controlled using up and the refilling of life's liquid is the routine of daily life. But uncontrolled loss, rapid or over time, can come about in many ways – stresses of work or an accident or more mystically through sorcery or a broken taboo, or from simply inappropriate emissions of blood or semen. Such liquids can lose their potency, or go bad or become thin and watery, and thus require tonics to impart added strength to them. Finally, old age is the time when the body gradually dries up; the normal emissions of blood and semen cease, the skin hangs loose and dry as leather. Tiredness takes over; the body is all but drained of its vital fluids.

The important point to stress here is that health is not a matter of being either well or ill, but rather a process or gradient on which the negative pole is death, whereas the positive pole is a state akin to the WHO definition. Most people live most of their lives somewhere in the middle. The point on the scale at which people fall sick varies culturally and personally: some refuse to go off sick until they are close to death, whilst others fall ill at the slightest shift away from what they perceive as well-being. Paradoxically, then, rates of illness can increase in step with the increasing availability of medicine.

definition of disease as a departure from a statistically set norm and of health as the body in homeostasis. 'Coping', 'adapting' become the key words within a concept of health that recognises there may be other more desirable goals than maximising personal fitness.

Just as chronic physical impairment can, depending on the way it is handled in the community, develop into a disability and a handicap, so too with illness. The extent to which an illness makes a scapegoat of its victim (or its supposed transmitter) depends on the cultural matrix and the non-medical interventions of the patient's kin and community. How people handle illness socially affects both the course of the illness and the welfare of the patient. Stigma is a notorious example of the way social barriers are imposed on the ill – whether it is leprosy in the past or today's HIV and AIDS. Indeed so severe can that stigma be that some analysts identify as 'sick societies' those in which culture so exacerbates the effects of illness that mortality is significantly higher than it might otherwise have been. Cultures are no more automatically good than communities are automatically supportive or united: habits, like feuds, die hard. Nonetheless there are practices and skills that increase the possibility of surviving an illness, and others that create the environment in which the distress of the victim (and the family) is minimised. It is these cultural practices that can determine – as much as medicine perhaps – a person's well-being, whatever the physical condition. In the absence of effective medicines (whether through poverty, temporary recession or the breakdown of government), these cultural practices and skills are the only remedy a community has.

Health policy, then, needs to support these cultural resources not just as an emergency back-up when medical resources fail, but also as an essential adjunct for maintaining the well-being of a person or a group in the face of illness. Local communities vary no less than societies in the specificity of their medical culture, including as it does immediate political and economic considerations. A prime example is the wide choice of religious groups, ranging from cults of affliction and diviners to charismatic churches and national associations of healers. The costs – as they appear to the sceptical outsider – are in subscribing to beliefs that are unprovable if not false, with subordination to a healer and bills for expensive therapies. The point for government, however, is not to back one or other (which would be impossibly invidious) but to ensure basic standards are maintained for the protection of the 'consumer', in the same manner as it regulates the medical market.

SOCIAL HEALTH AS 'TRADITIONAL' PUBLIC HEALTH

So what are the principles of 'traditional' or customary public health? In spite of the specificity of local cultures, certain common themes do emerge. The intention here is not to contrast some synthetic, notional 'traditional medicine' with biomedicine – health is a field containing a wide, overlapping range of definitions, theories and practices. But at the outset we must recognise that public health is not the invention of biomedicine; it is the primary outlook of most local medical systems. The community's health, 'the body politic', is the central concern, with the immediate illness of an individual being taken as a symptom of trouble in the community's well-being. In part this emphasis on the community or group simply reflects the obvious: [a] that an individual's illness affects the group, economically and often socially; [b] that the dominant voice is that of the other (well) members of the group and not that of the sick, not least because it is the well who have to care, and arrange the cure, for the sick; [c] that where there are no institutions on which to off-load the sick and few means of successfully living alone, the group, whether family or neighbours, has to take charge. This is not to argue that the group necessarily wants to, or that there is always such social harmony and cohesiveness that the group rushes in to help. There is simply no alternative. But long-term benefits are assumed: what you do for someone today will perhaps be done for you tomorrow.

Clearly in some contexts this primary, group-centred perspective is being undermined: huge conurbations are one obvious site where group cohesion is stretched, but so too are communities in crisis, whether that crisis has been caused by war or famine, drought or some other disaster such as an epidemic; the 'spare capacity' for reciprocity is suddenly reduced to zero. Absolute self-interest and a refusal to reciprocate may be necessary as a short-term strategy but in the long run the social costs are too high to sustain. The critical significance of the group-centred perspective suggested here is not an outsider's observation, but is part of folk aetiology explaining, among other things, the rise and fall of states, or the success and failure of households; it is, in short, common sense.

It is the group, not the individual, who gives meaning and legitimacy to an individual illness. The meaning may be generated by a healer along with the patient or the family but it still

BOX 2: PUBLIC HEALTH

Health, people often suggest, is primarily a personal matter. Yet there is a dimension of health that has always been public, located beyond the home, outside the family sphere. At different periods, in different places, this 'public health' has been limited (i) to caring for or controlling sick strangers who entered the community and perhaps to providing for those without family too, and (ii) to taking moral responsibility for community-wide epidemics. In a traditional context, epidemics are crises when those categories of people who do not normally die (notably young men and women in their prime) start to die in large numbers; with the community itself thus considered as a whole to be 'sick', its governors – elders, priests, the king (if it has one) – are held responsible, though they may then identify particular scapegoats as the immediate cause of the scourge. Subsequently, the domain of public health and the government's contribution to it have been extended to arranging for the common provision of good water, sanitation, even minimal housing for the needy, and now, later still, to monitoring the incidence of illness and to providing vaccinations and basic health education – in short, to protecting the community as a whole from common dangers to its health, however those dangers (and that health) may be conceptualised locally at the time. Thus the 'new' public health today extends the prevention of ill health to include seeking bans on a range of 'poisonous' substances in the public domain: limits are now placed on tobacco, alcohol and other addictive substances; on pollutants and pesticides; on foods found to be dangerous. The causes of accidents, in the workplace or elsewhere, are similarly targeted (accidents being seen no longer as chance events). This modern public health involves, just as it did in the past, not only government institutions but charitable bodies and popular pressure groups that function to mobilise the community in matters affecting its well being.

'Public health', by definition, is political, and has given rise to contrasting styles of public medicine. Cultures, for example, can be divided into (i) those that locate the sources of health primarily at the personal level, and therefore promote detailed health education and a greater health-consciousness for all; and (ii) those that locate the causes of ill-health outside and beyond the individual and hence expect others to protect the ordinary individual from these outside causes. For the latter, health concerns tend to be marginalised (and medicine to be seen as just another craft practised by someone else) whereas, for the former, health and self-medication are central to daily life, and the specialism of medicine, responding to this scale of interest and specific demand, is therefore highly rewarded. 'Public health', then, has rather different meanings across cultures, minimised by some as potentially a system of social control that threatens personal autonomy; by others, it is taken for granted as simply another service that is better provided in common, along with defence and security.

The argument here is not for a further extension of the new public health or the 'nationalisation' of the personal domain in medicine – for one thing, irrespective of whether it is desirable or not, such an extension is simply not a realistic option today. Instead our argument is that, under the present conditions of economic recession, newly resistant diseases and the growing costs of medical services, public health has to rediscover some of its earlier priorities. It needs to recognise, for example, (i) that in the midst of globalisation there are growing disparities in health both within and between communities; (ii) that it is possible to reduce the effects of these disparities by inputs at the social level; (iii) that those involved in 'public health' have to work with communities to identify the locally relevant social inputs, and then help to make it possible to implement them.

remains subject to group interpretation, in the patient's mind as much as in others'. Hence the fear of stigma, which is notoriously hard for the individual to shrug off in the face of public certainty. It is the group too that funds or participates in rituals of exorcism or purification, that welcomes back the recovered patient or labels them an adept of the deity associated with the particular illness. Even solipsistic meanings, such as paranoid delusions, have to conform to the cultural patterning to be acceptable – if they do not, it confirms madness and exclusion follows.

The process that gives meaning is the diagnosis, which sets out not so much to know what the immediate illness 'is', but to uncover its cause, and its connection to other illnesses and misfortunes. The group's interest in the process lies not in identifying just the current symptoms and labelling the illness, but a deeper issue (common to epidemiology and public health) of locating the underlying, connecting rationale. The quest for cause (often called 'divination' but concerned here not with the future but with tell-tale events in the past) looks at both the host (what the victim has done, who has s/he been with, etc.) and the external factors in illness transmission – the social fault-lines and ecology, the sources of conflict. This is not simply being mystical or personalistic, involving metaphors or magic, but rather pinning responsibility for the victim's manifest vulnerability on social or political factors. Just as few would doubt that illness resulting from exposure to toxic industrial waste is not merely an accident but the result also of a failure, say, of safety standards, a failure itself due in turn to ignorance or corner-cutting or the pursuit of maximum profits, so too 'traditional' diagnosis assumes that most causes of illness are ultimately preventable, once their rationale is understood. A chain of responsibility is invoked, and has to be plausibly established just as it is in a court of law during a suit for damages. Blaming an individual identifiable (on account of the illness caused) as a witch is not so different in principle from blaming Structural Adjustment, the World Bank or a tobacco company for increased mortality; all four would deny the intention to kill, though extra deaths may be deemed publicly to have been the consequences of their actions. And all four may consider themselves unjustly made into scapegoats by the community. The point here is that illness is a form of injury, and not an accident; it is in itself social in origin and essence. Giving meaning to that injury, then, has to be a profound, social exercise in trying to understand practically the reasons for otherwise unjustified events.

If it is groups who give meaning to events, so too groups can be the locus for treatment, especially if peace of mind or renewed hope is considered one of the essential signs of being cured. Just as in the absence of adequate paediatric medicine the only therapy is to treat the mother and ensure her care and confidence, so it is important to be able to treat the group, repair its conflicts and open up avenues that give it hope for its future. Thus to guarantee, for example, the fees so that the children of a person dying of AIDS can go to school transforms the 'health' of that household; so do laws that enable a widow to live in the family house, work the land and bring the children up, and avoid quarrelling with the in-laws. In short, being part of a group that is seen to be viable, even successful, is itself a tonic and an incentive, making the distress of illness that much more bearable, and possibly shorter-lived in consequence.

Central to this apparent altruism – that others' well-being is important to the sick – is the fact that the direst form of poverty is to be totally alone, without links of any kind to other people. In situations where banks or police or courts or other institutions are not part of the social furniture of everyday life, the only store of wealth, the only guarantee of security for self or property are other people. Amidst the dangers of riot, robbery or molestation, you are wholly dependent on your group, and others are dependent on you. For anyone brought up in the apparent security of the modern state, the experience of other people as the sole source of your being is often missing (as witnessed by UK former Prime Minister, Mrs Thatcher's remark that there is no such thing as 'society'). Given how vulnerable we all are (but especially those of us who live in a Third World vulnerable to prices as well as to diseases, insects and much else), that sense of dependence on others is at its highest in illness. If to some commentators there appears today a growing disregard for human life, it is that the lives of other groups have to count for less; in straitened circumstances, priority must go to one's own. The tragedy of death is the loss it represents to the living; a person who is one of yours is priceless – so hard to reproduce, so long in maturing, yet so quickly, irreversibly lost. The preciousness of people – and the poverty of peoplelessness –

is hard to overestimate in these circumstances; and the pets of the developed world, its cats and dogs and budgerigars, are no substitute.

Given how important groups are, wealth is thus not measured only in income or food or housing but in having a group (the larger the better usually) to be linked into. That link need not be a matter of friendship or even liking; family apart, the patron that exploits you, the client that blackmails, the distant relative who bewitches or the neighbour who harasses you, all in some way add up with others to constitute your group; without them you are no one, differing from the mad solitary along the roadside only in the way you can recognise your own desperate social poverty. Illness, then, threatens poverty of the worst kind; it is not just a matter of sentiment or a time for sympathy.

An essential element of any group is the dead who are ahead on the road of being, and the unborn following behind. We, the living, are responsible to them as they are responsible for us. As members of the group, each one of us has rights, and failure to observe the rights of others is as wrong as failing to be given our rights. Hence the sense of justice is central to the working of a group – there has to be equity if not necessarily equality, at least in the short term. Much of healing, then, is about the quest for justice: finding out who is responsible for the illness, and offering or getting redress. The healer is the mediator in this process, s/he is the judge who understands the rights of all and sets the sentence. Illness is no accident, but the rational outcome of a social process; it is, as we have said, an 'injury'. Good health, then, is closely akin to justice in its broadest sense; and public health is a system that not only ensures equity both within and between groups but also gives people confidence that there will be equity whenever a crisis arises. Hence the appeal to the dead or to deities as permanent arbiters in the struggle for well-being.

Plate 6.1 Laropi, Uganda, 1991. A young woman is possessed by spirits while being trained as an *Ojo*, spirit medium/healer. The importance of indigenous religious practice remains strong, even among the young.

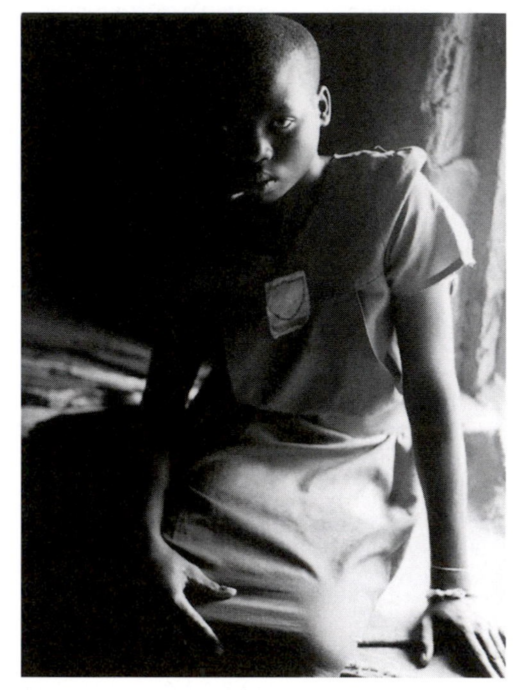

Source: Photographer, Tim Allen.

HEALERS AND THE SOCIAL HEALTH OF GROUPS

One of the key figures in ensuring that the social health of groups is maintained is the local healer, as the specialist called whenever the well-being of the group is threatened by a crisis. S/he in turn may call on the ancestors or on specific deities. Ritually his/her role is to keep the community free from pollution and cleanse it, should any taint come through or onto an individual. The healer has first to demonstrate the rationale behind the crisis. To do that, his/her understanding of both the tensions within the group and the anxieties of individuals is crucial in order to arrive at an explanation that can be generally accepted and thus form the basis for an effective ritual. In some societies there is a single such healer, but in others several healers offer these services, and serve to keep each other in check. Where there are many healers, the skills of divination and diagnosis are apt to be even more widely known, with a number of people not actively practising their expertise; again, such 'social knowledge' of healing helps to establish community-wide criteria of competence and morality. This professionalisation-from-below, based as it is on the many years a local healer has spent growing up in the community and the intimate knowledge neighbours have of his/her character is more effective a form of licensing than certificates showing exams passed or associations joined.

Healers of renown, of course, draw their clientèle from over a much wider area, as do healing shrines and temples. The 'medical communities' thus formed by systems of healing may not be quite like other spatially contiguous communities, but they share a common belief and practice and a sense of security that their well-being is guaranteed by the system they subscribe to. They may belong to more than one such 'medical community', in which case membership may well not be publicly acknowledged especially where a fundamentalist, dominant system seeks to impose a monopoly over the beliefs held in the community. Healers thus can be marginalised, even persecuted, within their own society – but with that marginality they combine a degree of insight (that is the envy of social scientists!) into how, beneath the surface, their society really functions. Marginal or not, healers have survived years of hostility from colonial or modern socialist authorities simply because people continue to insist on having their services. Similarly elders continue to maintain domestic religious practices, despite the risk of being accused of syncretism, in order to sustain the long-term well-being of their households on behalf of the rising generation. Healers and heads of households are not, of course, the only providers of therapeutic rituals. Charismatic churches and other ecstatic religious cults, for example, do so, and along with herbalists, midwives, barber-surgeons, as well as homeopaths and chiropractors, etc., constitute a significant part of a nation's medical culture.

Plate 6.2 A *nyanga* in Tonga, Zimbabwe. *Nyangas* are formally registered local healers who still form an important part of Zimbabwe's health care system.

Source: Photographer, Tim Allen.

A NATION'S MEDICAL CULTURE

Every modern state has what might be called a 'national medical culture', consisting of both the particular mix of ways it has for managing health and a certain set of prevailing values and beliefs that carry across the boundaries of particular 'medical communities'.

Just as a nation has its own political culture and its own religious culture, so I am suggesting we can also identify a medical culture, given form by government legislation and institutions but at the popular level informally expressed within this matrix by a wide variety of therapies. Within a nation there are several 'medical communities' each with its own practitioners with their theories and a clientèle of patients and their kin. Like the congregations of churches or the followings of political parties, these 'medical communities' are neither spatially contiguous nor necessarily stable, and like them are only intermittent in their activities, though their scope is particularly wide, embracing women and children as well as men. Distinct medically, they nonetheless interact economically and politically with other medical communities. Such medical pluralism is nothing new, though the range of therapies and their various adaptations to local circumstances has increased. What is happening, however, is that people are increasingly refusing to be confined to one 'medical community' with its particular medical theory but are choosing from the whole range of therapies on offer – assessing, for example, biomedicine as best for repairing a serious injury in a crash but a particular healer as excellent for treating the apparently 'possessed'. Integration occurs, then, at the level of medical culture in a way not possible at the political, religious or ethnic level. Such syncretism is not about blending methods of healing but, by keeping them separate, enables use to be made of each skill as appropriate. Syncretism requires social knowledge, that is to say the awareness people have of other people's different beliefs and practices. Individuals, no less than societies, can be successfully 'pluralist'.

Social integration, then, in matters medical is taking place not through governmental attempts at public health, not through 'medicine-as-social-control', but at the grass-roots, through consumer choice. Integration involves not just purchasing and trying out unaccustomed therapies but also giving credence to new notions of how illness may be caused. Hence acceptance of ideas about evil (and hence sorcerers and spirits) are spreading as the old cultural barriers come down. The apparent increase in sorcery accusations and their acceptance in some places now as evidence in courts of law reflects a new practicality. For in these modern, plural societies, it is practical reason rather than belief that becomes the rationale for therapeutic action: 'if it works, try it'.

Underlying this pluralism and the development of social integration-from-below is a fundamental shift that has taken place in many medical cultures over the last fifty years. Initially many nations experienced a colonial system of 'subsistence health' in which state health services for the general public were minimal and what services there were aimed largely at preventing epidemics and sustaining subsistence production. Curative medicine was mainly for the government's own use; the populace had its own means of sustaining health – just as people fed themselves through subsistence farming, so in a system of 'subsistence health' they were expected to cure themselves. With independence came the growth of a national, welfare-oriented health culture in which a much expanded programme of biomedical services, both curative and preventive, was taken out from the cities and into rural areas; the model was bureaucratic, the aim being to replace local therapeutic régimes with the modern, universal culture of world medicine. Now, some thirty years on, we have in its place today's more entrepreneurial model for the supply of medical services. The shift to medical entrepreneurship is partly a result of a global change both of attitude and in the economics of health. But it is due too to much deeper, much older, grass-roots attitudes about the place of the healer in society. This in turn is based on a recognition that attention to the individual sick person, or to the poor, is intrinsically less important than the struggle to secure wealth and good health; and that illness, like dirt, is ultimately best left to those who can be hired, whatever the cost, to clean it up. Healers, in this respect, are the launderers of our bodies; medicine is certainly necessary but that need not imply high status. The healer is more a provider than a patron, and the kin that call them to the patient are more customers than clients.

The notion of public health as a system of social control applies, at least historically, to certain developed countries but is often analytically irrelevant elsewhere. Social integration through medicine is occurring quite independently of the formal health authorities and in ways that is often at odds with biomedicine. But there is more to it than that. It is precisely because the culture of illness and medicine is rooted in the grassroots that epidemics can not only call into question the legitimacy of governments but can bring about widespread religious movements quite independently of the formal religious and political cultures. Medical culture, then, is not always so amenable to public health (or health education) and indeed is often highly critical of what governments have to offer. Any attempt at social health, I suggest, has to take this broader national medical culture into account state by state. At the practical level of policy-making, medical culture is not susceptible to the application of universal, general principles in the way biomedicine and conventional public health are.

The argument here has been that under the new conditions of globalisation there is an increasingly sharp division between those with ready access to biomedicine (and the health it can bring) and those with only restricted access. And second, that for the latter a crucial requirement for individual well-being is the health, broadly defined, of the social group. Merely to treat the individual's physical illness is especially inadequate in this context. Although social well-being has been on the agenda of WHO for nearly fifty years, little has been done. So the question remains: how to promote the health of the social group, especially in view of the fact that social groups (and the individuals that compose them) vary in how they define social health?

PUTTING THE 'PUBLIC' BACK INTO PUBLIC HEALTH

I have suggested that, given the likely deterioration of health for some in the new global economy, and given too the need to integrate those thus marginalised, we need to understand their ideas of health in order to formulate a workable compromise. I have suggested, too, that the notion of public health focuses on the group and is concerned with a wider well-being than simply physical health, and that to this end healers and other practitioners who can give meaning to misfortune and can cleanse both individual and group of their distress, are critical in ensuring the group's social health. Taken all together, the varieties of therapy – including biomedicine and conventional public health – constitute a medical culture peculiar to each nation. Such a medical culture has a certain autonomy, based on its being ultimately determined by people as consumers, and is not readily manipulable by the state or by formal religion. Indeed it can challenge both.

Government, as itself an expression of a nation's political culture, therefore has to work with this medical culture, and not against it; it is ultimately counter-productive to seek to destroy traditional therapies as if they were rivals, especially if the 'well-being of all' is the aim of government. Instead, if the marginalised are to be integrated as governments say they want, a compromise within the matrix of the wider medical culture has to be negotiated. Earlier we argued that besides the marginalised there are also those individuals and sections of the community who neither accept nor reject outright the global economy, but have worked out a compromise, just as other minorities have to in other contexts – a compromise that preserves both their own beliefs and their links to the dominant economy. It is a compromise of this kind that it is necessary for biomedicine to make within the wider medical culture of the nation.

It is the essence of medical cultures that they are the constantly evolving product of local circumstances. There will certainly be similarities with other medical cultures, but for policy-making it is the specificity of the local that counts – in short, any compromise is a matter of politics, and therefore to enunciate a prior set of rules here is pointless. It is important to remember too that solutions, far from being permanent, have to be but temporary adjustments in an ever-changing field: particularly in a plural society, people's cultural responses to crises have to keep up with shifting realities and unforeseen outcomes, which in turn modify experience.

Nonetheless, as a starting-point for considering culturally appropriate strategies of the sort suggested here, we can pick on three components:

1. *Patients and their kin.* What are the crucial practices (often unnoticed because taken for granted) that can rekindle the sense of well-being even in the face of distress? We have suggested here that, in contexts where both the well-being of the group is considered paramount and the sick individual's identification with the group is of the utmost significance, priority must be given to ensuring the functioning of the group. One possible way of doing that, already mentioned, is to guarantee hopes for a better future by ensuring the dependants of the patient have certain rights, whether it is to school fees for children or land and housing for widows. Similarly, it is imperative to enable the sick to keep in touch with their kin, whether through visits, mail, newspapers or radio; inexpensive transport becomes a matter of great importance, when there are sick relatives to see and funerals to attend, in efforts to keep the wider family network going as a support group. For the same reason, enabling the sick and their kin to have work enough to earn money is crucial, even if it is only enough to be able to offer gifts to guests on their visits. Such seemingly minor items reduce the stigma of sickness, and keep a patient functioning in the social manner that fits their real status as a responsible adult but is also one that gives pleasure too. Finally there may be a need to help patients and their kin create new groups that can offer additional mutual support. Such groups will need help with transport and other minor, if practical, items; but above all they need recognition of their importance, not just for the sick, but for the community as a whole.

2. *Healers.* What is it that healers offer to a particular community that results in such a public demand for their services? Could their undoubted skills be used more broadly, for other tasks within the community? In medical cultures a key element is the provision of a meaning for each misfortune and that an acceptable meaning, with all its ramifications, is essential for a cure (defined generally as a return to well-being); in which case, healers have to be recognised as providing that meaning. Given the political potential of healers, governments are sometimes anxious to control healers, whether through associations or a programme of formally incorporating them as subordinates in a primary health care scheme. But unless healers both retain their autonomy and have the opportunities to develop new understandings in their own way (and this does not exclude learning about biomedicine), they become mere agents of government, thus destroying the special, extra-social quality of their insight and with it the peculiar authority they have to mediate between misfortune and the community. Since professional control of healers usually comes from below, through public opinion over the long-term, local committees representing the community are an alternative way of giving formal expression to the need for local control over health matters. But control is not the real problem. In the new conditions of civil strife and moral crisis, the task is to find ways of working with healers in re-cementing communities riven by modern conflicts or stricken by misfortune. We need to be creative and positive about all the social skills available to a community, and not carp about control or cripple people's trust that their own institutions, their own professionals can develop responses that build up well-being. Laser surgery may not be their métier, but healing the social wounds of war is; the ability to reassure and re-inspire people, to link them once again to their roots, cleansed and at peace with themselves, needs to be not just highly valued but encouraged to develop further in ways that fit the times.

3. *Ministries of health.* Ministers need to stand back and look at the nation's entire medical culture, of which they themselves are

but a part. They need to learn how to recognise its changing dynamics and how to work with it, and accept its various contributions to the nation's well-being, medically and socially. For through people's use of the medical culture – and the experience, common to us all, of misfortune – there is generated a degree of social integration and support that is rarely found in other social arenas. Conscious then of the medical culture as a distinct field for co-operative action, government may be called upon in ways it had not foreseen – for example, to set up a council or ombudsman's office to intervene and arbitrate when there are disputes and malpractice. Malpractice, like the faking of drugs, will occur in the global economy, but if the public is explicitly entrusted with noting and reporting such matters, then the kind of 'health police' that once bred so much resentment and resistance becomes unnecessary. Nonetheless, nothing absolves government from also doing what it does best: improving the accessibility of ordinary biomedical services and ensuring that the basic material conditions for health – water, drainage, housing, schooling – are established and maintained. An emphasis on social health is not a substitute. Social health focuses on the cultural conditions for health, and makes good use of people's particularities rather than just the universals of biomedicine. It puts the 'public' back into public health.

For any of these strategies to work, there has of course to be a government that is not fighting, either metaphorically or literally, its own people. In short, some of the prerequisites for effective public health are political, and not medical or social. But in the absence of those political prerequisites, what then? Can there be health despite politics? In practice, people know there has to be: otherwise, pessimism becomes an illness; like self-pity, it incapacitates. Hence the final paradox: amidst all the illness and death, celebration becomes a crucial therapy – whether it is carnival or a funeral wake or simply an occasion for singing arranged by torture-victims for themselves. It is an act of defiance, a show of solidarity with others, as well as fun. Far from being inappropriate, such joyfulness re-affirms, if only briefly, the common capacity for happiness and humour. This is not to deny the sorrow and pain of ill-health, nor to down-play the imperatives of public medicine. It is vital nonetheless to recognise people's capacity, among themselves, to make room for well-being, for moments of joy, amidst the suffering and the struggles of everyday life. For in the final analysis it is not health itself but only the means to health that can come from WHO or from governments.

NOTES

1 The proportion spent on mental health within a nation's whole health budget varies widely, with, for example, the United Kingdom spending 16 per cent and Japan only 6.2 per cent; truly comparable data are difficult to obtain, however. In the established market economies, mental illness accounts for 15 per cent of male life-years (DALYs) lost, but only 9.3 per cent of women's, according to the World Development Report 1993, *Investing in Health*. The proportion world-wide of estimated life-years (DALYs) lost through mental illness is 6.6 per cent for women, and 7 per cent for men, with alcohol and drug dependence responsible for the excess of mentally ill men over ill women (women outnumber men 2:1 in depressive and stress disorders; the ratio is 1: 5 for dependence-related illness).

2 'Globalisation' can enable a local disease to escape from its usual habitat and become a pandemic, with devastating results world-wide. Via trade routes on land and sea, the Black Death spread from Central Asia to Europe, and during its seven years (1346–53) killed between 30 and 50 per cent of the populations affected. Similarly, cholera was able to break out of India in the early nineteenth century and spread through other parts of Asia and into Europe; often carried by troops as well as by traders, it gave rise to a series of seven pandemics. Perhaps the most devastating was the influenza pandemic of 1918–1919 which in a brief space of time killed over 21 million world-wide, in the aftermath of the First World War. These rapid pandemics are spectacular, but lesser-known epidemics, when diseases have been introduced from abroad to non-resistant populations, have been as locally devastating to particular communities (such as those indigenous to the Caribbean and the Americas).

3 Some 62 countries are currently identified as being seriously affected by mines. Angola, Mozambique and Somalia, for example, have between them some twelve million mines requiring clearance; Angola has already some 20,000 people who have lost a limb. Cambodia has been even worse affected. Mines are designed not to kill but to maim, embedding into the body plastic fragments that are hard to x-ray. Laid randomly or dropped from the air, mines are now extremely difficult to detect safely. It is estimated it may take up to fifty years to clear all the mines and enable farmland to be worked without fear. Until they are cleared, casualties will be a major problem for the health services; but the longer-term consequences for the general health of the disabled and their families have yet to be estimated.

III

REPRESENTATIONS AND CULTURAL COMMODIFICATION

7

FINDING THE RIGHT IMAGE

British development NGOs and the regulation of imagery

Henrietta Lidchi

INTRODUCTION

This chapter will attempt to investigate the culture of development but by a different method. To invoke the question of culture in the context of development has most frequently been to question whether the practice of development elicits a 'culture clash'. In other words to assess whether the success of development projects or programmes is affected by a cultural disjunction and, if so, in what manner: how such conflicts arise, how they can be solved and avoided in the future. This chapter will not attempt to contribute to this debate. Its preoccupation with culture is located elsewhere. Questions of 'culture' will, in the following pages, be explored in relation to issues of representation. In this brief analysis 'representation' will mean the construction of meaning in and through images of development as produced by British NGOs.

The chapter will accord pre-eminence to the 'image of development' – in visual and verbal terms – by addressing issues of context and content. By delineating why the 'image of development' became an issue of concern in the 1980s, it will demonstrate that it is possible, and desirable, to submit representations of development – visual and other – to cultural analysis. The chapter will begin by considering why questions of image rose to a position of prominence in the 1980s for, most particularly, British development NGOs. Following from this it will consider an advertising campaign produced by the British-based development NGO, Christian Aid. Using tools known to cultural theory, most particularly semiotic theory, it will show how images of development can mobilise and represent complex ideas about development process and practice in seemingly ingenuous ways. Finally, this analysis will allow us to recast development NGOs as profoundly cultural organisations which respond to forces external to their own sector while articulating quite distinct understandings of development which gain popularity at specific historical junctures.

SURFACING: HOW THE 'IMAGE OF DEVELOPMENT' BECAME THE OBJECT OF CONCERN

It may be difficult now to recall with what passion and persistence the question of imagery was discussed amongst development practitioners in the late 1980s and 1990s. The history of development is a comparatively short one, the largest and most prominent development organisations in this country – Oxfam, Christian Aid, Save the Children Fund – have institutional histories that do not extend beyond this century. Conceived as small social justice organisations which emerged as a response to the need for peace and reconstruction in Europe rather than elsewhere, they saw their role in development secured in the 1960s. Yet within this limited historical timespan, the question of image has been a preoccupation for an even shorter time.

The first to address the question of representation and imagery properly was Jørgen Lissner in a thesis entitled *The Politics of Altruism* (1977). Radical for its time, this book effectively delineated the parameters of a debate that would subsequently emerge in the aftermath of the Ethiopian famine in the mid-1980s. Briefly, Lissner's argument was based on the following premises: (1) that development NGOs in the North (NNGOs) were harbouring a destructive internal conflict between fundraising and education, (2) that this was symbolised in the images and messages these discrete groups of professionals produced, and (3) that the image of development fundamentally impacted on development practice: *negative* images of development encouraged *negative* development practice and vice versa. In the *Politics of Altruism*, Lissner characterised the activities of fundraising and education as conflictual and competing. Fundraising, in his view, was a primarily economic activity dedicated to raising money, with no further goal than ensuring the long-term institutional survival of the organisation. Development education, on the other hand, was a morally oriented pursuit, committed to increasing global awareness of development and conscientisation of Northern audiences. Fundraisers had short-term goals: the ends (money) always justified the means (advertising); the best methods were quite simply those with the largest profit margins. Development educationalists were more visionary and 'means'-oriented, their work was a practical application of ideals of solidarity with the poor, and they were, furthermore, particular about the type of messages and images they produced (Lissner 1977: 145–7; 1981).

These differing priorities were rendered most visibly manifest, for Lissner, in the images of development these different groups produced. Fundraisers were crude behaviourists content to flatter or shame the donor, either by showing how cost-effective their donations could be, or by dwelling on the despair and devastation in the South (Plates 7.1 and 7.2).

Plate 7.1 Disasters Emergency Committee East African Appeal, press advertisement, June 1980.

Source: Reproduced with kind permission of East African Emergency Committee.

Moreover, because there was nothing 'sexy' about development in advertising terms, fundraisers sought to elicit donations by making full and constant use of a narrow range of images – utilising, in the main, distressing portraits of malnourished young children – 'starving baby images'[1] (Lissner 1977: 189; 1981). For Lissner, whose sympathies lay squarely with education and social justice, it was simply a question of opting to show the 'truth'. Starving baby images, he held, were neither true nor accurate pictures of reality overseas, but a reflection of the laziness of fundraisers who chose to feed Northern prejudices for profit. Lissner defined these images as *negative* – demeaning, lacking dignity and untruthful. In this manner Lissner gave voice to an argument which stated that *negative* images were the product of a power imbalance between those *representing* – the NNGOs, the North – and those being *represented* – the poor, the South.

> The public display of an African child with a bloated kwashiorkor-ridden stomach in advertisements is pornographic, because it exposes something in human life that is as delicate and deeply personal as sexuality, that is, suffering. It puts people's bodies, their misery, their grief and their fears on display with all the details and all the indiscretion that a telescopic lens will allow.
>
> (Lissner 1981: 23)

Lissner's drawing of an analogy between negative imagery and pornography at this early point was particularly prescient because he linked knowledge to power through the visual image (see Adamson 1991 and 1993 for later, similar, accounts). Such an argument was not mainstream to development thinking, but drew much from work on photography in other disciplines.

Lissner advocated that imagery should be regulated, most particularly photographs. In his analysis, he tacitly acknowledged what Edwards (1992: 7) later stated: that all photography works on the basis of spatial and temporal dislocation. A photograph is of the past, it freezes a moment which has ceased to be; yet it functions in the present, transforming the 'there-then' into the 'here-now'. Moreover, its immediacy and realism can be replayed an infinite number of times: 'the photograph repeat[s] mechanically what could never be repeated existentially' (Barthes 1984: 4). It is this combination of factors that allows photographs to stand for, and symbolise, historical events, whilst failing to disclose within the frame the conditions under which these events came to exist. The existence of a photograph does not permit the viewer to distinguish whether what he/she is looking at is the result of a 'candid' shot of the subject, or one that it posed (Nichols 1991: 150–1; Barthes 1977). Temporal dislocation is coupled with spatial dislocation. Photography frames and shapes the moment, exposing it to historical scrutiny. Furthermore it travels. Photography can make what is spatially distant, and what may never be personally encountered – famine, for instance – familiar. Consequently, the most basic characteristic of the documentary snapshot is that it simultaneously appropriates an image and decontextualises it. The camera, then, rather than being seen as a *medium* which represents the 'real', can be best characterised as an *instrumentality* (Burgin 1986: 43). Instrumental in the sense that not only does it construct meaning but that, as a tool of representation, it is intrusive: it has the power to observe at close proximity, and to remove the image absolutely from its original context.

Plate 7.2 Christian Aid, poster advertisement, April–May 1990.

Source: Reproduced with kind permission of Christian Aid.

For Susan Sontag the camera is a sublimation of the gun, something that is 'loaded', 'pointed' and 'shot' at subjects, capturing their image and exposing their reality for the purposes of consumption (cited in Pinney 1992).

Consequently it is often argued that photography combines voyeurism and control, because visual images are taken *by* the powerful *of* the powerless; the *subjects* of the photograph are transformed into *objects* by virtue of being 'shot'. So photography can produce the colonised and the powerless as fixed realities: entirely knowable and visible, but equally 'other', irreconcilably different: the *objects* of desire and derision (Young 1990: 143). The representer/viewer combines both epistephilic (deriving pleasure from knowledge of/over) and scopophilic (deriving gratification from looking at) desires. Photography constitutes subjects not only as *objects* of knowledge, but also as fetishistic *objects*, docile and visible bodies.

All these reservations underlay Lissner's critique of *negative* images. For him, images of starving children demeaned the *subjects* of development because they represented them as being devoid of dignity: they transformed into *objects* of representation, and by implication *objects* of development. So, for Lissner, representation – the image of development – was decisively linked to intervention – the methodology of development. The wrong type of image could elicit the wrong type of development. Consequently Northern NGOS (NNGOs) were in a crucial position. Northern audiences were particularly susceptible to 'negative' and 'exotic' images of the South because they had no chance to 'double check' their 'veracity', nor did they have a ballast of the normal with which to compare them (Lissner 1981: 24; Adamson 1991). When NNGOs reproduced charity images that dwelt on disaster and despair, or promoted a dependent view of the South, they conspired with the media to undermine their own capabilities to increase either global awareness or democracy (Lissner 1981). So Lissner advocated consistency between images and practice. By reflecting a dated charitable approach based on a modernisation view of development, fundraiserers constituted poverty as an inescapable fact of the 'global order'. But development organisations, he believed, ostensibly favoured a more recent structural view of poverty causation and advocated social change and global social justice. This agenda was reflected in poorly resourced, infrequently produced, development education material. Fundraising targeted the sin of omission – the failure of the rich to give to the poor – promoting an idea of charity, but education sought to address the sin of commission – the manner in which the rich North constantly appropriated wealth from the poorer South – advancing the more desirable view of potential social justice. Development education had an agenda driven by moral values, not profit margins (Lissner 1977: 159–60, 167–71, 173; Whitaker 1983: 160; Arnold 1988: 14).

So Lissner's account described Northern NGOs as sites of internal conflict. In his fatalistic narrative, fundraisers were seen as a parasitic but progressively more dominant force in NNGOs (Lissner 1977, 1981; Whitaker 1983: 177–8). NNGOs were in turn characterised as lacking the will to work through the tension between institutional growth and a vocal commitment to global development and social justice. Lissner's portentous warning – of at best an impasse, at worst an erosion of principle – did not forecast the possibility of a major alteration in the state of affairs. Yet the extraordinary events that surrounded the Ethiopian famine in mid-1980s ushered in a cultural revolution which proved tensions delineated in Lissner's account to be far more blurred.

THE STORY BREAKS: IMAGING THE ETHIOPIAN FAMINE IN 1984–5 AND ITS CONSEQUENCES

The Ethiopian famine of 1984–5 was a revolutionary moment for development NGOs in two significant respects. First, in the realm of development practice. For development NGOs the events of 1984–5 marked a 'coming of age' denoted by their unprecedented leap in income and prestige[2] (*New Internationalist*, 1992: 19; Poulton 1988; Gill 1988; Gordon Drabek 1987a, 1987b). The crisis in Ethiopia seemed to be *the* example of 'development in reverse' that NNGOs had long been warning governments about. Its effect was to secure a loss of faith in the ability of official aid institutions and their programmes to relieve or alleviate poverty in the long-term.

> Mass death by starvation in Africa was the most emphatic proof possible that the development era had been a washout, that most official aid served to impoverish communities rather than to enrich them and that such reassuring notions as the 'trickle-down' theory of development were a cruel deception ... Only the private agencies emerged with any credit from the catastrophe.
>
> (Gill 1988: 169)

Throughout the famine, it appeared, NGOs had remained true to their principles and steadfast in their priorities. The events in Africa in the mid-1980s had the effect of convincing many that an NGO contribution was more effective, professional and caring than anything governments and the larger aid organisations could hope to provide. Thus, NGOs emerged as the new hopeful instruments of development, representing a new model of what development could mean and how it should be conducted[3] (Chambers 1986; Sewell 1988: ix; Gordon Drabek 1987b: vii; Poulton and Harris 1988: 4). In a situation of dire need NGOs had promised and delivered salvation; they represented and carried through a more human/e alternative.

Secondly, in 'cultural' terms, the events surrounding the Ethiopian famine of 1984–5 had profound effects. It forced NNGOs to consider their image in the singular – how NGOs looked and were identified – but most particularly images in the plural: how they had represented their 'partners' in development. Questions of representation moved to the fore – no longer a question of gilding the lily – now a principal concern.

The consequences of the Ethiopian famine were to invest confidence in NGOs and ensure their survival, through exponentially increasing their income and popularity. This had been achieved through a barrage of powerful if extremely disturbing imagery: that of starving masses and doomed individuals. Yet the fate and fortune of the development NGOs could not be divorced from the power of the media. The recognition of this caused development organisations to reassess their responsibility in representational terms, since they came to the unsettling realisation that popularising the right images might ensure their institutional survival, increase their 'political' leverage and promote their kind of development by mobilising popular support.

The Ethiopian famine gained global recognition in October 1983 when a report filmed by the late Mohamed Amin – then Visnews' Africa Bureau Chief – and filed by Michael Buerk – of the BBC – was screened on the Nine O'Clock News. Up to that point, despite repeated warnings, detailed official accounts, a Disasters Emergency Committee[4] appeal and news reports failed to move the famine to centre stage. It was judged to be marginal and largely 'unnewsworthy'. That Amin and Buerk worked together was largely a product of good fortune and sound journalistic instinct. Buerk had, in fact, been into Ethiopia in July of 1983 to report back on the success of the Disasters Emergency Committee appeal, but had only filmed in the southern part of Ethiopia, which was comparatively lush and suffering to a lesser degree. In October 1983, Amin and Buerk focused on the northern towns of Korem and Makelle, the epicentres of the famine. They were unprepared for the scale of the human distress they encountered and emerged in a profound state of shock (Amin 1989). This was a notable feature of the visual and verbal images that were subsequently broadcast on the BBC's evening news of 23 October 1984.

> Dawn, as the sun breaks through the piercing chill of night on the plain outside Korem, it lights up a biblical famine, now, in the 20th century. This place, say workers here, is the closest thing to hell on earth. Thousands of wasted people are coming here for help. Many find only death ... Death is all around ... Korem, an insignificant town, has become a place of grief.
>
> (Michael Buerk's commentary quoted in Harrison and Palmer 1986: 122)

The footage is said to have hushed newsrooms producing tears and unprompted donations from habitually hardened news staff. Carried by 425 of the world's major broadcasting agencies, in Britain, it jammed Oxfam's switchboard for three days. Even

British tabloids such as *The Sun* joined the fray with a two-inch headline 'RACE TO SAVE THE BABIES' (28 October 1983). An enormous aid operation ensued, fuelled by popular support. Aid workers were now swamped by clumsy news teams eager for a story. It was the biggest news item to come 'out of Africa' in the 1980s (Harrison and Palmer 1986: 110).

Extraordinary though these events were, they did not secure the 'cultural revolution' in NGOs. This was left up to the pop star – Bob Geldof – who saw the Amin-Buerk footage. He responded by recording the single *Do They Know It's Christmas?*, a record that sought to 'save the world at Christmas time', under the name 'Band Aid' (Hebdige 1988: 216). Geldof's energy fuelled a media circus around the famine, in which he emerged as the self-appointed voice of the people. As a concerned, independent and active citizen, he considered himself to be empowered and entitled to confront politicians and governments on the level of their commitment to the South. Geldof's abrasive approach opened a surprising number of doors, to the consternation of more established and experienced development professionals (Burnell 1992: 122–3; Harrison and Palmer 1986: 131; Philo and Lamb 1986: 26).[5]

The Band Aid saga culminated in the legendary Live Aid Concert of July 1985. Staged simultaneously at Wembley Arena and the JFK Stadium in Philadelphia, it was sixteen hours long, beamed via 13 satellites to 120 nations to an estimated 1.6 billion people, a third of the world's population. It was the 'biggest [philanthropic] music concert in history' (Philo and Lamb 1986: 26). Frenzied performances were punctuated with appeals from an exhausted 'Bob' bullying people to part with their money 'now' and the by-now iconic footage of starving people, including, most particularly, the Amin/Buerk report (Harrison and Palmer 1986: 131–3; Hart 1987: 1). Substantial quantities of money poured in, during and after the event to the tune of US $60,000 a month in 1985 (Allen 1986: 32). The Band Aid Trust eventually raised £144 million world wide[6] (Burnell 1992: 203–4). Live Aid secured populism a place in charity fundraising, and spawned a succession of similar events, but for development NGOs it created a precedent both alarming and seductive. It showed how people could be mobilised behind development. The crucial question was: at what price?

For many who had worked in the development field, particularly those who had striven against public indifference, the success of Band Aid warranted investigation, in part because it seemed that the media and the NNGOs had jumped on a bandwagon with little or no thought of the message they were broadcasting. The media and the NNGOs had dehistoricised, depoliticised and trivialised the complex and life-threatening issue of famine, by making it an issue of money and food. By opting for money over truth they had privileged a *negative* image of Africa. Once more constituted as 'other' – the 'Dark Continent' – it was lastingly inscribed in the minds of millions as a timeless space where a biblical famine – an event alien to the modern industrial West – could unfold without resistance (Nash and Van Der Gaag 1987: 30). Ethiopia, which in turn was synonymous with Africa, was depicted as a country, poor to begin with, brought to its knees by famine, and needing outside assistance to feed itself, on a scale without historical precedent (Horgan 1986: 18).

Northern audiences, whose popular belief it was that Africa was poor and underdeveloped, now had proof. African nations were visibly passive nations constantly threatened by the possibility of a 'natural disaster' and always in need of assistance, and dependent on Western goodwill (Nash and Van Der Gaag 1987: 1; Nyoni 1988–9: 9; Hart 1987, 1989). Both the media and the NGOs had made full and frequent use of images of starving Africans, most particularly women and children (Kaida-Hozumi 1989: 26; *Dialogue*1988) (see also Plate 7.1). They had popularised and used a predominantly negative image of Africa and Africans. Unnamed and unidentified, Africans were mostly photographed when they were powerless to refuse, African subjects were used without reservation to incarnate, on an individual scale, the timeless mass of starving Africans (Stalker 1991). For a respected Southern partner who prepared a Zimbabwe report on news coverage of the

Ethiopian famine, these images were disturbing:

> Images represent people. People will fix images to the faces depicted, regardless of whose personal image it is ... Images should not be applied en masse, but to individuals; to people with dignity, with an identity. When images lose their identity, they become a way of looking down on people ... It seems to be that if you respect somebody you want to learn their name ... you want to know them as individuals. But the way that ... African people [were portrayed] is as if they were not people at all.
>
> (Nyoni 1988–9: 7)

African *subjects* were represented as the passive recipients of aid – *objects* of development – who had no voice, no identity and no contribution to make during the crisis. The West, in contrast, was constituted as being full of active subjects, development workers, fundraisers, journalists or world citizens (Hart 1989; Bailey 1989). In an argument echoing Lissner's, these negative images were said to be both counterproductive and untrue. Untrue, because the peoples of Africa, it was stated, were active and resourceful, constantly engaged in diverse and effective strategies to circumvent the possibility of starvation. Counterproductive, because negative images were self-fulfilling prophecies. They attracted the wrong type of development and development agencies, those which encouraged dependence rather than 'empowerment', 'dialogue' or 'self-reliance' (Nyoni 1988–9: 7–8).

A number of countries, thirteen in all, participated in the *Images of Africa* survey, which attempted to understand why these images were produced (*Dialogue,* 1988). They concluded that neither NGOs, the media, nor Band Aid had made a concerted effort to address and broadcast the positive indigenous efforts to allay the crisis. NNGOs had produced educational materials somewhat after the event. Less resourced, less attractive and less widely circulated than their fundraising counterparts, these more enlightened images and messages had not reached the same audience (Nash and Van Der Gaag 1987). The *positive* image of development where subjects participated, were self-reliant, and self-determined had simply not been available to Northern publics (Simpson 1985: 21; *Dialogue,* 1988; Nyoni 1988–9). They concluded that although NNGOs had had the power to portray the crisis in a different light, they had not done so. They had been swayed by other more persuasive forces.

On the basis of this exceedingly brief précis of the critiques that emerged as a consequence of the events of 1984–5, credence might be given to Lissner's prophecy, but in effect the impact of this moment was revolutionary. The events were to move the question of image to centre stage and to reconcile the agendas of education and fundraising by making the images that both these sections produced synonymous with institutional image and development practice. From the late 1980s until the present day, the more visionary NGOs are guided by the belief that: '[T]he problem of images and perception cannot be separated from the methodology of intervention' (Nash and Van Der Gaag 1987: 77).

Moreover there had, at some level, to be a recognition of two most obviously positive results: (1) the money had saved lives and (2) the events resulted in a mood of concern about the developing world not manifested in over twenty years (Black 1992: 265; Burnell 1992: 12). So the events of 1984–5 gave rise to the widespread view in NGOs that image and practice were causally linked. This engendered several transformations: (1) in development organisations who now viewed broadcasting the right image as a priority; (2) in the images themselves – with noticeable increase in the production of *positive* images or 'balanced' messages; (3) in representational practice, the creation of formal or informal guidelines on representational practice, which sought to impose constraints on the content and regulate the production of images; and (4) a reconsideration of the value of a more populist approach – NNGOs were loath to lose the constituency and support they had gained from Band Aid and Live Aid.

POSITIVELY THE TRUTH: NEW TURNS IN THE IMAGING OF DEVELOPMENT?

Having considered the historical context in which the question of imagery was brought to the fore, we can now look at how such a debate impacted on imagery itself. We are therefore moving from context to content. In the following section we shall undertake a reading of a *positive* image. There are three reasons for this. First, to show how the codes and conventions of images of development – as of all other images – are designed to 'make sense'. Second, to gain an insight into the reason why positive images and negative images are, in effect, two sides of the same coin, and how they are underpinned by a 'realist' understanding of the nature of representation. Third, how this adherence to the criterion of 'realism' prevents new modes of representation emerging.

The reading attempted below will draw a great deal from the work of the French cultural theorist Roland Barthes and the tools that he developed from Saussurian linguistics. Barthes, drawing from Saussure, asserted that the relationship between linguistic signs was essentially arbitrary so that the meaning of each utterance – *parole* – could only develop when it was activated within a predetermined and static system of meaning – *langue* (see Barthes 1967; Nichols 1981; Burgin 1982). Barthes used the concept of a *sign* to unravel the layers of meaning. The sign is divided into its constitutive components, the *signifier* and the *signified*, to separate expression from meaning. The difference between these two components is clear, the 'substance of the *signifier* is always material (sounds, objects, images)' (Barthes 1967: 112), whereas the *signified* 'is not "a thing" but *a mental representation of the thing*' (1967: 108; my emphasis). These elements cannot exist independently; they are defined in *relatum*. *Signification* is the arbitrary process which binds them and produces the *sign* (Barthes 1967: 113). To give a simple example, there is nothing intrinsically active ('go') about the colour green and static ('stop') about red, or preparatory about amber. These colours signify these states only when used in relation to one another in a system of traffic signals. Their meaning is contextual and relative. The same kind of analysis can be utilised in relation to images to unpack levels of meaning. For Barthes visual images – photographic or otherwise – could never be 'messages without a code'. They constituted a language in themselves and in their wider circulation within society (Barthes 1977: 17). Barthes isolated different levels of meaning, using the concepts of *connotation* and *denotation* to explore the complex articulation of meaning. He defines the realm of *denotation* as the first order of meaning, which is focused on analogy or description; to clarify denotation reference must be made to the system of signs (Barthes 1977: 16, 18). *Connotation*, on the other hand, operates as a second order of meaning; it is the way in which the image is understood; it is 'the manner in which the society ... communicates what it thinks of it'. To elucidate the connotations, reference must be made to the 'rules of social life, of history, of social practices and usages ... to what Barthes has called the "collective field of the imagination of the epoch" ' (Hall 1972: 66). Both *denotation* and *connotation* will be used in the following analysis.

The images discussed are not fiercely contemporary, but those produced by the British NGO Christian Aid. The poster image and the press advertisement (Plates 7.3 and 7.4) were allocated to the weeks running up to the 1990 Christian Aid Week. As the 'priming' April communication materials for Christian Aid Week (its major fundraising week), they occupied a prominent space in Christian Aid's fundraising calendar in 1990. Choice here is guided by the fact that these advertisements were quite deliberately designed to be positive. It was, in a sense, the first step towards new strategies of representation for Christian Aid (Lidchi 1993).

Christian Aid was, and is, well known for its distinct methods of communication and its adherence to new development ideals, foregrounding partnership, empowerment and participation as principal development goals. These materials, although not ultimately typical of Christian Aid's fundraising strategies in the long term, were definitively 'of the moment'. They received widespread approval from other members of the British NGO community who spoke of these images as positive and as making the right kind of statement. This was partly attributable to their perceived political nature. The 1989–90 series of Christian Aid appeals, of which the poster

Plate 7.3 Christian Aid, press advertisement, April–May 1990.

Keep the health service going.

At 9 am, six days a week, Elizabeth, like many other health workers in Bangladesh, cycles off to the local villages.

She carries out an immunisation programme, teaches villagers about sanitation and the dangers of unclean water, and often has to treat infected wounds.

She keeps a careful record of the progress of pregnancies. By detecting abnormalities early, the maternal mortality rate is being lowered considerably.

Health workers like Elizabeth spend time with the mothers of sick children, showing them how to prepare nourishing food.

Providing funds to take care of the health of the poorest is a vital part of Christian Aid's work in the Third World.

Please give us the money to keep the health service alive and well.

GIVE NOW, RING 01-200 0200 OR POST THIS COUPON

To: Christian Aid, PO Box 100, London SE1 7RT.

I would like to keep the health service going.

I enclose my gift of £200 ☐ £100 ☐ £50 ☐ £25 ☐ £10 ☐ £.... ☐

I enclose a cheque/PO ☐ Or

Please debit my Access/Visa/American Express account no.

Please send Covenant Form ☐

Signature

Name

Address

Postcode

Christian Aid

CHURCHES IN ACTION WITH THE WORLD'S POOR.

Source: Reproduced with kind permission of Christian Aid.

and press advertisement formed a significant component, had a theme of bringing Third World problems closer to home (Plate 7.4).[7] The 1989 September poster and press advertisement stated 'Make Water Public' (in reference to the privatisation of water in the UK), whereas the Christian Aid Week statement was 'The World is Our Community, This is our Charge' (1990 was the year the Community Charge was introduced in Britain) (Plate 7.4).

In the following analysis the poster (Plate 7.2 on page 89) will be used to illustrate visual signification, whereas the press advertisement (Plate 7.3) will be referred to when it is important to emphasise and amplify certain points, particularly the interrelationship between image and text in the navigation of meaning. Such a reading is at one level personal. Meaning, as semiotic analysis permanently reminds us, can never be fixed. Yet these images aspired to general relevance and broad appeal, so both as motivated and public signs, they must be read with a belief in their general level of signification.

The striking aspect of the poster image (Plate 7.2) is its documentary appearance. It is almost as if the photographer just happened to be there as the woman was riding by, early in the morning on her way to work (we know she goes to work early from the press advertisement). She took him by surprise, he 'froze her in time'[8] (Hall 1972: 79). The photograph has a literal content; it reproduces mechanically, repeatedly and literally the moment that was fortuitously caught. This is amplified by the quizzical look on the faces of the men, seemingly interrupted, the temporarily immobilised cow, and the fact that the bike and its rider are slightly out of focus.

This realistic style and the image's usage encourage the impression that it is simply *denotation*, that it depicts reality, and this validates it as evidence: an objective document. To have taken the photograph, the photographer must 'have been there'. It seems to be an authentic record of 'what happened', an unprompted, representative visual image of the everyday in the South and Christian Aid's work there. The *connotation* is that this photograph was the result of a 'lucky find' – for the photographer – a fortunate *trouvaille* (Barthes 1984: 33).

The viewer is then, perhaps, struck by the linguistic message, the poster's very obvious caption and explanation 'Keep the Health Service Going'. This linguistic message locates the visual image and navigates the reader through it. It ostensibly exists purely at the level of *denotation* because, although it does not describe the photograph, it seems only to amplify it, the only noticeable construction being the pun (this is more evident in the press advertisement).

However, conversely, the linguistic message of 'Keep the Health Service Going' is illustrated by the visual image; the woman ('Elizabeth') represents/embodies the health service which is denoted by her bag with the cross on it (connoting the Red Cross and its medical services). The bicycle connotes mobility and dynamism. For the British viewer in 1990 there was the added *connotation* drawing parallels between the underfunding of the British Health Service and the similar

Plate 7.4 Christian Aid advertising schedule, 1989–90.

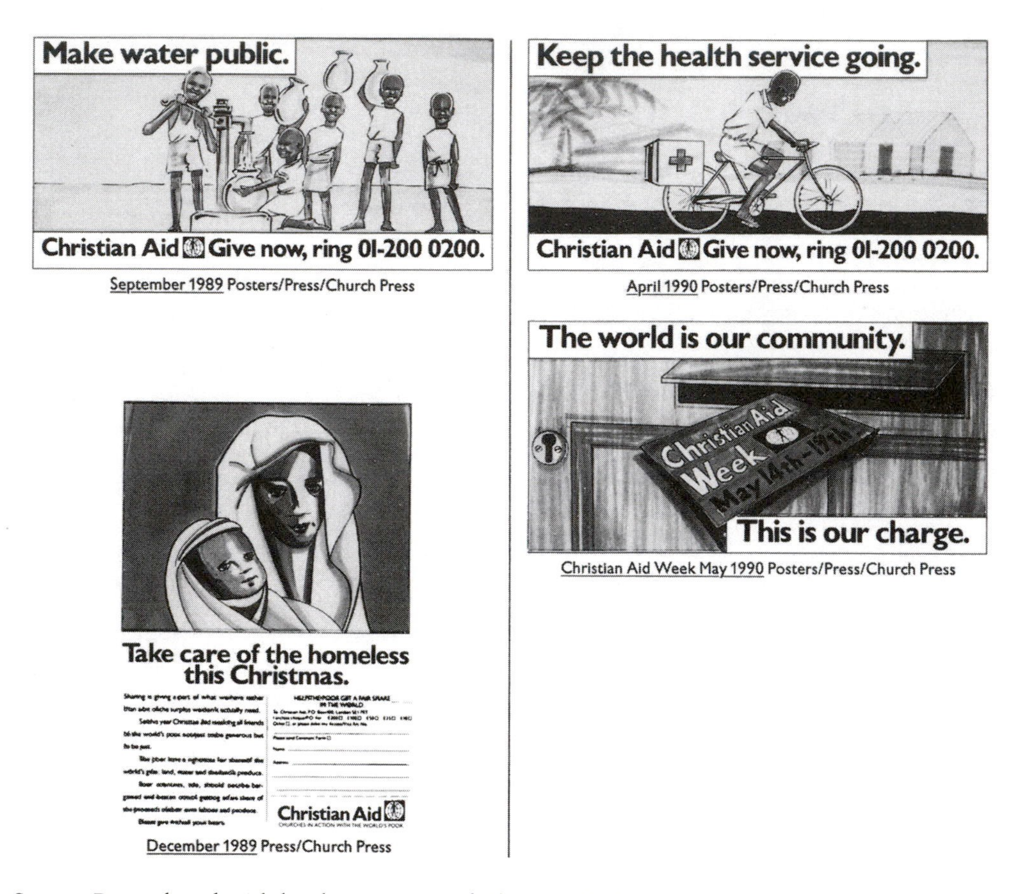

Source: Reproduced with kind permission of Christian Aid.

plight of public health services overseas (affected by structural adjustment programmes). This is, therefore, also a humanitarian and universalist message, a testament of similarity, of 'common struggle', over and above difference. All world citizens, we are given to understand, should have access to adequate health care, and all those who have it within their power to do so should ensure that public provision of health services are sustained, as Christian Aid does.

In the poster the short verbal message leaves room for a richer, polysemic visual image. First of all one is struck by the fanfare and intensity of colour. Strongly contrasted with the sepia tones, or black and white coding of negative images, and their habitually desiccated and desert environments, this is an upbeat rural tableau where the scenery is green and lush. The hut, the cow (complete with 'pats'), the young woman in a white, black and red (the three colours of Christian Aid's logo) *shalwaar kameez*, the white flowing scarf, the yellow flip-flops, the 'sit up and beg' bike, the old umbrella (black with a red handle) and the rudimentary medical equipment (one bag with a green cross suffices to contain all that she needs) all serve to *connote* spatial location. It achieves this in particular through evoking 'local colour' – what Barthes (1971) might call 'Indian-ness' – and, perhaps, a feeling of 'community' (is she smiling at those inside or outside the frame, or both?), but most especially through invoking 'authenticity' (the photographer and the subjects were there and, what is more, Elizabeth has mud on the tires of her Phoenix bike).

It denotes a different reality, a distinct type of health care; the health worker's 'medical technology' is contained in the bag with its green cross, and her transport is the bicycle. This could have the positive connotations of 'primary health care' – grassroots, appropriate and participatory – or the negative ones of 'rudimentary' health provision – technologically backward and ineffective. This largely depends on the reader of the image. The latter reading, however, would be somewhat contradictory, since the image's tone is clearly joyous, and the connotations of success and dynamism are conveyed by both visual images and verbal messages.

What makes the poster compelling, what might be called the *punctum* (Barthes 1984: 27–60; Graham-Brown 1988: 3) – that 'something' which draws the viewer in and requires his/her attendance, and which presumably distinguished this image from others during selection – is clearly the central *subject*: 'Elizabeth' (identified and depicted alone in the press advertisement) despite the fact that she is not centrally placed. This glowing image depicts her in a neat and clean suit, with a whiter than white scarf casually draped over her shoulders and wearing a beaming smile. She connotes health, a certain set of standards and purposefulness. She knows where she is going. There may also be an element of feminine 'liberation', the *connotations* of mobility, professional status and public participation defining themselves in opposition to consensual notions (in the North) surrounding Muslim/Asian women as domesticated and docile.

The selection of this image was clearly not accidental. Those who selected it did so, probably on the basis that it was oppositional and non-stereotypical. This circumscribes one's ability to read it pessimistically. Elizabeth *is* a squeaky clean, conscientious worker, so we cannot help but admire and warm to her. Moreover, for the more informed, the visual image and the verbal message combine to constitute her not as the *object* of development – helpless and despairing – but as its glowingly empowered *subject*: independent, competent and self-determined. These *connotations* can be read against the text of the press advertisement, which focuses quite explicitly on her diligence. In a number of respects she symbolises many of the key aspects of NGOs' new development thinking: the active citizen/participant who through her own self-realisation empowers others.

'Elizabeth' is notable by being a *named* subject and incarnates partnership while satisfying the need for a human touch in advertising. She rises above sheer statistics; she is as a genuine person, embodying the triumphs and tragedies of the developing world. Elizabeth is what Barthes calls the 'mythical signifier' (1989: 134–5), part analogy, part motivation; she purposefully hails the viewer in the name of 'Third World/ health/development/empowerment' and makes him/her take note of her as a self-determining subject. This naming device, therefore, encourages a convenient slippage between representation and intervention. There is an inference that by being the identified/iable subject of representation, Elizabeth is concomitantly, the identified/iable 'partner' in development. She is clearly not an object of pity, and although these visual images might elicit donations, the British donor is not misled into thinking that Elizabeth is exclusively dependent upon his/her magnanimity.

However, Elizabeth's status is somewhat uncertain. On closer observation, it is clear that she is not one of the 'poorest of the poor' (compare her with the old man). She could as well be a 'do-gooding' charity worker as a more politically motivated, empowered and self-realising subject. Ambiguity permeates at another level, left open by the text of the press advertisement (Plate 7.3). Care work is habitually feminised, so no revolution there: the particular roles being accorded to Elizabeth in maternity, pregnancy, nutrition, immunisation and sanitation work are hardly controversial. So though this image might be, and was, hailed as 'positive/political', it focuses on areas which NGOs might normally consider to be precisely those traditional arenas of charity concern. This image is advantageously ambivalent, since its message could as easily be read as a humanitarian message with charitable referents or as an exhortation to be in solidarity with the poor. There are also some intriguing and unanswered questions. Elizabeth's identity, for instance. This is a Christian name in a predominantly Muslim country. What does this imply about Christian Aid's selection of beneficiaries, or representatives?

In this short analysis, it has been shown how an ostensibly clear image such as the one of 'Elizabeth' can have many levels of meaning. Such a remark may not appear particularly astonishing given that advertisements are highly motivated representations. Although an analysis of content is useful, it can only go so far. To get beyond the construction of meaning, to evaluate the worth of individual representational documents, their content, signification and relevance, one needs to excavate their 'hidden histories' (Edwards 1992: 12). To gain greater benefit, reference must be made to other contexts, the context of image production, the motivation of those who produce them, the reasons why they attain the status of truth. We can only touch upon these briefly here. In certain respects this has already been delineated in terms of an emerging discourse where image and intervention were seen as inescapably intertwined. But there are other useful contexts, first that of photographic production. Despite appearances, this photograph's existence was not entirely 'fortuitous':

> This picture ... was part of a series of pictures. I mean, there was the health issue and ... we'd agreed that we would ... take local community health workers ... and X [the person from the advertising agency] had been looking at old ... stuff and he'd picked up ... the bike, you know, cycling around. ... He thought that would be a good idea, to have one of them cycling around due to the red cross, which wasn't such a great idea ... So I shot loads of pictures of ... these health workers riding around all over the place ... We ... thought we'd got it and then we were walking back from one village and we passed this little scene and I thought, that looks ... almost archetypal ... the ... old man and the cow and the little grass hut and ... the banana tree. ... And so he said, 'Okay, well look, get on your bike, cycle past this one ... say three or four ... [times] ... backwards and forwards on a bike'.
>
> (Author's interview)

Though the setting of the image was partly due to luck, the image was largely constructed – scripted and framed – before it was finally executed and selected. If this was a representation of reality, it operated purely in a 'generic' manner, in the sense that it is true that health workers *do* ride on bikes through the Bangladeshi countryside, but in this instance this 'truth' was somewhat deliberately 'caught' by the camera. It might be referred to as a *technical truth* (Adamson 1991), something which in all probability could happen, rather than actually did, unprompted.

Secondly, this image is deeply rooted in the culture of Christian Aid. This image defined itself specifically in relation to previous communication images and messages created by Christian Aid, it was evaluated in relation to these, and in event proved to be a stepping stone towards a new type of fundraising which ultimately led to a revision in communications strategy. Christian Aid, as its name implies, is an ecumenical organisation, with strong links overseas which it fosters through a network of 'partners' who are mostly church-based. In 1990 the guiding document at this moment was the social-justice-oriented pamphlet *To Strengthen the Poor* (M. Taylor 1987). Throughout the 1980s Christian Aid acquired a well-deserved reputation for its bold and innovative style of advertising and communication, which favoured depictions of single, and mostly inanimate, objects in its press and poster work. Strong messages drew on a sophisticated discourse of development and justice in the form of religious metaphor/allusions and word play.

This enabled Christian Aid to develop 'political' but non-confrontational messages in conjunction with iconographic and symbolic, black-and-white representations of woks, bread, ploughs, rocks, arms and feet but also the ubiquitous Christian Aid Week envelope, red with the logo 'slim jim' in a diagrammatic world.[9]

The 1989–90 series (Plate 7.4) was only the second to feature identifiable people, the first and only in the series to use full-colour posters. These images of Elizabeth were therefore different and *positive* because they distinguished themselves from previous Christian Aid communication strategies, but equally contrasted with *negative* images used by other NGOs (as defined in previous sections).[10] They did not in any way reify the desperate condition of the poor, with the connotations of 'all past and no future' (J. Taylor 1987: 50). The positiveness of the image was signified by its active nature and up-beat tone. By choosing a woman as main subject, they furthermore showed a leaning towards reversing certain other stereotypes about participation and development. 'Elizabeth' was the image of an empowered 'participating ... hardworking, industrious and self-determined' subject (Hart 1989: 14). The image was clearly a response to the prevalent and persuasive argument that the practice of development – intervention – was reflected and influenced by the images produced of it.

So by giving the image a series of contexts through which it can be framed and analysed, one aspect about positive images, at least, is brought into focus – their intentionality. Positive images are not, in any sense, closer to the 'truth'. They were/are very deliberate, highly motivated answers to the truth claims and immediacy of negative images. Departing from the proposition that representation and intervention were linked, they took as their specific task to counteract certain assumptions about development, developing countries and development agencies. It is this connection made between image and the practice that creates a 'realist' impasse in images of development. Development NGOs rely heavily on documentary photography. In large part this is due to the assumption that such photographs can provide transparency, an unmediated window on the world, fixing 'reality' or 'truth'. There is also the assumption that *positive* images are somehow closer and more representative of the truth than *negative* ones. Subsequent to the mid-1980s a great deal of effort and energy has been expended on ensuring that the photographer, the lens or the processing do not allow power to corrupt the integrity of this initial image and therefore its essential truth.[11] Yet the underlying supposition is that a photograph is somehow linked to truth, and can, therefore, provide evidence of it. Such assumptions are unfounded for Nichols (1991). For him, like Sekula, the deceptiveness of documentary or realist photography lies in its deliberate ingenuousness (his argument takes place in relation to film, but is equally applicable to photography):

> The [only] ... 'objective' truth that photographs offer is the assertion that somebody or something – in this case an automated camera – was somewhere and took a picture. Everything else, everything beyond the imprinting of a trace, is up for grabs.
>
> (Sekula 1984: 57)

So the root of photography's persuasiveness lies in its rhetorical strategies and stylistic form. Nichols' contention is that representations combine argument and evidence, merging representations *of* the world with those *about* the world. As such, photography in the realist or documentary mode functions both as a *reflection* of reality and as a *discourse* on it. Though the viewer may engage in the structure and meaning of the image, what s/he recognises, and inevitably responds to, is its 'realism' (Nichols 1991: ix–x, 177).

Nichols' critique undermines conclusively the polarities created between negative and positive images. For him, they cannot be oppositional entities which, respectively, mobilise falsehoods or truths, since he contends that *all* representations are quotations from, rather than samples of, reality and consequently hold the ambivalent status of both argument and evidence. By favouring positive images over negative ones, NGOs denied positive and negative images the same level of motivation because positive images were not, ostensibly, concerned with money but primarily with ethics (therefore better). So this proposition must be reassessed. It emerges that the discourse of accuracy and truth tells us more about how these representations are being used to validate intervention – development practice – than about their ability to depict accurately or truthfully (Hacking 1983: 146). So when considering NGO representational practices, the most pertinent question becomes: why is a realist, or documentary mode, of representation judged the most appropriate for development NGOs?

A possible answer to this question might lie in the fact that documentary photography invests the discursive assertions of NGOs with immediacy, poignancy, and most importantly, authority. The audience is encouraged to witness subjectively and believe the analogue of overseas reality with which they are presented. An objective depiction functions as evidence of NGO experience; it demonstrates that these development organisations 'were/are-really-there' and acts as conclusive proof that they 'were/are-really-doing-something'. Documentary/realist representations work simultaneously as the basis for knowledge and the validation for the correctness of NGO discursive assertions, but equally as inspirational images: images that require some form of active response. They seemingly anchor and limit the conditions of truth and, therefore, meaning.

There is a final aspect worth considering. Through privileging the documentary photograph, NGOs present photography as primarily a medium not an instrumentality, thereby covertly dissociating representation from power. The recognition of power is made in institutional terms, in the form of guidelines. The guidelines subsequently drafted after 1984–5 by Oxfam (1987), SCF (1988, 1991) and the NGO-EC Liaison Committee (1989) emphasised the need for greater accuracy and truth but in tandem with substantial verbal contextualisation, dialogical production and appropriately controlled methods of reproduction. So it is not a question of 'false consciousness'. Development organisations know that representations fuse power with knowledge, but this is not something which is conveyed to their audiences, who are presented with ostensibly unmediated realities.

CULTURAL CONSEQUENCES

The outcome of the events of 1984–5 initiated a fundamental reconsideration of the content and process of production of visual images amongst British development charities. The most immediate result was a move towards positive imaging, producing images which stressed the strength, dignity, and self-determination of the human subject in the face of adversity, rather than images which objectified and dehumanised the subject of development (Montague 1988; Hart 1989). But the reflections around imagery put into question the quality and the quantity of NGO images as well as their techniques of production and reproduction. This served to link, decisively, the practice of representation – the image of development – with the practice of development or intervention. But this operated according to a realist premise, positing that 'out there' there was one real truth or reality ready to be captured and conveyed in visual images. Yet, as we have seen, all visual images – *positive* or *negative* – articulate something very different from dry truth or reality (Chaney 1988).

To privilege imagery is to conceive of development NGOs in cultural terms. First, in that they do possess, an institutional culture, they perceive development as a practice which defines problems and prescribes methods for achieving certain delineated goals. As organisations they have a certain way of seeing the world, producing discourses on how the world is and how it should be. These understandings may change over time and/or differ across the gamut of organisations. Nevertheless, as a group of organisations involved in a common project, development organisations actively articulate sophisticated understandings of development practice, which allows them to be seen as a moral and intellectual community of practitioners. Second, development NGOs are cultural institutions. They do not confine the understandings and beliefs they hold to themselves. Indeed it is in their interest not to. They are actively engaged in creating meaningful images and messages on the subject of the developing world for the purposes of dissemination and conversion. Only by reaching a wide variety of people and ensuring that they take on board the challenge and the 'need' for development can development NGOs ensure their ultimate survival. As prolific cultural producers they search constantly to circumscribe, and hopefully fix meaning, primarily through the regulation and use of documentary photography. Indeed, one might speculate the extent to which sophisticated ethical discussions such as those taking place in the late 1980s were a function of financial buoyancy.

CONCLUSION

In this chapter, the culture of development has been discussed in the light of representation or imagery, in the knowledge that to do so is to resituate the debate about culture and development. The issue of imagery is of relatively recent emergence and has been accorded importance because the images of development were and are said to reflect and refract onto development practice. This chapter sought to elucidate both the genesis of this debate, its context of emergence, and how representations could be understood in their own right. Although it might argued that the question of representation has undergone its most public moment in line with current financial restraints and that development NGOs occupy the public consciousness to a lesser extent than they did in the mid-1980s, questions of culture remain. The history of development is, in many ways, a short one, far outstripped by the valiant and constant attempts to create moral communities and cultural consensuses around meaningful truths.

NOTES

1 This was an image which was developed quite early on by Northern NGOs to denote the raw urgency of 'need' in the developing world (Black 1992: 195; *New Internationalist* 1992) (Plate 7.1).

2 In its report in 1987 The Charities Aid Foundation (which gathers statistics on giving in Britain) concluded that between 1977 and 1986 the underlying trend of giving to international aid was steadily rising, despite a dip in 1982. The most striking aspect was the jump from just over 11 per cent of the total voluntary income in 1984 to 22 per cent in 1985 (Rajan 1987: 88). This increase can be seen also in relation to specific voluntary agencies. Oxfam's income in the period spanning the years 1981–2 to 1990–1 leapt from £16.26 million to £69.223 million (Burnell 1992: 292; Oxfam 1991). By 1988 of the total $51 billion aid received by developing countries, $3.6 billion was provided by voluntary agencies, of which Britain contributed $239 million (Burnell 1992: 24).

3 This new development thinking advocated a more equitable and active role for Southern partners in development, with methodologies that spoke of the rights and entitlements of the poor, and their role as partners, participants, empowered and self-reliant, ensuring that the poor could *have* more, and *be* more (Boyden and Pratt 1985: 13).

4 The Disaster Emergency Committee (DEC) appeals are group-held. At this point the main actors were Christian Aid, Save the Children Fund, Oxfam, the British Red Cross, and the World Wildlife Fund for Nature. Today the DEC includes twelve charities.

5 In January 1985 Geldof visited six capital cities in the Sahel, accompanied by the world's press. He met Mother Theresa in an airport lounge, and had specially arranged flights with the compliments of the Ethiopian Government to areas previously out of bounds for 'security' reasons (Harrison and Palmer 1986: 131; Philo and Lamb 1986: 26).

6 The Band Aid Trust was wound up in 1991 (Benthall 1993: 84–6, 235).

7 Christian Aid had four stable points in the fundraising calendar represented by posters: September, Christmas, April foreshadowing the Christian Aid Week poster, and Christian Aid Week in May.

8 The photographer was a male and a regular Christian Aid photographer.

9 Christian Aid's logo has since been changed.

10 Christian Aid produced the educational leaflet outlining its view on images in 1988 *Images of Development, Links between racism, poverty and injustice* (Christian Aid 1988).

11 The formal guidelines produced by Oxfam (1987), SCF (1988, 1991) and the NGO-EC Liaison Committee (1989) seek to regulate all the stages in photographic production and reproduction.

8

REPRESENTATIONS OF CONFLICT IN THE WESTERN MEDIA

The manufacture of a barbaric periphery

Philippa Atkinson

INTRODUCTION

The reach and influence of the media in the modern world is so pervasive that it shapes, unconsciously, the way we think about foreign cultures with which we have little other contact. The media exerts an enormous power also on these cultures, in its effects on areas as diverse as foreign policy formulation, demand for exports or tourism revenues, or even visa applications to the West. In times of war and disaster in other countries, the media in some sense 'plays God' in terms of its influence on the extent and nature of Western responses. This was demonstrated during the Ethiopian famine of the mid-1980s, when the broadcasting of particularly emotive television images of starving children prompted a massive humanitarian response, leading to the setting up of Band Aid and Comic Relief (Benthall 1993: 84–7).

The media has always played a major role in the formation of ideas about foreign culture, first through the press, and, since the advent of film technology, through the more powerful medium of television. This has served to 'bring into our living rooms' images of previously inaccessible disasters and conflicts, contributing to the much heralded globalisation of human experience, and increasing the difficulty for the West of ignoring or dismissing such far away events. In recent years the ways in which the media represents such disasters and conflicts has come onto the agenda of those seeking to analyse Western roles in and responses to such events in the rest of the world. The media is under scrutiny in terms of its impact on shaping our cultural interpretations of distant wars, and through this, its impact on broader policy agendas. Its importance lies in its indirect influence in determining the level of humanitarian engagement on the part of the West, and the nature of Western political or military intervention.

At the level of national news, the media plays the role of creating a public forum for debate, and is important in the promotion of diffuse ideas and discussion within the national democracy (Curran 1996). International coverage, particularly of far-off conflicts, is naturally more limited however, by both space and interest, and thus has a far less positive role. While the media as an abstract institution may be neutral in that it has no intrinsic interest in portraying foreign conflicts in any particular light, in practice it is influenced by a wide variety of different factors. The lack of localised knowledge to inform media coverage of distant events and the tendency towards sensational short-term reporting serve to bias reporting, or at least to reinforce existing biases.

There are no mechanisms to guide the style or content of media coverage, and there is no assurance of balance or accuracy in reporting. That the media *should* have some responsibility for ensuring the integrity of its product, it is argued, is because of its interactive role in both reflecting and helping to shape public opinion.

It has been suggested that, in the past, intellectual arenas helped to legitimise colonial interventions through their debates on natural selection and cultural relativity necessitating the destruction of 'barbaric' civilisations by progressive ones (Lindquist 1997). Today, it is argued that the media is playing a similar role in terms of the manufacture of a 'barbaric other', or is at least colluding in the process. Its superficial and biased reporting serves to obscure, in part deliberately, the political and geopolitical realities of local conflicts, by focusing on the unhelpful explanations of ethnic hatred or competition, and on the savagery of those engaged in fighting. This reinforces the political agenda of those in the West whose actions may be in some way contributing to the perpetuation of these 'peripheral' wars (Duffield 1997). It is important in helping to conceal the exploitative aspects of Western or free-market commercialism, which is a directly causal factor in many distant wars, as it was in the days of the colonial project. This is part of a broader post-Cold War trend of the establishment of the supremacy of the Western liberal ideology and market-driven culture.

The media that is discussed in this article is the Western media. The aim is not to suggest that it is a homogenous entity, but to highlight its general tendency to reinforce the values associated with Western culture, and thus to perpetuate misunderstandings about the nature of foreign conflicts. This is partly for institutional reasons, such as the structure and market-driven nature of the Western media, and partly because the media can only be a reflection of the ideologies and understanding of its own culture. The media tends thus to reflect the concerns of its Western audiences, and their own internal fears and questions. This is shown clearly in the differential reporting of wars closer to home, such as in Bosnia or even Israel, where an attempt is made to understand and relate to some of the local realities, as compared with reporting about African wars, where images of barbarism and anarchy abound, and where the causes of conflict are portrayed as incomprehensible.

The chapter is in three parts: the first, a discussion of the structural factors that encourage superficial reporting of distant conflicts; the second, an analysis of the new barbarism discourse and its impact on policy; and finally a discussion of the implications of this type of media coverage on Western understanding of and responses to such events.

MEDIA INSTITUTIONS

The media is dominated by Western agencies, which in the 1980s were producing and transmitting over 90 per cent of the world's news (Harrell-Bond and Carlson 1996: 104). Some media corporations, such as Agence France Presse, are government funded, and their output thus reflects to some degree the policy and agenda of that state. Most are private corporations controlled by individuals or small groups of businessmen with particular biases or traditions, which thus project their news accordingly. The extent of First World domination militates against the development of a more southern-orientated perspective, and contributes to the lack of depth in reporting of foreign cultures. Western political considerations sometimes have a direct role in influencing media coverage. The coverage of press agencies present in Iraq during the 1991 Gulf War was closely monitored by the Allied forces, in an example of direct manipulation of the media by Western political powers. This demonstrates the difficulty for the media of retaining independence when powerful political agendas intervene in a war situation (Benthall 1993: 209).

The way stories are made and transmitted furthers the tendency towards superficial coverage. Producers are increasingly subject to market forces, driven to put out eye-catching scoops that will attract new customers, or at least keep existing ones. There has been much discussion on the impact of increased competition on the quality of traditional press and television, and the trend towards tabloid-style journalism is continuing. The effect of this is to underscore existing tendencies of superficial and short-term coverage. Disaster stories, for example, will go in and out of the news depending more on the nature of competing news stories, or the extent of existing coverage, rather than on the basis of their intrinsic interest or newsworthiness.

The trend towards directing news to the 'lowest common denominator', by using simple stories and easily understandable images, is combined with the nature of news reporting, as fast-moving and predominantly space-filling, to produce a situation where sensationalism rules, and little in-depth political analysis is developed. News producers have to be aware of all stories and potential stories as they are happening, and choose which to cover for various reasons, but above all what they believe will be of interest to the maximum number of people. Their main driving force is to fill space, in seconds or in column inches, and their own understanding of the issues they are covering is, by necessity, limited. Even where specialist reporters are employed, such as by the *Economist*, or for in-depth television documentaries, the writers or producers may only be covering the story for that week or issue. Cost cutting at many newspapers and television channels has greatly reduced the number of dedicated reporters, so, for instance, Africa Editors at the daily newspapers have disappeared. There are notable exceptions to this trend of course, and individual journalists such as Robert Fisk demonstrate the depth of analysis that can be gained through specialisation. These cases are, however, rare.

The use of traditional story-telling techniques by journalists also discourages more detailed assessment of particular situations. The setting of a disaster news story in the familiar language of story-telling helps to render a far-away event more accessible to Western viewers or readers. Good and evil may be symbolised in some form, and the West, or the US may be set in the role of hero or mediator intervening (Benthall 1993: 188–91). Complicated events can thereby be transformed into easily grasped 'soundbites'. The portrayal of wars as ethnic conflicts may be more easily understood than complex political or economic analysis, and famines are explained by drought, so complicated issues of politics again can be avoided. This method of journalism, of a packaged image and message, satisfies both the necessity for timeliness and interest, and the belief that consumers are incapable of understanding anything deeper. It is self-perpetuating, as, for example, once the story of ethnic conflict is established for one war, it becomes the easiest explanation for both the producers of the media and the consumers, and can be applied to other situations, becoming a convenient point of reference. It may also be in the interests of other parties, such as governments or aid agencies, to portray conflicts in a simplified way, in order to make them more accessible to Western donators, for example.

Television imagery is especially powerful, and can establish or set ideas in a particularly insidious way. Television has gradually replaced the printed word as the main source of news for the majority of people, and especially for news concerning overseas affairs (Benthall 1993: 201). Although standards of technical professionalism are high, and responsibility for stories is diffused among more people than in newspapers, the fast-paced nature of television news as a medium for transmitting images means that, even more than with the press, fashion and novelty are the determinants of what becomes news. The power of the imagery of television means that film portrayals of starving children in wastelands become entrenched in the minds of viewers, who may develop their ideas of foreign cultures purely on the basis of such images. Many people now view Africa as a continent devastated by war and famine. The 'heart of darkness' image, used in film and quoted time and again in headlines, encapsulates this perception.

Superficial reporting and journalistic prejudices in terms of simplification or misrepresentation may result in the portrayal of positive or negative imagery, but both are potentially damaging in that cultures are in some way misrepresented. A common image of diverse Third World cultures is that of happy peasants tilling the land or engaging in simple crafts, working hard but honestly, with a backdrop of sunshine and blue skies. This image is often presented through the tourist industry, which seeks to attract Westerners to visit countries supposedly unsullied by the negative aspects of industrialisation. Authenticity is a major selling point, and cultural phenomena such as dances or dress-styles are shown as if they are unchanged over many generations. While there may be little in such images that is inherently disturbing, the portrayal of falsely romantic images strengthens the stereotyped views many hold of Third World countries as simple and backward, and acts to obscure important modern realities, such as political repression or the drugs trade. The tourist trade can help to support repressive political regimes both materially and symbolically, and in some countries such as Turkey and apartheid South Africa opposition groups have campaigned to discourage tourists from visiting.

The imagery that is used, particularly in relation to Africa, is, however, becoming increasingly negative. The tendency for coverage to relate exclusively to disaster stories is becoming more marked. Many African countries will not figure at all in the media, either television or press, unless affected by war or natural disaster. Coverage of other stories tends to be limited to South Africa and Kenya, countries in which the British public is assumed by editors to have some interest. The continent is seen by the media and the public as a mess of ethnic wars, environmental degradation and famine, with little hope of joining the modern world of capital accumulation. That African culture is in some way to blame for its problems is gaining ground as an explanation. This negative imagery is both created and sustained by media portrayals that fail to convey either the complexity of the political problems faced by many African countries, which in all cases have deep historical roots in colonialism and subsequent exploitation, or the reality of local efforts to challenge the problems they face.

THE NEW BARBARISM THESIS

The trend towards simplistic analysis of Africa's wars and disasters has been examined by some writers in terms of current Western theories of culture and race. Paul Richards, an anthropologist, terms as 'New Barbarism' the tendency to portray African wars as anarchic and primitive, and argues that careful analysis is necessary to reveal the rational motives behind actions that on the surface may appear barbaric (Richards 1996). The New Barbarism thesis, developed originally by the influential American academic and journalist Robert Kaplan (Kaplan 1994), suggests that African wars are motivated by fixed ethnic and cultural realities, and are thus beyond the reach of Western understanding and help. The title of Kaplan's article 'The Coming Anarchy' encapsulates the emotions of fear and dismissal prevalent in the reactions of many to images of war in Africa and other peripheral areas such as Afghanistan. Kaplan blames 'cultural dysfunction', 'loose family structures' and 'communalism and animism' for Africa's current problems (1994: 44–76), helping to reinforce stereotypes of Black Africa as a dangerous and unpredictable place.

This type of superficial analysis supports policy strategies of disengagement and isolationism popular particularly within right-wing Republican circles in the US. By focusing attention on internal cultural factors, broader geopolitical trends are obscured, and isolationists are able to argue that Western intervention can have little positive impact on these wars. Thus both the causal role of the West is denied, and at the same time its potential to intervene constructively is lessened. Media coverage of the wars in Sierra Leone and Liberia has contributed greatly to the New Barbarism thesis, focusing as it has on the so-called primitive and bizarre methods of the fighters, and by failing to provide any analysis of the causes of the conflict. It has thus helped to perpetuate the failure within the international community to tackle the realities of the conflicts there (see Atkinson 1996). With little appreciation of the complexities of the internal and external factors that have helped initiate and sustain the wars in Liberia and Sierra Leone, reflected in the superficial and sporadic coverage in the international media, it has been difficult for either multinational institutions or interested national governments to formulate informed and effective interventions.

Two recent examples of reporting on the Liberian war will serve to illuminate these tendencies. Cannibalism in Liberia made the front page of the *Observer* newspaper in April 1996 when an out-of-date interview with a child fighter was recycled during the intensification of the war in the capital Monrovia (*Observer*, 14 April 1996). The fighting in the streets of the city was totally unrelated to the story printed, which referred to events in the countryside months, if not years previously, and little attempt was made in the article to analyse the current crisis. In June the country reached the main six o'clock news of the BBC World Service Television, with the showing of a scoop 'cannibalism' film of fighters removing the heart from a dead man (11 June 1996). There was no new news about the war in this story, and the teenage fighters in the film appeared to be playing to the camera. The newsreader, however, looked appropriately shocked as he related the story to the world. These examples illustrate the kind of false and negative imagery that can result from the media's tendency to sensationalism, and the lack of any serious analysis. Other reporting of the fighting during April and May came mostly in the first week of the crisis, a 'quiet' news week in early April, and focused on the evacuation of aid workers and other expatriates. News coverage of the war in Sierra Leone has similarly focused on atrocities committed by rebel fighters, with little attempt to understand the reasons for the war.

Mark Duffield develops Richard's critique of New Barbarism in his analysis of developmentalism, and what he calls 'New Racism', a trend in commonly-held perceptions of Third World cultures that is subtly informed by post-modern ideas of multiculturalism (Duffield 1996a: 179–86). The perception of foreign cultures and conflicts as fundamentally different from our own experience leads, Duffield argues, to interventions by the West that are greatly constrained by their failure to see or understand local realities. The conception of multiculturalism allows outsiders confidently to describe African experiences of war as savage or primitive, because it holds that such things are innate, rooted in African history and culture, and therefore unchangeable. Duffield suggests that this type of analysis works to disguise or obscure the actual power relations that obtain in war situations, and which are what influences the outcomes of different policy choices. The glaring omission in New Barbarist stereotypes is the lack of meaningful analysis of the modern African state, which plays a pivotal role as the arena where competition over access to power and resources takes place, and where the nature of links with the international economy are determined. The failure to understand the obtaining social, political and economic relationships in African wars produces policies which may unconsciously work to make the situation worse. Many modern Western interventions in African wars are characterised by this lack of political commitment and understanding.

The apolitical nature of policy analysis may be highly dangerous to those caught up in war and disaster, as it prevents an appreciation of how policies can work to strengthen those who do hold the power, and who are fighting the war (Duffield 1996a and 1996b). Food aid policy for example, is rarely, if ever, informed by any understanding of the local political importance of large-scale provisions of food. Control of aspects of food aid distribution has been helpful to many warring groups, in the Horn of Africa, in Liberia and Sierra Leone, and in the Hutu refugee camps in Eastern Zaire. In the last case, control of registration of beneficiaries was delegated by UNHCR and WFP to Hutu 'traditional' leaders, many of whom were implicated in the organisation of the genocide in Rwanda. These leaders later took over actual distribution of food in some camps, which gave them in effect life and death control over the refugees, and helped them to prevent the return of the Hutu population to Rwanda (Anderson 1996). The conventional understanding of the genocide, as reported by the media, as the latest manifestation of ethnic hatred, was in part responsible for informing the decisions that were taken. As long as policy makers fail to analyse wars in terms of the obtaining power relations, in this case, the bid by extremist Hutus to retain or regain control of the Rwandan state, such mistakes will continue. The issue of the growing use of humanitarian relief by the West as a substitute for political intervention (Duffield 1996a; Macrae 1996) is an even more fundamental policy failure informed by the same thinking.

Some commentators would argue that the use of superficial images to suggest the inherently barbaric nature of African wars is in fact part of a more deliberate strategy to underplay the role of the global economic and political system in sustaining such wars. In order to avoid a serious examination of the role of international business and political relationships in fuelling war in peripheral areas, it is convenient for the West to interpret conflict as based on internal, sociological factors. Media imagery reflects this concern, reinforcing stereotypes of the distance between the civilised West and the barbaric fringes of the world, where the rules by which we guide our lives are not supposed to apply (Duffield 1996a). The images projected are similar in content to those of the nineteenth century when representations of 'primitive' civilisations as savage and brutal helped in the justification of colonial interventions (Lindquist 1997). This imagery is used today to help disguise the inter-relatedness of the conflicts around the world, and the integral role of peripheral areas in the world economy as suppliers of commodities to the West and buyers of arms products. It further allows for the delivery of humanitarian aid as the main Western response to conflicts, to be carried out in an apolitical manner, with little concern about the political impact of the often large amounts of resources that are thereby supplied.

IMPLICATIONS

The media both helps in forming such perceptions, and is itself informed by them. 'New Racist' understanding of foreign cultures and conflicts impacts on policy formation at a number of levels. Western policy makers are influenced directly by the media, and indirectly through the lobbying from the different actors who are themselves informed by media representations. Decision-makers within aid agencies, for example, are subject both to pressure from the public of charity-supporters, as both financial supporters and political activists, and to the decisions of their main donors, the Western governments and government agencies that control aid budgets. UN bureaucrats are less directly affected by public opinion, but play an important role in actually providing information to the media themselves. Government decision-makers are guided by diplomatic and strategic demands, which involve economic interests, cultural or historical ties, and global political relationships, as well as, to some extent, by public opinion. All of these various groups and individuals are affected by media representations, and in turn, help shape the perceptions of those groups which make up the media.

The power of the media lies thus in its influence on the groups of people whose impressions of foreign culture have an actual impact on events in Third World countries through the policy strategies chosen by the West. Its responsibility lies in its role in the provision of information and analysis, as acknowledged by CNN for example in its claim to be 'informing the world'. The growing pressure of the marketplace at all levels within media organisations and the institutional constraints thereby imposed create many difficulties in the fulfilment of these obligations: the tendency to sensationalism and superficiality; the random forces that determine which disaster news will reach the screens or the front page; the political considerations that influence the way news may be presented – all act to prevent a free flow of informed commentary on far-away events. Underlying these problems, the concepts of New Barbarism pose the danger of biasing the news that is reported through a fundamental misreading of the realities of the power relations at work in conflict situations in Africa and elsewhere.

The importance of humanitarian relief in helping protect the lives of those affected by war has been remarked on by many, including Benthall, who suggests that aid flows have now become the main cash-crop in many countries facing crisis (Benthall 1993). The growing domination of the media by large cable-television corporations and newsgroups subject to market pressures in an unprecedented way makes it even less likely that such crises will be portrayed in an accurate manner, and gives such organisations unprecedented power over the amounts of money that may be raised in an emergency, and the types of policy that may be pursued by the West. Little space remains within these organisations for the individual writers and photographers who have traditionally brought news of disasters and wars to our living rooms, and little time can be found for today's news journalists to study or appreciate the particular aspects of particular situations. Some attempts are being made, mainly by aid agencies, to work against these trends and improve the quality of reporting on disasters. Some agencies, such as Save the Children, have adopted guidelines to monitor their own use of images in disaster appeals, in attempts to avoid contributing to the negative imagery of helpless Third World populations reliant on the West. Bernard Kouchner, founder and former director of Médecins sans Frontières, has argued that humanitarians must form a close alliance with television journalists, whose images of disasters play such a crucial role in shaping perceptions and raising funds (Kouchner 1991). By working together it may be possible to improve the understanding of journalists, and thus the quality of information reported.

While the danger of misrepresentation could be somewhat lessened by improvements in methods and standards of reporting within Western media organisations, the fundamental issue of misconceptions of the causes of such crises will only be addressed through fundamental change. The most important mechanism for this change will be the support, financial and symbolic, of Third World media. The development of Third World media can play a crucial role in the evolution of a deeper understanding of Third World culture and issues internationally, as well as having an important impact at the national level. Access to, and production of, more and better quality information and comment within the Third World would work to empower local populations at a number of levels. Their own understanding of the issues faced in their societies, and their ability to express this understanding in the international arena, would greatly benefit from the development of local discussion. As mentioned above, within a national democracy, the media can play a vital role in raising and discussing issues, and in airing diverse opinions held by different groups. 'Preventative' journalism may be particularly powerful in conflict situations as in Africa in helping to inform and unite people in opposing the war, or in resisting powerful groups. Richards has discussed the potential use of radio in conflict resolution activities in rural Sierra Leone (Richards 1996: 159), and it has been argued that local radio in Burundi played an important part in discouraging participation in ethnic cleansing in 1995 (Kayanja 1996).

Organisations such as PANOS are playing an important role in this development, by working with existing media in Third World countries and supporting the development of new skills. It has been suggested recently that plans made in the 1970s to develop a 'new world information and communications order' spearheaded by UNESCO must be regenerated to redress the present imbalance (Harrell-Bond and Carlson 1996). Under the UN Special Initiative on Africa, US$7.8 million has been allocated for communications for peace-building over the next five years (Kayanja 1996). The Internet may prove a valuable tool in this endeavour, both to assist in increasing the availability of information overall, and as a medium for communication of southern ideas and analysis to a wide audience. In the light of the continuation of the current crises in Africa and the growing importance of humanitarian aid in the West's responses to disasters and conflicts, it is vital that such initiatives are activated and supported. The inability of the Western media to portray accurately the realities of far-away conflicts, and the dangers presented by misrepresentation, make the development and support of southern media an urgent task.

9

SEX TOURISM

The complexities of power

Jan Jindy Pettman

INTRODUCTION

Sex tourism is an industry, a set of social-sexual relations, and a site for the exercise of different kinds of power relations. Exploring sex tourism means asking questions about relations between sex and power, men and women, First and Third Worlds, and relations across national, racialised and cultural boundaries. The increase in sex tourism is linked directly to processes of global change, most noticeably to aspects of globalisation and restructuring which both facilitate expanding international travel and tourism, and generate increasing inequalities within and between states.

This chapter begins by looking at the different discourses around prostitution and the politics of sex more broadly; then traces the rapid increase in tourism as experience and as big business since the 1960s. It then investigates the relationship between prostitution and tourism summed up as 'sex tourism'. It looks at two of the most popular sex tourist destinations, Thailand and the Philippines, to explore complex linkages between militarism, development and sex tourism; and the connections between representations of sexuality and cultural difference. Here male/female and 'First World'/'Third World' disparities in power and dominant group men's racial fantasies of 'others' materialise in the sexual use made of the bodies of Southeast Asian women and children (Pettman 1997).

A SEXUAL POLITICS OF PROSTITUTION

'Prostitution occupies a significant position at the intersection of feminist [and other] debates about the relationship between power, sex, sexuality and work' (Sullivan 1995: 184). While there are differences over time, place and class, discourses of sexuality and femininity frequently distinguish between good women and bad women, with virtuous wives and prostitutes the emblems of this difference. The good women work for their families, and provide sexual services for their husbands, as a labour of love. So it is not sex as such that defines the prostitute, but rather the payment for sex, and providing sex for more than one man. The lines may blur. A woman who pursues her own sexual pleasure or does not appear under the control of a man may be labelled a prostitute – an accusation which in many cultures severely reduces the woman's marriage prospects. But using prostitutes is rarely held against men. There are different constructions of 'the prostitute', varying over time and place, but often including representations of prostitution as immoral; as sexual oppression; as economic exploitation; or as economic necessity or opportunity. Prostitution as immorality sees the prostitute as a fallen woman, or prostitution as a social evil in which the prostitute is caught. Prostitutes are regarded as selling their bodies; though many prostitutes reply that they are selling only sexual services, and that their identities are not confined to or limited by the work they do. Social stigma attached to prostitutes has been recharged in recent constructions of them as carriers of AIDS, with the threat of more surveillance and control; while the dangers to prostitutes from clients are rarely taken seriously.

There is much debate on whether marriage and good morals are undermined by prostitution, or validated and protected by it. Social attitudes and the cultures of prostitution vary (Davis 1993). So in Thailand, for example, there is wide tolerance of men's use of prostitutes, where some 4 to 6 million Thai men patronise a brothel at least once a month (Dunn 1994: 10). This is seen by some women as less threatening than their husband taking a second wife. Feminists differ in their views on prostitution (Shrage 1994). Some see constructions of the good woman and of deviant sexualities as part of the repressive regime of sexuality, and call for an end to criminalising prostitution. Other feminists view prostitution as female sexual slavery, an extreme case of women's condition, where they are expected to service men's needs and fantasies, and often lack either economic independence or bodily autonomy to secure their own lives (Barry 1985). Women are seen as sexual objects, there for (heterosexual) sex. So debates about prostitution are, often, debates about gender relations, sex and sexuality more generally.

Prostitution is also often understood in terms of exploitation, or of opportunity, in terms of women's work and status more generally. In this reading, women's economic dependence on father or husband and lack of alternative employment or adequate social security force women into the sex industry. This is especially so where rapid social change, rural impoverishment and displacement force young women into towns and cities in search of jobs. Prostitutes are then seen as victims of poverty or development, as well as of patriarchy.

Prostitutes, especially those organised and politicking as sex workers, challenge the victim imagery of those who argue sexual oppression or economic exploitation (Sullivan 1995). They assert that prostitution is work, and that many prostitutes make a rational choice in terms of job alternatives and the returns available from their work. A fierce debate rages between those who wish to assert distinctions between forced and free prostitutes, though both unequal gender relations and economic inequalities undermine any simple view of coercion or choice.

There are moral panics and political campaigns around prostitution, especially concerning those seen as most vulnerable, in child prostitution and trafficking across state borders, and in terms of international sex tourism. But prostitutes work under very different conditions in any one country. It is usually assumed that the client or sex buyer is male, which is overwhelmingly but not always the case. Far more men than women go as tourists to sex-destination states like Thailand and the Philippines, mainly from Western states and Japan. It is also often assumed that the prostitute is female, usually a young woman or girl. However there is a trade in boys and young men's bodies, too, though mostly for men's use. Some older women supplement their incomes providing services, including sex, to itinerant or local men; and working-class and poor men may use prostitutes whose earnings are somewhat above their own. So the gender, age and class dimensions of prostitution need interrogating, to see whether they compound or subvert dominant power relations.

It is impossible to pursue these debates without acknowledging wider power relations, which include gender relations and culturally shaped notions of masculinity and femininity; business and other interests in prostitution; and the role of the state in criminalising, circumscribing or regulating prostitution. Where prostitution is illegal, 'protection' abounds, often in the form of pay-offs and direct or indirect participation of state agents in the industry.

AN INTERNATIONAL POLITICS OF SEX

In many states, significant numbers of prostitutes come from racialised, minority or 'foreign' backgrounds. Racialised sex carries with it meanings informed by colonial relations and contemporary unequal international relations, where difference is sexualised, and the exotic and the erotic are often associated.

The demand for 'other' women for paid sex sees an international trade in women from those countries which also function as sex tourist destinations. So 286,000 Filipinas and 50,000 Thai women entered Japan under the euphemistic 'entertainment' category between 1988 and 1992 (David 1992). In Australia there is evidence of the importation of Thai women to service the 'non condom' prostitute demand, in a situation where local sex workers have successfully organised for safe sex (Brockett and Murray 1993). In the case of sex tourism, however, the client goes overseas in search of sexual adventures, seen by some as a form of sexual imperialism, or sexploitation.

Thanh-Dam Truong (1990) analyses international tourism by putting together a political economy of women's labour with sexuality. She notes:

> The intersection of prostitution and tourism cannot be understood as a patchwork of discontinuous events resulting from individual behaviour, or simply as a synchronic expression of sexism and racism. Instead, it must be placed in the context of the operations of relations of power and production in the field of air travel which preceded its development. The emergence of tourism and sex-related entertainment is an articulation of a series of unequal social relations including North–South relations, relations between capital and labour, male and female, production and reproduction.
>
> (Truong 1990: 129)

Tracing the exercise of power in the construction of sexual subjectivity, she asks about the relationship between constituting sexual subjects and economic relations, especially in terms of the sexual division of labour. Prostitutes' labour contributes sexual services, usually to men; but also contributes to the global production of leisure services, and to business and state wealth. This approach leads to asking how it is that particular developing countries become integrated into the international division of labour through the provision of leisure services, and how sexual labour becomes integral to this process (Truong 1990: 100).

INTERNATIONAL TOURISM

Since the 1950s, there has been an enormous increase in the numbers of people crossing state borders for tourist adventures (see Table 9.1). In 1950 there were some 25 million international tourists; 165 million in 1970; 459 million in 1990; and over 500 million now. This dramatic upsurge was spurred mainly by the increasing leisure time and incomes in the rich states, together with technological developments, fuelling the growth of airline, accommodation and other tourist services into one of the biggest and most lucrative transnational industries in the world.

Tourists are often defined as leisured travellers, seeking short or longer term escape from

Table 9.1 Arrivals of tourists from abroad and receipts from international tourism

Arrival of tourists from abroad (Day visitors excluded)				*Receipts from international tourism (International transport excluded)*		
Years	Total Thousands	% change over previous year	Index 1950 = 100	Total US$ million	% change over previous year	Index 1950 = 100
1950	25,282	–	100.00	2,100	–	100.00
1960	69,320	10.61	274.19	6,867	12.58	327.00
1965	112,863	7.90	446.42	11,604	15.20	552.57
1970	165,787	15.52	655.75	17,900	6.55	852.38
1975	222,290	8.08	879.24	40,702	20.34	1,938.19
1980	286,249	1.12	1,132.22	105,198	26.23	5,009.43
1985	329,538	3.29	1,303.45	117,374	4.36	5,589.254
1990	459,212	6.56	1,816.36	264,714	21.22	12,605.43
1995	561,027	2.78	2,219.08	380,693	10.17	18,128.24

Source Adapted from from World Tourism Organisation, Madrid

home or ordinary life; though in the case of sex tourism they are supplemented by work-related transnational mobility, particularly of military personnel and business, professional and technical workers. Wealth, age, gender, nationality and other factors determine who can have leisure time and afford the fares and other expenses. The huge multinational industry of airlines, tourist agencies and hotels is based largely in the rich and tourist-sending states. The industry sells a range of destinations and services, including package holidays for specialist markets, for example in sex tourism, or eco-tourism. International tourism offers commercialised hospitality and commodified culture. The promise of 'excitement, of authenticity, of the exotic and of the extraordinary' (Sharpley 1994: 32) is offered in a safe, well-organised environment.

Tourism is also a development strategy for tourist-receiving states, especially those seeking foreign currency to service mounting international debt repayments. Growing liberalisation of international trade and often falling prices for traditional tropical state exports combine with pressure from international financial institutions like the World Bank and IMF to increase export earnings and open economies to transnational business and finance. Thailand declared 1980 its Year of Tourism. Foreign arrivals in Thailand in 1960 numbered only 81,340, but in 1986 2.8 million arrived (Hill 1993: 136). Tourism is the largest single source of Thailand's foreign exchange earnings (see Table 9.2). But there is a massive leakage of tourist earnings to Western and Japanese-owned industry interests, estimated in Thailand's case to be around 60 per cent of tourist money spent in Thailand (*New Internationalist* 1993: 19).

Tourism is viewed by some as an economic opportunity and a business vital to national development, and is sponsored by governments and their advisers and backers. Others see tourism as a form of neocolonialism, as dependent development, as exploitation and distortion in political economy and in social and cultural relations (Brohman 1996). This is especially so in 'First World' – 'Third World' tourist exchanges, where cultural difference and even poverty are 'staged' for rich state tourists.

INTERNATIONAL CULTURAL POLITICS

Tourism provides experiences to be consumed. International tourism offers escape, adventure, daring, doing something different, being out of place. Notions of cultural difference and authenticity become part of the experience. Travel agents and airlines, tour guides and brochures provide images of other places and people – the latter usually marketed in contrast to the home place/people.

> What the brochures do not usually show are details and pictures of a destination's level of industrialisation, its problems of poverty, over-population and pollution ... the tourism industry markets and operates a sanitised authenticity.
> (Sharpley 1994: 147)

National identity and cultural difference thus become part of advertising and selling international tourism. These differences signal boundaries in power relationships (Swain 1995: 261), between 'First' and 'Third World',

Table 9.2 Tourist arrivals and income in Thailand, 1985–90

Year	Visitors (millions)	% growth	Tourist expenditure (million baht)
1985	2.438	3.9	31,768
1986	2.818	15.6	37,321
1987	3.483	23.6	50,024
1988	4.230	21.5	78,859
1989	4.809	13.7	96,386
1990	5.299	10.2	115,700

Source: Tourism authority of Thailand in Hall, C. M. (1994) *Tourism in the Pacific Rim: Developments, Inpacts and Markets*, Melbourne: Longman.

Note: Baht 25.35 = US$1.00

rich and poor, male and female, and often older and very young, in 'a peculiar and unstable combination of sexuality, nationalism and economic power' (Leheny 1995: 369).

Gendered imagery and representations of women have long been used to mark the boundaries of national, racialised and cultural collectivities (Pettman 1996). Women's roles as symbols of the nation, as group reproducers and cultural transmitters, subject them to forms of body policing which often include restrictions on their sexual relations with 'other' men. Women who provide sex for foreign men can thus be accused of national as well as family or community betrayal. On the other hand, some governments urge their citizens to be friendly to tourists, which might include providing some of them with sex. However, this does not prevent others from looking down on those who do.

'Complex gender and race hierarchies in a regional political economy drive sex tourism' (Swain 1995: 251). Tourist industry deployment of difference is gendered, racialised and culturalised. The tourist is often imagined and sought as male – an explorer, adventurer, engaged in action, discovery. The tourist destination is often gendered female, associated with nature, or with passive or erotic female sexuality. These images replay and recharge colonial and racialised images of 'the other', and so reproduce 'First World'/'Third World' difference yet again. They also function to associate border crossing with sexual temptations or transgressions.

'Third World' national, racialised and cultural differences are frequently marked as feminine and sexualised in the smiling Thai or Filipina air hostess, for example. In the case of sex tourism, sexual 'delights' are often explicitly advertised, utilising 'cultural' promises. These images connect with those already circulating in newspapers, magazines, films and advertising. 'The Asian woman' is packaged and sold internationally, with particular constructions of sexuality that suggest female availability and male adventure (Enloe 1990). So it is necessary to 'ask how cultural ideologies about race, nationality, ethnicity, class, age and so on, intersect with those about gender in shaping the commercial sex industry in our contemporary world' (Shrage 1994: 142). This includes the ways in which different masculinities are matched or appealed to in terms of images of 'other' women, enabling travelling men to purchase and play out their racialised sexual fantasies.

The utilisation of gendered and sexualised difference to 'sell' states and cultures as tourist destinations and the close associations between the entertainment and hospitality industry with the provision of sexual services makes it difficult to establish the relationship of prostitution to international tourism. Many come to look but not to touch, or at least not to do. But others come to do things they might not do at home, taking advantage of being unknown and unresponsible in the tourist place. They are able to persuade themselves – with the help of the tourist industry – that the sex is culturally sanctioned; or that they are simply 'helping out' the prostitute, or the economy (Pettman 1996).

MILITARISATION, DEVELOPMENT AND SEX TOURISM

Why do some states become sex tourist destinations, while other states 'send' sex tourists (Enloe 1990)? Why do some states permit, or encourage, sex tourism? Why do so many women, and children, come to work in that industry in states like Thailand and the Philippines?

Seeking answers means looking at the history of each state. Relevant aspects include indigenous prostitution; indeed local demand for prostitutes usually exceeds that for sex tourism, including in Thailand. In most places, too, colonial prostitution generated localised international sexual politics (though Thailand was not formally colonised). Colonial power relations were played out in the use of local women for sex by coloniser and military men; while coloniser women were 'protected' from sexual contact with local men. Sexual use was a form of, and reward for, colonial domination. This was frequently justified through ideologies of cultural difference which saw colonised women as promiscuous, immoral, or already degraded by their own men and culture. The good woman/wife and the bad woman/whore became a racialised boundary too.

In colonial relations, sex across racial lines

without responsibility or recognition was a by-product of colonial authorities' anxiety to secure the racialised boundaries of the colonial relationship, and to reserve for white and legitimate coloniser children rights of citizenship. This was especially important in terms of family-less or transient colonial officials, soldiers and traders, or where male settlers far outnumbered settler women. Where sexual relations across the lines were forbidden or frowned upon, illicit sex added danger and fed fantasies which are replayed in contemporary sex tourism.

Another key dimension in the development of sex tourism has been foreign military involvement. Not coincidentally, both Thailand and the Philippines had massive foreign military presences, and developed huge base prostitution industries (Enloe 1990; Sturdevant and Stoltzfus 1993). During the Vietnam War, Thailand became the major rest and recreation destination, servicing some 700,000 US military personnel between 1962 and 1976. Prostitution flourished too around the Subic naval base and Clark air force base in the Philippines. Careful negotiation and management of military sex by local, state and US military authorities ensured sex for military men (but not military women?), which was assumed necessary to satisfy the male sex urge. It was also seen as building cohesion through shared 'adventures', as military men's reward and guard against loneliness in foreign parts, and as defusing the dangerous homoerotic tensions of close military living. Racialised sex played a part in marking out national identity, shaping military men's attitudes to local culture and to local women. At the same time, police and other officials and businessmen developed huge vested interests in the foreign sex industry.

The winding down of the US military in Thailand, and more recently the closing of the US bases in the Philippines, threatened massive dislocation in the sex industry and activated the search for alternative sources of foreign currency. Tourism was already identified as a growth strategy, as noted above. Development policies and programs and recent restructuring also generated social and economic changes, including rural impoverishment, displacement, growing indebtedness and rising consumer expectations. Demand for more jobs and money occurred within a changing international division of labour marked by increasing feminisation of the labour force. Young women make up most of the new workers on the global assembly line, in export processing zones for example. The growth of hospitality and service sectors also heightened the demand for young women's labour, which was frequently sexualised. Families developed income earning strategies which entailed sending a daughter to town or the city; often encouraged by networks of agents and recruiters. Other women or girls sought escape from oppressive or abusive family relations. They were supplemented, in Thailand, by numbers of illegal workers and refugees from Burma, whose migration status and fear of deportation made them a ready source of super-exploitable labour (Asia Watch and Women's Rights Projects 1993).

SEX AND CULTURE

There is an international political economy evident in sex tourism. There is also controversy over the role of culture as permissive or determining causes of young women and girls going into prostitution. This raises complex questions about gender relations and especially attitudes to prostitution 'before', and how these were affected by colonial, militarised and now tourist sex. There are questions too about the effects of sex tourism on those in the trade.

In Thailand, for example, the long history of local prostitution and growing numbers of prostitutes following the abolition of slavery in 1905 encouraged a large local industry and an acceptance of men's use of prostitutes (Hantrakul 1988). Buddhist culture may be seen as devaluing girls and women, less worthy in themselves, and as female not able to achieve merit by becoming monks. Men are more associated with spiritual pursuits and women with less respectable business or material interests. But cultural or 'Buddhalogical' explanations of social change in terms of women's traditional status are inadequate, though gender relations and family values clearly play a part (Tannenbaum 1995).

Many Thai prostitutes see themselves as

good and dutiful daughters, doing a job that allows the family an escape from poverty, or to acquire more possessions. Against the 'dishonour' of sex work is the understanding of these women as supporting their families. One study found that nearly all prostitutes surveyed in Bangkok sent money home (Phongpaichit 1982). Prostitution then becomes something women and girls do, not something they are. Some women gain status from high returns, exercising choice and more control; while others see little of their earnings and work in despotic or dangerous conditions. Sukanya Hantrakul suggests that the birth of a daughter is now more a cause for celebration in Thailand (1988: 132); though for 'use value' rather than for her own sake. Here we see a political economy of social and cultural change. 'Culture' is a dynamic resource in making sense of emerging social relations, including sexual and gender relations. Culture and political economy are mutually constituted.

AN INTERNATIONAL BODY POLITICS

In the numerous discourses around sex tourism (van Esterik 1992), prostitute women's own experiences and what they make of them are rarely visible, apart from testimonials collected for political campaigns. We need to ask where different women are located in gendered power relations at family, local, state and, increasingly, transnational levels. There is a long history of mobilisation against prostitution and especially against international traffic in prostitutes, for example the International Convention for the Suppression of the White Slave Trade in 1910, the Convention for the Suppression of the Traffic in Persons and of the Exploitation of the Prostitution of Others in 1949, and more recently in terms of child prostitution and pornography, and international sex tourism (Reanda 1991). Dichotomies are mobilised in some discourses, distinguishing for example between child and adult prostitutes, and between voluntary and forced prostitution. But the boundaries are often blurred in practice.

The End Child Prostitution in Asian Tourism campaign (ECPAT) estimates that there are some 1 million child prostitutes in South and Southeast Asia. Both Thailand and the Philippines are identified as major international child sex destinations. The Philippines in particular has sex-patriates who have settled in the country for child sex. They often use 'local culture', local poverty or fear of AIDS to rationalise their activities.

The age of consent for girls in Thailand was 13 until 1987, and in the Philippines was only 12. But as a result of local and international pressure, the age of consent in Thailand is now 16. The UN Convention on the Rights of the Child defines children as under 18. Now, in the Philippines, children under 18 engaging in prostitution are assumed to be the innocent party, and those using them are vulnerable to prosecution (Hodgson 1995). But there is little to suggest that poorer, illegal or debt-bonded girls, or boys, gain more choice or personal power on becoming legally adult.

The age distinction in sex tourism is also undermined by the existence of so many child workers in other forms of employment who also suffer from exploitation and abuse, including sexual abuse. A recent International Labour Organisation report estimates that there are about 250 million children aged 5 to 14 who are put to work in the 'Third World', some 61 per cent of them in Asia. So child sex tourism is part of a wider picture of impoverishment and need for families or children themselves to seek some income through the children's work (ECPAT 1996: 7).

So, too, the free/unfree or voluntary/coerced distinction is hard to maintain where the choices are loaded, whether through power and pressure in the family, or poverty and lack of options for families or the prostitutes themselves. Sukanya Hantrakul suggests only 10 per cent of Thai prostitutes are forced or tricked into the industry (1988: 121). However, reports on women and girls trafficked into Thailand from Burma, for example, document horrific forms of debt peonage, forced confinement, rape and deprivation (Asia Watch 1993). At the same time, victim images of prostitutes deny the many who do make something of their lives, often in very grim circumstances, or present themselves as responsible and generous contributors to their families.

It is necessary then to look at personal stories and local and workplace configurations of power. There are clear differences between the tertiary-educated Bangkok escort and the Burmese girl trafficked into a Thai brothel, whose situation is compounded by the illegality of both her work and her status in the country.

Working conditions for other girls and women, especially for refugees, minorities or illegal immigrants, are often dangerous, and involve harassment and intimidation. The illegal status of prostitution in both Thailand and the Philippines make labour organisation for sex workers' rights especially difficult. Can we then see prostitution in the wider context of the rights of women, and of workers?

RESISTANCE AND WAYS FORWARD

Sex tourism, and especially trafficking in women, are the focus of international campaigns that locate their arguments often within the context of women's rights and the violation of other human rights. Representations of prostitutes as victims reflect some women's experiences and perceptions, and may enable mobilisation of moral outrage in international campaigns. But some critics, including organised sex workers, suggest these images mirror those of the sex trade itself, and help perpetuate the racist distinction between 'First World' sex workers who have control over their bodies and work, and 'Third World' prostitutes who are constructed as passive victims yet again. There is a difficult politics and ethics here in terms of who speaks for the sex tourist worker, and questions about the connection between representations and materiality – what happens to women's and children's bodies, and how is power organised, experienced and resisted in the industry?

There are many local and international players in debates around sex tourism in Thailand and the Philippines (van Esterik 1992; Pettman 1997). The End Child Prostitution in Asian Tourism campaign (ECPAT) was formed in 1991 in response to evidence of increasing child prostitution and the involvement of Western paedophile sex tourists in the region, especially in Thailand and the Philippines. Today there are some 250 ECPAT groups world-wide. ECPAT was a key contributor to the first World Congress against the Commercial Sexual Exploitation of Children in Stockholm in August 1996. Now ECPAT has moved from being a campaign to being a non-governmental organisation (NGO), and has changed its name to End Child Prostitution, Child Pornography and Trafficking, in accordance with the recognition that child sex tourism is a global problem which requires a global response.

Other women's and human rights NGOs, including Education Means Protection of Women Engaged in Recreation (EMPOWER), the Third World Movement Against the Exploitation of Women (TH-MAE-W) and GABRIELA, a Philippines' coalition for women's rights, all campaign against sexual abuse andexploitation within international networks. Their public education campaigns target local corrupt officials and practices, and sending states, and include protests and attempts at public shaming of those men who come to buy sex from locals.

NGO campaigns have succeeded in persuading international travel agents' associations and airlines to take a stand against, and issue guidelines to discourage, child prostitution in particular. They have also put pressure on receiving state governments. These campaigns were significant in securing recent legislation in Thailand which increased penalties for prostitution, and also shifted the focus from sex workers themselves to their customers and others involved in the trade, including parents who sell their children. Much still depends, though, on whether there is the political will and official capacity to enforce this legislation.

NGO campaigns go beyond addressing specific practices or sites. Their critiques reveal the complexities of power in which women's and children's bodies are caught. So, for example, GABRIELA locates its analysis in the context of dependent development, militarised political economy and unequal gender relations. Such critiques reveal an international political economy of sex. They call for attention to women's and children's experiences, and caution that sex tourism and child prostitution are not only moral issues. They demonstrate that analyses of sex tourism must be located within wider interrelations of cul- ture and gender relations, development and restructuring, state power and the globalising political economy.

10

THE CITY AND IDENTITY

News frames and the representation of London and Londoners in the *Evening Standard*

Jenny Owen

INTRODUCTION

In this chapter I want to explore some of the ways in which a newspaper might be said to construct identity; in particular, how London's sole evening newspaper, the *Evening Standard*, conceptualises the identity of the city and its inhabitants. In Media and Cultural Studies there exists an influential strand of argument which has suggested that in the late twentieth century we are witnessing an increasing 'interconnectedness of cultures brought about by the transnational flow of images, commodities and peoples' and that this has led to the 'formation of a global culture, dominated by transnational corporations and increasingly Americanised and commercialised' (Gillespie, 1995: 15, citing Mattelart *et al* 1984 and Schiller 1969, 1973). This global culture, it is argued, has been further disseminated by the digital/electronic revolution which threatens to engulf and destroy local cultural autonomy.

Recent studies however, suggest that 'any tendencies towards cultural homogenisation' are matched by 'simultaneous tendencies towards the fragmentation, pluralisation and diversification of markets, cultures and peoples' (Gillespie 1995: 3) and that 'local transformation is as much a part of globalisation as the lateral extension of social connections across time and space' (Gillespie 1995: 15, citing Giddens 1990: 64). For example, Gillespie's ethnographic study of young Punjabi Londoners analyses their consumption of television programmes such as soaps, news and adverts, and demonstrates the willingness of audiences (in this case from an ethnic minority) to use popular cultural products like television programmes to shape new identities.

MELTING POT THEORY AND THE CITY

In the United States these questions to do with the relationship between global and local cultures and identity formation have long been understood in terms of a debate popularised by the use of the metaphor of the 'melting pot'. This metaphor and the use of terms such as 'acculturation' can be traced to a group known as the Chicago School which was made up of 'urban sociologists as well as anthropologists' like Robert Park and Ulf Hannerz. These scholars investigated 'how it could be that ethnic groups remained distinctive in American cities – and to what extent they did so through time. In other words, they were concerned with continuity and change in ethnic relations' (Eriksen 1993: 19).

Park 'regarded the city as a kind of ecological system with its own internal dynamic, creating diverse opportunities and constraints for different individuals and groups' (Eriksen 1993: 19). At the same time, in his view, the city also contained several distinct 'social worlds' which were based on class and race or ethnicity but which would nevertheless be subject to a degree of mobility. For this reason wrote Ulf Hannerz 'in an assessment of the Chicago School' the 'typical race relations cycle' would 'lead from isolation through competition, conflict and accommodation to assimilation' (Eriksen 1993: 19).

In short then, the main thrust of Park's work was that 'every society is a more or less successful melting-pot where diverse populations are merged, acculturated and eventually assimilated, at different rates and in different ways, depending on their place in the economic and political systems' (Eriksen 1993: 19). Melting-pot theory was subsequently challenged as having never taken place – at least as far as African-Americans and Native-Americans were concerned.[1]

My concern however, is not to become embroiled in the debate about whether or not the melting-pot metaphor should be viewed as an accurate description of urban life in the United States. Instead, what is important about melting-pot theory is its focus on the relationship and dynamics between various ethnic groups in the city and the question of how identities develop over time. Equally, my aim is not to focus on the consumption practices of audiences or readers (as is the case with Gillespie); instead, the point is to consider the representation of a city and its people and what those representations might mean in terms of how identities are constructed.

Finally, this is not a study of the global influences on a London newspaper. Although the starting point is the assertion that London is a global city[2], this is not to say that the city is merely the 'centre for the production and diffusion of a "Western" mass culture'. In fact, 'its diversity of peoples, its ethnicities, its subcultures, its alternative cosmopolitanisms, its representations of both core and periphery [are] an instrument for changing that culture' (Paterson 1995: 73).

REPRESENTING LONDON AND LONDONERS

In particular, I shall focus on the extent to which the representation of London and Londoners in the *Evening Standard* can be understood in terms of a consensus or dominant ideology about the city. For example, this is a consensus which typically frames all discussion about London and its people in terms of arguments about urban decay, crime and loss of 'community'.

It is not just the mass media who articulate the idea that urban living must be hell. The social historian Roy Porter, for instance, writes of contemporary London in terms of a 'new urban order' where 'in place of the employed, self-sufficient and respectable working classes who abounded from the time of the guilds to the 1960s "affluent worker", a new outcast London is coming into being, poorly integrated into the discipline of work, family and neighbourhood, into common values, and lacking expectations of a better economic future' (Porter 1994: 370).

In addition, for Porter (1994: 354) London's ethnic mix is a fact not to be celebrated but one to be explained in terms of 'Britain's imperial chickens' coming home to roost. Indeed, he remarks that 'given the scale of immigration' in the 1950s and 1960s, it was 'remarkable' that racial tensions were not greater. These racial tensions he associates with 'poverty, unemployment, rotten housing and a growing bush war between blacks and the police', but not with racial discrimination or hostility on the part of whites.

My concern therefore, is to try and analyse the role of the *Evening Standard* in reinforcing this vision of London, where the city becomes a 'synonym for fear and violence, a symbol of the nation's decline, a sort of madhouse where chaos threatens and loonies rule', and the present is contrasted with an idealised portrait of life 'before the fall' (Phillips 1991: 117).

Clearly not everyone agrees with Porter's diagnosis. Novelist Mike Phillips for instance, resists the idea of the city as a symbol of decay and loss of community and argues that London is a 'living machine whose focus is identity', a place where 'the national project is reassessed' and where 'citizenship is divorced from racial origins, and you can't tell an Englishman from an Indian, or an African or a Chinese' (1991: 121–2). In other words, for Phillips, London is a cosmopolitan city whose inhabitants reflect its role, past and present, on the global stage and where identities are being reworked in the crucible of the city.

CULTURAL FRAMES AND THE LONDON *EVENING STANDARD*

So, what kind of newspaper is the *Evening Standard*? Although its politics are generally 'right-wing', it is not a newspaper which can be dismissed as simply an organ of the Conservative Party. Indeed, despite its traditional hostility to 'Old' Labour and the Left, in the 1990s it has emerged as rather 'liberal' on social issues – particularly homelessness – and has maintained a constantly critical approach to the Conservative government's (1979–97) record on public services, especially public transport.

Indeed, I shall want to argue that although the *Evening Standard* frequently presents London in terms of cultural frames which reinforce the consensus about the city's decline, this is not the whole story. Indeed, part of what the *Evening Standard* achieves is indeed a reassessment of the identity of the nation's capital city, and by extension the nation itself; and this reassessment takes place within a paradigm which both engages with the multiplicities of identities that Londoners possess, as well as addressing all inhabitants as Londoners.

There are, of course, solid business reasons why the paper is active in the reassessment of identity. The paper must appeal to a large number of people who either work or live (or are merely passing through) the greater London area; these people do not buy the paper for its politics: they buy it because it tells them about the city where they live or work or are visiting, and they want something to read as they commute across the city. The paper on the other hand has no real way of distinguishing between its readers and, given that its aim must be to maximise its market share, then these readers (and potential readers) must all be addressed as citizens on equal terms; the paper cannot afford to discriminate against or define who is and who is not a 'genuine' bona fide Londoner; if you are here, then you are one of us is the principle. Identity is to do with presence in the here and now, not the colour of your skin, your accent or your place of birth.

To some extent this reflects the history of London; unlike many other capital cities London grew up as a result of commerce and trading rather than as a site of strategic defence; and in the eighteenth century it became, albeit temporarily, the largest city in the world – a 'monster city' – reflecting the size and importance of the British Empire. London was also a city founded by immigrants (the Romans) and sustained and enriched by continual waves of immigration (both national and international).

For example, in the nineteenth century the East End was known for its Jewish immigrants; and in the twentieth its Bengali. At one time, as Roy Porter acknowledges, 'Finsbury was Little Italy and there were large German communities in Fitzrovia, Camden Town and Kentish Town' (1994: 354). After 1960 significant areas of north London were settled by Greek and Turkish Cypriots; Asians, from the 1950s onwards; while black people (largely from Africa) had first arrived as slaves and servants and then as freed slaves in the eighteenth century (although Caribbean immigration in more recent times is associated with the 1950s and 1960s).

My decision to focus on this particular newspaper also relates to its historical role as the sole London-wide 'local' newspaper and its complex (and often tempestuous) relationship with Westminster, the Greater London Council (GLC) – when it existed – and local government. The *Evening Standard* had, for instance, played a vocal part in the vociferous, Fleet Street-led campaign against the GLC in the early 1980s;[3] however, in the 1990s, in the absence of a city-wide government, it has provided an increasingly important service as a forum for Londoners to conduct a public debate about the future of their city.

Media sociologists argue that the news media select events for reporting according to notions of newsworthiness or news values; some of these values are more overt than others and might include 'general values about society such as "consensus" and "hierarchy"; journalistic conventions; nature of sources; publication frequency and schedule' (Fowler 1991: 13); in other words news is socially constructed (Cohen and Young 1973; Galtung and Ruge 1973; Glasgow University Media Group 1976, 1980, 1982, 1985; Hall 1973, 1980, 1982).

The concept of the news frame is also a valuable method for interpreting the ways in which the media makes sense of the world for its audience; and I shall want to consider the 'cultural frames' routinely used by the *Evening Standard* to shape its journalism and address its readership. News frames highlight the extent to which news is culturally constructed as opposed to 'natural' or self-evident. For instance, one study of news reporting in war-torn El Salvador has suggested that news was only explicable when the cultural frames which shaped the reporting were taken into account. In the case of El Salvador these frames presented the country in terms of a 'culture of violence': reporters were encouraged to think in terms of a news frame in which Salvadoreans were presented as 'cruel and ignorant' (and the violence was therefore 'natural'); while the US and its allies were presented as a force 'struggling to develop a civil society amongst the violence of outlaw extremists – a civilising force in a primitive land' (Pedelty 1995: 89, cited in Owen 1996: 164).

TIRED OF LONDON/TIRED OF LIFE?

So, what are the news frames which are employed by the *Evening Standard* to 'make sense' of London and its people? Boswell famously said that 'when a man [*sic*] is tired of London, he is tired of life'; and so it is that the London of the *Evening Standard* is represented as a place of huge contrasts, and the paper frames much of its output in terms of the value of these perceived differences. The magazine, for instance, which appears with the paper every Friday, exploits the capital's local colour, its ethnic diversity, its sense of fun, its place at the centre of the nation's cultural universe; as well as positioning itself as interpreter of the experience of the 'lighter' side of London life.

CASH AND CURRY

On 2 February 1996 a feature entitled *CASH AND CURRY* pondered the popularity of 'tacky suburban tandoori' restaurants. The article informed the reader that in '1955 there were ten Indian restaurants in London, but by the end of the 1960s there were 800 and by 1995 1,200 excluding those that are solely takeaway'; which makes the Indian restaurant 'Britain's most popular venue for eating out' and makes 'chicken tikka masala ... London's most frequently ordered dish'.

MIND YOUR MANNERS

By contrast, on the next page a feature called *MIND YOUR MANNERS* explored the world of the Lucie Clayton's Young Londoner Grooming Course. Lucie Clayton's was once the finishing school for the likes of Jean Shrimpton, Shirley-Anne Field, Joanna Lumley and Shakira Caine but is now 'full of unpretentious, normal girls, aged between 17 and 25 ... who want to avoid making fools of themselves' and come from 'Essex, Hong Kong, Croatia ... and Tesco'. ' "A Tesco checkout girl saved up for a whole year to do the course" said Kate Smart, vice-principal of Grooming and Finishing. "She loved it" '. In other words, 'normal' London girls are as likely to come from Hong Kong or Croatia as Essex.

A LONDON NATIVITY

Similarly, the Christmas 1996 edition of the magazine, in a feature entitled *A LONDON NATIVITY*, offered readers a selection of photographic images – a list of names on a board outside a barristers' chambers in Lincoln's Inn Fields entitled 'No room at the inn'; two prostitutes outside a strip joint in Shepherd's Market entitled 'Shepherds watch their flocks at night'; Barbie and Ken (plus plastic baby) living in a cardboard box, entitled 'Unto us a child is born'; and a young female character (Tiff) from the BBC1 soap opera *EastEnders* entitled 'A star in the East'. These images were:

> symbols for a modern Nativity with a London angle. For what is the Nativity but a story of remarkable incongruities – the Son of God born in a humble cow shed – and what is London all about but the very same thing? We refer to the kindly action of the stranger on a fetid Tube train; the hint of grace in the underpaid, overstressed bus driver who nevertheless waits at the stop for the lumbering pensioner.
>
> (*Evening Standard* 20 December 1996: 10)

THE *EVENING STANDARD* AND LONDON'S TRANSPORT SYSTEM

The *Evening Standard* also offers a more traditional journalistic approach to life in the city. Throughout the 1990s it has campaigned vociferously on the subject of London's transport problems – a typical report described how 'construction workers and white collar London Underground managers' were 'being sent on a £750-a-time psychology course to teach them how to be nice to one another'; this despite the fact that the Jubilee Line Extension project was '£200 million over budget and two months behind schedule' (9 February 1996); and I shall argue that it is transport in particular which is used by the paper as a framing device to address and construct its readers as Londoners.

For example, the paper campaigns to stop the abolition of the London Routemaster bus (the one you can jump on or off while it's still moving): the red London bus or Routemaster is a globally recognised icon of Londonness and its proposed abolition is tantamount to the destruction of a Londoner's 'civil rights'. The paper also systematically attacks the government for its lack of investment in public transport and London Underground for its inefficiency, but it is its lyrical approach to London's transport history, which is more characteristic and revealing.

HISTORY, SEX AND THE NORTHERN LINE

History is used to recreate the past, both as a pedagogical tool for contextualising contemporary developments and as a method of exploring the nature of London life and what it means to be a Londoner. For example, feature writer Simon Jenkins, in a piece entitled *Sex, Henry James and the Northern Line*, describes how the Northern Line 'was born 100 years ago this spring. Its parents were dynamism and greed, with sex on the side'. The Northern Line, that most London of transport jokes (the 'Boredom Line' on account of its slowness), was the 'brainchild of Charles Tyson Yerkes, a Philadelphia tramway tycoon and buccaneer' who had fled Chicago and his creditors with $15 million and arrived in the city in 1900 at the age of 63 to exploit the electric railway boom. 'With him came not his wife but his mistress Emilie Grigsby'.

By 1905 Yerke's finances were in disarray and in December 1905, the Northern Line and Miss Grigsby, 'bleeding him of money, an exhausted Yerkes died reputedly in her arms at the Waldorf Hotel'. Jenkins remarks: 'I like to think of the scent of Miss Grigsby's boudoir wafting up the lift shaft at Chalk Farm, for which it helped pay. She lived on unmarried, until 1964'.

Eventually Jenkins contrasts this tale with current government failure to produce a coherent transport policy for London, cuts in expenditure on the Underground in 1997, and implications of corruption surrounding the Jubilee Line extension: 'ordered by Lady Thatcher against the strong opposition of London Underground to help her friends, the Reichman Brothers'. It is, however, his use of London's history to invoke a sense of the capital's character and the character of Londoners (an American venture capitalist and his mistress) which you remember as a reader and not the political invective.

TUBE TALK

In a similar vein the magazine also carries a weekly column called 'Tube Talk', which is a forum for readers' anecdotes and snippets of arcane and often amusing information about the capital's transport system. 'Tube Talk' addresses the reader as a Londoner on the grounds of a shared experience of the horrors of London travel. The cultural frame is redolent with the notion of the idealised 'plucky' Londoner (who first emerged during the Blitz) and the dominant sentiment is that London transport may be hell but, being Londoners, not only can we laugh about it, it is what makes us what we are.

For example, on 1 November 1996, 'Tube Talk' discussed the re-run on satellite television of the 1970s sitcom 'On the Buses' (described as having a 'certain seaside-postcard vigour') as well as a complaint from two women readers 'about the growing problem of people picking their noses on the Underground'; the response to the latter being a discussion of other anti-social practices – in particular the 'objectionable habit of males on the Underground: their aggressive style of leg crossing' which is described as 'almost as annoying as sitting next to somebody whose knobbly elbows occupy not one half but six tenths of the armrest'. This is described as 'a damnable encroachment that any true Londoner can immediately spot'.

Transport is both a metaphor for the failings of the city's politicians and a shared experience which unites Londoners; but in contrast with the 1980s there are relatively few Loonie Labour Council stories, these having been replaced by tales of European bureaucracy gone mad, the crumbling National Health Service and falling standards in state schools. The European Union is an easy target: an editorial entitled *SHELLFISH BEHAVIOUR* (29 January 1996) described a 'new directive from Brussels, the Welfare of Animals Transport Order' which 'demands rest breaks and stress relieving showers for mussels – yes, mussels – in transit across Europe', but this is tame compared with the heady days of baiting the GLC.

POVERTY IN LONDON

In fact the paper expends greater energy highlighting poverty and homelessness than criticising Labour councils or European bureaucrats. In January 1995 it ran a week-long series entitled *THE BETRAYED – THE TRAGIC WORLD OF LONDON'S FORGOTTEN PEOPLE*. In a four-page centre spread the grim reality of London's 'remarkable incongruities' is exposed: beneath a black-and-white photograph of mourners standing at a graveside are the words:

> They had just buried the baby ... Noticeably there was no mother present, why we do not know but are permitted to guess.
>
> Nearby a grave digger watches and waits. When the minister leads his little group away, he shifts back the planks and sheets of rusting corrugated iron and shovels in a few spadefuls of earth to cover the tiny box. He doesn't close it because this is no normal grave.
>
> This is a communal grave ... You can get a lot of babies in a pit 20 feet long, three feet wide ... Where is this sad place? Somewhere in some awful recess of the Third World ... ?
>
> This is happening here, now, London, England, 1995!
>
> (*Evening Standard* 9 January 1995)

The piece continues. London 1995 is compared with London a hundred years ago and the comparison is not a favourable one: statistics show that twice as many babies die in Tower Hamlets than in the health authority which caters for south and west London; while the poverty and deprivation of the East End is contrasted with the enormous wealth of the City, 'the greatest financial centre in the world', and the 'glitz and glamour of the West End'.

The thrust of the piece is that the East End has been betrayed; although this is also the language routinely used by far-right, racist groups like the British National Party (BNP), which has some support in the area and which (rather inevitably) becomes the focus of the rest of the article. For the BNP it is the immigrants who are the source of the area's problems; for others, like Sister Christine Frost, who belongs to the Order of the Faithful Companions of Jesus, or the Reverend John Lines, the betrayal has a political and economic source (the government and the big developers). Meanwhile the piece ends with a quotation from Joseph Conrad, who:

> once described the East End poor as an 'unsmiling sombre stream not made up of lives but of mere unconsidered existences' ... Is this the inevitable fate of the individuals behind those grey, tragic faces? What is to be said of a great city where unconsidered exis-

tences are destined to end, unconsidered, in the pit at the East London Cemetery?'.[4]

Most recently the newspaper has attacked Tony Blair, the leader of the Labour Party, for his assertion that he never gives to beggars. Brian Sewell, the *Evening Standard*'s art critic and roving social commentator, complained that Blair 'conjures for our fears the threatening image of the aggressive beggar on the streets of London – a ragged version of the dapper highwayman'; and concluded that 'London is not the place for the zero tolerance of beggars' (8 January 1997).

Later that same week a centre-page spread entitled *GRIM FACTS ABOUT LONDON'S HOMELESS* used case histories (like the man who lives in a bus shelter but had four separate Christmas dinners on account of the efforts of public spirited Londoners) to describe the reality of London's homeless, the wider social and economic implications, and the efforts of the voluntary sector to combat the problem (10 January 1997).

CONCLUSION

In conclusion, the *Evening Standard*'s role in the construction of the identity of London and Londoners would seem to be both complex and contradictory. At one level it communicates a sense of London as a city on the brink of political and economic disaster (at times with good reason); at another there is a celebration of the capital's cultural life, its ethnic diversity, its vibrancy, the pluckiness of its citizens. Equally, although the paper uses London's history to excoriate contemporary political blundering and corruption, this history is also used to describe what is distinct about London and Londoners. Londonness, for instance, is in part defined as a kind of resilience to the hardships of city life – in particular commuting – and partly as something moral (hence the attacks on homelessness and poverty) and public spirited.

The cultural frames which define and shape the *Evening Standard*'s approach to London and London identity are therefore to do with a specific grasp on the city's past; and while it is not the case that this history is free from the ideology of the city as an urban hell, and the city dweller as alienated and alienating (or indeed, right-wing ideologies of various kinds), the paper nevertheless represents an attempt to reclaim the idea of urban life from the agenda of crime, decay and loss of community as well as to construct an identity for the Londoner which includes rather than excludes.

As Mike Phillips has suggested, and a reading of the *Evening Standard* would seem to confirm, London is indeed a 'living machine ... creating its own climate, its own ecology and its own internal structures. The product is change: the assembly and remodelling of identity' (1991: 121).

NOTES

1 The greatest challenge to melting-pot theory was mobilised by Nathan Glazer and Patrick Moynihan in their influential and controversial study *Beyond the Melting Pot* (1970 [1963]), (Banks 1969: 69). In this work they documented the extent to which various ethnic groups (the Irish or Italians) continued to manifest an 'ethnic' identity long after this identity would be recognised as 'Irish' or 'Italian' (Banks 1969: 71) and were much criticised for their failure to distinguish between ethnicity and race.

2 By the nineteenth century London had become the centre of a massive empire, and while that empire is no more, its influence continues to impact on the city. Thus today London is both one of the world's largest financial and media centres, as well as being a multi-ethnic, cosmopolitan city of approximately 8.4 million people.

3 One typical *Evening Standard* editorial opined that 'Surely no-one – but no-one – deserves Mr. Ken Livingstone. His rule at the GLC goes straight into the cruel-and-unnatural-punishment class' (24 July 1981, cited by Holingsworth 1986: 80).

4 The following day the series on London's 'forgotten people' focused on the plight of a family of Asian origin living in a squalid local authority flat and desperate to be rehoused.

IV

CULTURE AS EXPLANATION

11

CULTURE AS IDEOLOGY

Explanations for the development of the Japanese economic miracle

Roger Goodman

THE 'DISCOVERY' OF THE JAPANESE COMPANY SYSTEM

In the early 1980s there appeared a genre of books which suggested a close connection between Japanese culture and Japan's post-war economic miracle. It says much about the thirst of the outside world for the secrets of this economic success that several of these books could be found on book stands at airports. Until the 1980s, much of the western world had either refused to take Japan's seemingly inexorable economic growth seriously or else had explained it away in terms of unfair trading and 'dumping', both of which practices the western world could outlaw, thereby returning Japan to the fold of developing economies.

This particular genre included works such as William Ouchi's *Theory Z: How American Business Can Meet the Japanese Challenge* (1981) and Pascale and Athos' *The Art of Japanese Management* (1981), but perhaps most illustrative of the style was the 1982 reissue of Miyamoto Masao's seventeenth-century samurai classic, *The Book of Five Rings*, with the new subtitle 'The Real Art of Japanese Management' and the following front-cover splash (in the Bantam Books edition): 'Now the secret of Japanese success in business can be yours. It lies within the pages of this age-old masterpiece of winning strategy'.

The underlying premise of the genre can be summed up relatively simply: Japan had developed a form of capitalism that was different from the ideas behind 'western'[1] capitalism; Japan exhibited consensus, harmony, affective relationships, hierarchy and groupism that were very different from ideas of individualism, class conflict and putative egalitarianism. (Table 11.1 sets out – in very attenuated fashion – some of the assumptions thought to underlie the values of 'Japanese', as opposed to 'western', capitalism.)

These Japanese 'cultural values' were felt to be clearly identifiable in the style of the Japanese company, most specifically in the relationship between management and workers. This was best exemplified by the so-called Three Jewels (*sanshu no jingi*) of the Japanese employment system: life-time employment (*shūshin koyô*), seniority promotion (*nenkō jōretsu*) and company unionism (*kigyō kumiai*). Life-time employment meant that once one was employed by a company, one could expect to remain in that employment, even in times of recession, until retirement age. Seniority promotion was the system by which long and loyal service would automatically be recognised. Finally, company unionism meant that workers in the company would be part of a single union which was

Table 11.1 Opposing values said to underlie 'Japanese' and 'western' capitalism

'Western' Values	'Japanese' Values
1. Individualism	Groupism
2. Psychological independence	Psychological dependence
3. Horizontal interpersonal ties	Vertical interpersonal ties
4. Egalitarianism	Hierarchy
5. Contractual-ism	'Kintractual'-ism
6. Sense of rights	Sense of duties
7. Logicality	Ambivalence
8. Rationality	Emotionality
9. Universalistic ethics	Situational ethics
10. Racial heterogeneity	Racial homogeneity

specific to that company and which would negotiate on their behalf for non-wage benefits such as health insurance, housing assistance, company-sponsored recreation, as well as pay deals. Such unions would have no interest in members who were not members of the firm. As Whittaker (1990: 322) says: 'The fate of workers is thus closely connected with the company, and socialisation into the "corporate community" reinforces this'.

The consequences of the system – the *empowerment* of shop-floor workers to make decisions in their own sphere (such as stopping the production line) without need to refer to management; the *commitment* of workers to the company as instanced by some four million workers voluntarily participating in quality-control circles in their own time and for no pay; and worker *participation* via numerous suggestion schemes and the *ringi seido* system, where those at the bottom of a company can channel policy initiative up to senior management – seemed to explain Japan's so-called economic miracle: few days lost in strike action (see Table 11.2); fewer days taken as vacation even when eligible; an openness amongst workers towards the introduction of new technologies without fear of losing their jobs; high-quality products at a low price; and the fastest growing economy in the world.

The literature of the 1980s highlighted a number of other features integral to the Japanese workplace which were regarded as particularly effective in producing a loyal workforce with a strong sense of identification with their company. The pay differential between the newest recruit and the chairmen of the largest companies was in Japan a fraction of that in the US and the UK.[2] Workers rotated between jobs, thereby rendering demarcations meaningless and giving individuals a much greater awareness of how their particular job fitted into the work of the company as a whole. Management would help the unions to organise and collect dues and union chiefs would be expected to proceed into management. Perhaps most importantly, although the customer in Japan might be king, they would be closely followed in the priorities of the company by the worker, only then the management and, a poor fourth, the shareholder who was expected to maintain a long-term interest in the company in return for very low dividends.[3]

The organisation of space and time further reinforced the sense of the 'company-as-family', as the model was widely called. There was an almost total lack of symbolic distinction between blue- and white-collar workers: all wore the same clothes; parked in the same car parks; ate in the same dining rooms and used the same toilets. In white-collar environments, open-plan offices meant that workers worked as a group rather than individually. Indeed the lay-out of the office hardly differed from one company to another and illustrated perfectly the symbolic hierarchy of the employees. However small the office, it would always include space for those known euphemistically as the *madogiwa* ('by-the-window tribe') – who had been found wanting when it came to taking on higher responsibility but who retained their jobs and their

Table 11.2 Total days lost by annual labour disputes (1,000 days)

	Japan[a]	USA	UK	Germany	France	Italy
1980	998	20,844	11,964	128	1,511	16,457
1981	543	16,908	4,266	58	1,442	10,527
1982	535	9,061	5,315	15	2,257	18,563
1983	504	17,461	3,754	41	1,321	14,003
1984	354	8,499	27,135	5,618	1,317	8,703
1985	257	7,079	6,402	34	727	3,831
1986	252	12,140	1,920	28	568	5,644
1987	256	4,469	3,546	33	512	4,606
1988	163	4,364	3,702	42	1,094	3,315
1989	176	16,996	4,128	100	800	4,436
1990	140	5,926	1,093	364	528	5,181
1991	92	4,584	761	154	497	2,985
1992	227	3,989	526	1,545	359	–
1993	–	3,998	–	–	–	–

Note: Based on data from each country. Labour disputes, as a rule, involve a protest action.
[a] Japanese figures include days of disputes involving a protest action and a factory closure.
Source: Bank of Japan, *Comparative International Statistics*, 1994.

Figure 11.1 Short-termism in the west: what managers cite as their top priority in percentages.

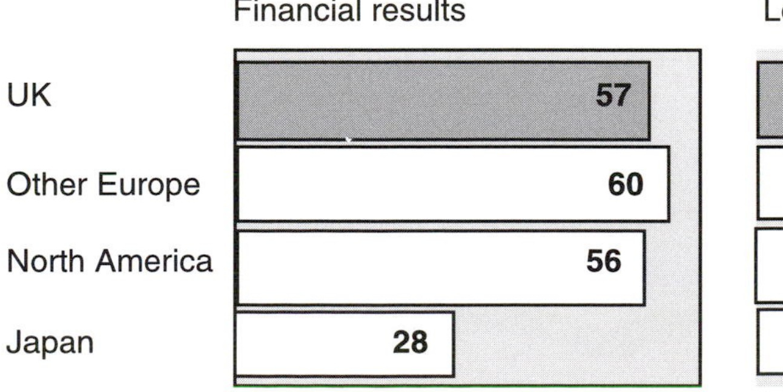

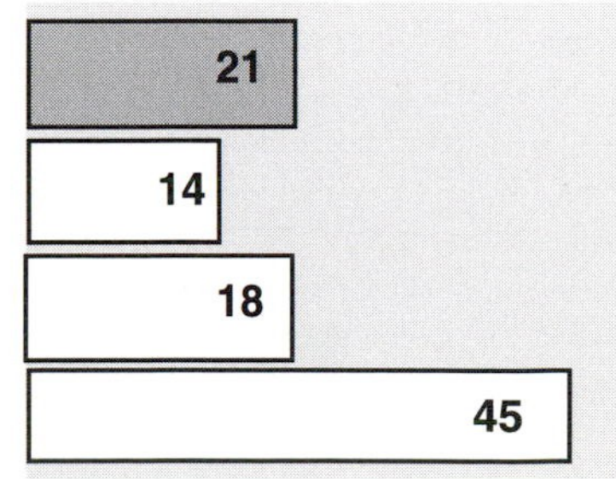

Source: *Observer*, November 1996

salary increases – and an area for consensus-building, decision-making discussions (*nemawashii*) over green tea or coffee (see Figure 11.2).[4]

Workers would be expected to spend time with colleagues outside normal working hours. Those who insisted on returning home to their wives and families would be tarnished

Figure 11.2 Typical Japanese office set-up.

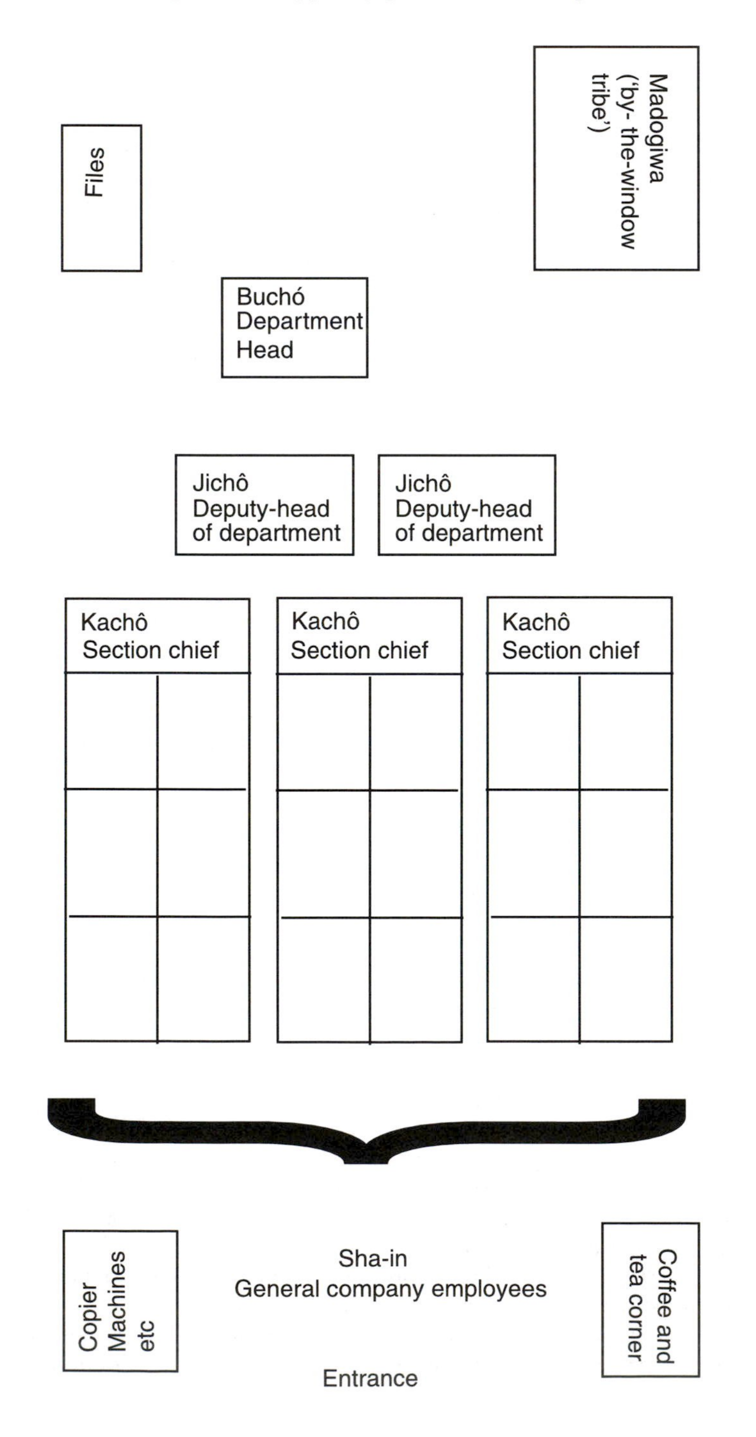

with the tag *maihomuizumu* (my-homeism), the use of the English 'my' accentuating the association of individualism and selfishness with western-ness and repeated in the expression *mai pesu* (my-pace) for those who insisted on working at their own – rather than the group's – pace. Workers, as Atsumi (1980) pointed out, could easily distinguish between this enforced socialisation with peers and bosses (which they called *tsukiai*) and time spent voluntarily with friends and kin: the three categories of relationships – *tsukiai*, friends and kin – were not mutually exclusive but certainly cognitively separate. *Tsukiai* relations, moreover, did not limit themselves to work-related matters. One's manager was both entitled and expected to take an interest in one's personal life. If one was not married by the expected time (*tekireiki*) – around 24 for a woman; 27 for a man – then one's manager should become as concerned to correct this as one's parents. In many cases, he (rarely she) would arrange meetings (*omiai*) with prospective marriage partners.[5]

THE PURPORTED 'ORIGINS' OF JAPAN'S UNIQUE COMPANY CULTURE

Above we have outlined – however sketchily – the basic features of the Japanese employment system that captured the imagination of western entrepreneurs and Wall Street gurus in the early 1980s. Many authors – and their readers – used the syllogism that since the Japanese company was organised in the way described and since Japan's economy was doing so well, then there must be a direct relationship between the two. Interestingly, however, the explanations proffered were not new; indeed, much of the discussion was outlined some twenty years earlier in James Abegglen's *The Japanese Factory: Aspects of Its Social Organisation* (1958), but at that time the Japanese economy had yet to take off and hence few (outside the academic world) could see the potential interest in the Japanese example. In the 1980s, however, those who bought their books at the airport book stands wanted not only to know how the system worked, but where the idea had originated and how, if at all, it could be copied. Unfortunately for them, the roots of this model seemed to be founded in feudal Japan, in particular in the organisation of the local community and of the household. This made copying highly problematic.

There are a large number of published texts which set out in detail the relationship between traditional culture and economic development in contemporary Japan and which inform and share assumptions with many of the works written in the 1980s on the Japanese company. We draw here only on the best-known ones which are available in English.

In *Japanese Society*[6] (1970), the social anthropologist Nakane Chie explains the whole structure of Japanese society in terms of dyadic parent–child hierarchical relations. According to this theory, individuals find it difficult to conceive of themselves as individuals outside a group or a hierarchical context; everyone sees themselves as in a junior or senior position to someone else in a particular group. For most adult men, of course, the group is the company within which they work.

Nakane draws on the work of earlier anthropologists for explanations as to *why* individuals cannot see themselves outside of a group context. In part, she says it is related to the fact that Japan is a wet-rice society, following a form of subsistence which requires co-operation between a large number of people – i.e. no one can grow wet-rice by themselves (see Ishida 1971) – and, in part, it is related to the topography of Japan – small, isolated hamlets in a mountainous environment, which required individuals to work together to survive, or large crowded conurbations, where harmony had to reign in order to prevent the breakout of anarchy.

Moreover, as others pointed out, the fact that Japan is an island helps people think of themselves as separate from – and in opposition to – other people.[7] This island mentality, combined with a long period of seclusion from the outside world (the *sakoku jidai*) between 1637–1854, when no one could leave or enter the country, has resulted in the developmentof an unusually homogeneous popu- lation and a 'unique' culture.[8]

Hierarchical relations are perpetuated by traditional child socialisation practices, which seek to develop what the psychologist Doi Takeo (1973) calls *amae* or the 'dependency' of the child on the parent. This is a manner of relating which is reinforced by co-sleeping, co-bathing, child-carrying and non-verbalisation between parent and child (see Lebra 1976) and which Doi believes is transferred in later life to one's superiors in the outside world, including the workplace. According to the above authors, however, two traditional influences have been particularly powerful in developing the Japanese-style company: the Japanese household system and Confucianism.

The idea that the modern Japanese company is a natural development of the traditional Japanese household (*ie*) is a common thesis. The two institutions share a sense of their past (and respect for their founders) and an obligation to their future. Ideal inheritance is patrilineal primogeniture but its explicit rules of gender and age can be overturned if the good of the company or household demands it. In both cases, the individual is expected to subordinate their personal interests to those of the household or the company and to show unwavering loyalty and respect to their seniors, especially the Head. The Head in return is expected to act benevolently towards his (occasionally her) juniors and to take responsibility for all their actions. A good Head is one who can create an environment where consensus develops rather than one who makes authoritarian decisions. Sharp distinctions are made between those inside the household or company and those outside, both in language use and in behaviour. In short, some argue, one should think of the *ie* – at least as it existed in the pre-war period – as an occupational group, and of the Japanese company as based on (fictive) kinship. According to this thesis (see, for example, Murakami 1984), the modern company emerged from the already existing occupational structure of the feudal *ie*.

At about the same time as Ouchi, Pascale and Athos and others were publishing their best-sellers on Japanese companies, the Confucian explanation for Japanese economic development received perhaps its best-known expression in English in the work of the economist Morishima Michio entitled *Why Japan has 'Succeeded': Western Technology and Japanese Ethos* (1981). Drawing on Max Weber's thesis that there was a connection between the development of capitalism and Protestantism in Europe, he outlined a similar role for Confucianism in Japan, with its emphasis on education, respect for authority, hard work and productivity. In the decade following Morishima's work, the 'Confucian model' has been used to explain the successive economic successes of Japan's neighbours – South Korea, Taiwan, Singapore and Hong Kong.[9] Underlying this thesis, as Dore (1987: vii–viii) points out, is a very different idea of the person: start from the idea of original sin and this assumes society will be motivated by self-interest; start from the idea of original virtue and this suggests that the importance of bonds of loyalty, community-belonging and the sense of responsibility that accompany them will be given greater weight.

In anthropological terms, the arguments which explain the development of Japan's economy in terms of its culture can be best described as functionalist, essentialist and static. Underlying them is the assumption that land=language=people=culture and that this equation explains Japan's economic miracle. In methodological terms, these arguments suffer two major weaknesses: they confuse ideology with reality, and history with the invention of history. It is to these issues that we now need to turn.

THE 'REALITY' OF THE JAPANESE WORKPLACE

The 'three jewels' of the Japanese employment system are not, in reality, enjoyed by all its workers. The employment system is characterised by a core–periphery distinction which works in two directions. There is a strong distinction made between those who work in the major companies and those who work in sub-contracting companies which serve these major companies. Put simply, the bigger the company the better able it is to realise the promise of the 'three jewels'. The basic pay of the biggest companies is twice that of the smallest; bonus payments (which constitute up to one-third of all pay) are around five times as great. Significantly, Japan has a relatively small number of large companies and a very large number of medium and small companies. Even the large companies are not very large by international standards; Toyota has only around 60,000 direct employees compared with GM's more-than 800,000, though Toyota's production, in comparison, is about half. In part, this is because of the efficiency of its subcontractors.

The second core–periphery distinction exists *within* the big companies where many workers – seasonal, temporary, part-time – are also excluded from the benefits of the 'three jewels' system. As a result, those who do receive – or expect to receive – the benefits of life-time employment, seniority promotion and company unionism constitute only around 25–30 per cent of the total workforce. The vast majority of these are men.[10]

Kamata Satoshi's (1982) account of life as a seasonal worker in a Toyota factory shows a very different side to the life of the Japanese workers from that offered by Ouchi, Pascale and Athos and others. He describes how workers are locked into their dormitories at night in compounds with high fences surrounding them; how visitors of the opposite sex are allowed entry on pain of losing the opportunity of doing overtime; how there is constant surveillance and rigid demands to conform. Moreover, he describes how the assembly line is gradually speeded up and how overtime is extended as the workforce is pressured to meet production targets. In particular, Kamata notes the lack of job security for seasonal workers and the paltry (by the standards of core workers) bonus of 4.5 days' worth of extra pay at the end of six months.

Kamata's account is clearly problematic; as a left-wing journalist undertaking six months covert research in order to demonstrate the power of the capitalist owners, his objectivity must be brought into question. It is particularly interesting, therefore, to see the extent to which he internalises the values of the company-as-family and, by the end of his six months, is a model worker who shares, with his fellow workers, the view that it is competition in the automobile industry and not management inadequacies that is the source of harder and longer working hours. He even develops an *esprit-de-corps* with his fellow workers that includes their 'benevolent' foreman.

A further problem with the company-as-family model is the notion that company workers – even core workers – somehow 'naturally' do not express themselves in a conflictual manner versus the management. Eyal Ben-Ari (1990), for example, suggests that workers actually do express conflict but in forms somewhat different from those normally used in western capitalist societies. Conflict is both symbolic and ritual in Japan and generally centres on key moments in the course of the annual spring wage negotiations (a tradition called *shuntō* – Spring Offensive – which dates back to the 1950s). It is during this 'Offensive' that core workers symbolise their power *vis-à-vis* the company management without needing to resort to actual strike action. Simply the threat of strike – or occasionally the shortest of strikes arranged so as to avoid inconveniencing customers or company – is sufficient, according to Ben-Ari, not only to defuse conflict by allowing workers to let off steam but also to convey a series of important messages to the management: we are élite workers attached to this company; we can demonstrate worker solidarity even if only choose to do so once a year; we accept the idea of our general dependency on the company; but we take on the role of 'not-us' in order to allow us to express demands that our social relations with our superiors do not normally allow us to express; hence we still belong to the company and having worked out an advance strike schedule, we know that you managers will indulge us in this performance because you recognise it is in your long-term interest to do so.[11]

As Ben-Ari is at pains to point out, these actions only apply to those workers who are members of company unions, which in turn are interested only in defending the interests of their members. It is more their potential power as core workers, *vis-à-vis* the company, which allows them to express conflict symbolically, than the idea that culturally they are averse to the notion of conflict *per se*.

THE DEVELOPMENT OF THE COMPANY-AS-FAMILY MODEL

If there are doubts raised about the generalisability of any description of the contemporary Japanese workplace which focuses on the principles of the 'three jewels' to the employment conditions of the majority of the Japanese workforce, then there are even bigger questions over the development of the 'three jewels' system and the idea that there is a direct historical link between the structure of pre-industrial family/occupational relations and the contemporary Japanese company. To put it simply, early factory management in Japan was not benevolent nor were worker diligence and loyalty unquestionable; in British terms, early factory conditions are best described as Dickensian (Hane 1979).

Rodney Clark's (1979) description of the development of the Japanese company perhaps most clearly shows how the idea of the company-as-family developed out of the early industrial form of labour relations. Japan's early industrialisation was based around the textile industry. Most of those employed were young girls from poor farming households. They worked in factories where they were placed in the 'care' of older male owners. These girls worked, at best, for very low wages; in many cases they were virtually indentured, especially in northern Japan, where famine meant that parents were forced to 'sell' their daughters to mills where they had to work out their contracts. The work was tedious, unhealthy (there were high rates of both tuberculosis and accidents) and very long – often thirteen or fourteen hours a day. Foremen, who were paid according to production rates, applied as much pressure as they could. Abuse – of every description – was rife, and the workers, in many cases little more than child labourers, had no one to whom they could turn for help.

The first real signs of worker protest were recorded in 1886 but it did not become effective until the 1890s, when employers, in a rapidly expanding market, began to face labour shortages and increasing labour mobility as workers took their new skills to competitors. To confront this, employers began to improve labour conditions: to offer their female workers housing, food and *okeiko* (training in the traditional arts needed to improve marriage prospects).

When the new industries – metal working and engineering – that involved the employment of an increasingly skilled male labour force came to the fore at the turn of the century, employers made the shift from daily wages to offering career prospects, seniority promotion in return for loyalty, welfare schemes and bonuses, all as a means of keeping their workers. In the case of both female and male workers – though particularly the latter – employers developed a new rhetoric to legitimate and justify their position. This rhetoric was founded on the idea of the company as family, as a unique Japanese form of occupational grouping that arose naturally out of feudal (*Tokugawa*) tradition. As Clark puts it:

> The metaphor of the family, besides harking back to Tokugawa tradition, was perfectly adapted to interpret employment practices forced on employers by the labour market. The notion of family-as-firm was also consistent with one of the central political concepts of the Meiji period, a concept widely supposed to have remote historical antecedents, but one which was in fact a new garment of old threads: that the Japanese nation itself was a gigantic family with the Emperor at its head. Such a well-connected and plausible doctrine as familism, therefore, was able to assert itself over the brash and contentious theories that might have proceeded from some obvious facts: that there were enormous differences in the way managers and workers were treated (and) that industrial relations were sometimes very bad.
>
> (Clark 1979: 41)

The efforts by employers in the 1890s to have the Japanese company seen in terms of a family have been repeated *ad infinitum* over the past one hundred years so as to mask the glaring inequalities that still exist. Dean Kinzley (1991) outlines the details of the Kyōchōkai (Harmonisation Society) which in 1919, with government sponsorship, did much to 'reinvent' the rhetoric of 'natural' management–employee co-operation in Japan; Sydney Crawcour (1978) describes the work of the Zensanren (The All-Japan Producers' Union) in responding to increasing union strength in the late 1920s and early 1930s. It was in the course of this campaign, Crawcour says, that the 'traditional' spirit of Japanese labour relations took shape as orthodox doctrine. And once again, in the immediate post-war period, faced by increasingly left-wing union agitation, the Japanese government together with the American occupation authorities (facing the Cold War) acted to encourage the growth of company unions in order to serve the interests of the company as a whole and not only its workers. In each case, the discourse called on the structure of the traditional Japanese household (*ie*) as the model for industrial relations in the Japanese company, and in each case it was what Kinzley (drawing on Hobsbawm and Ranger 1983) has termed the 'reinvention of a tradition'.[12] It was this reinvented tradition that western commentators, suddenly alerted to Japan's new economic strength, 'discovered' in the early 1980s, almost one hundred years after it first appeared.

CONCLUSION: CULTURE AS IDEOLOGY

If the current organisation of industrial relations in the Japanese company is simply the product of efforts by employers to retain their workforce, then what, if anything, is the influence of culture in the Japanese workplace? In a sense, culture is reduced in this theory to a mere ideology manipulated by élites to mystify their workers and thereby maintain allegiance and loyalty in order to boost profits. In similar vein, Brian Street (1993: 25) has argued in his paper, 'Culture is a Verb', that culture is an active process of meaning-making and that there is a tendency in organisational studies to lose sight of culture as something an organisation *is* and to treat it as something an organisation *has*. As Sue Wright (1995: 83–4) in discussing Street's work adds: 'Culture then becomes a tool of management, trying to introduce a new set of preferred characteristics throughout the organisation's operations'.

When, as in the case of Japan, the manipulation of the idea of cultural influence draws not only on 'history' and 'historical practices', but also on Japan's physical topography and traditional reliance on a subsistence economy, then it is very difficult for anyone to challenge its apparent explanatory power. 'Culture', therefore, is a dangerous word unless it is understood as a shorthand for the convergence of moral and spiritual ideologies with political, historical and economic forces in ever-continuing processes. Otherwise, attempts to account for Japan's system of industrial relations in terms of 'culture' that ignore these other inter-relating factors and its dynamic nature are in danger of legitimating, perhaps even helping to construct, that system as much as actually describing or explaining it. Hence it is that we earlier described such types of explanation as functionalist, essentialist, static and ahistorical.

However, Japanese workers are not powerless. Indeed, they constitute the world's most highly educated workforce – in itself perhaps a better explanation for the country's economic success than its system of industrial relations. Not a few indeed are anthropologically and sociologically literate. As we have seen, many, especially those among the core workforce, have been able to recognise the 'constructedness' of the 'tradition' of the Japanese company-as-family. But as the Japanese economy has continued to expand over the past fifty years and as their quality of life has improved along with it, they have been prepared to conform to the dominant cultural explanation. If anything, it is now management in Japan which has the problem of sustaining the 'tradition' which it initially created. This explains the extraordinary lengths to which management went during the 1990s, while Japan endured its longest recession since the 1930s, not to lay off any of it core workers and thereby shatter the illusion of the 'three jewels' of the employment system. Japan may have survived the recession but, according to many commentators, the ideology which management invented in order to keep loyal workers and workers loyal, almost proved to be its undoing.

NOTES

1 I write 'western' and the 'west' with a small 'w' to emphasise that I do not subscribe to the view – prevalent in much of the literature on Japan – that there exists a Western Culture or that The West is a monolithic entity that can be directly contrasted with another monolith called 'Japan'.

2 Tasker (1987: 300) cites a Nikkeiren report survey that in Japan the 'average company president receives 7.5 times the salary of a new recruit, against 30–50 times in the US and Europe'.
3 Shareholding in Japan operates in general on a system of companies cross-holding shares in each other within large conglomerates, thereby ensuring the ability of management to operate with a very long-term perspective without fear of take-overs. For the difference between Japanese and other countries' perceptions of time, see Figure 11.1.
4 For more on Japanese forms of consensus-building, see March (1991).
5 Surveys of workers have regularly shown that most would prefer a manager who became involved in their personal affairs to one who would not.
6 The original title in Japanese translates directly as 'Human Relations in a Vertical Society'.
7 This is known in Japanese as *shimaguni konjō* (an island mentality) and is often identified as a characteristic shared with the British.
8 For a full overview of the major arguments on the development of Japan's 'unique' culture, see Befu (1993).
9 The irony of the 'Confucian model' for the economic success of the East Asian economies is that until the late 1950s Confucianism – with its anti-individualism, and its respect for authority and the *status quo* – was seen by both western and indigenous observers as the main *obstacle* to the development of advanced economies in the region.
10 For a good overview of the core–periphery distinction in the Japanese work force, see Chalmers (1989).
11 When workers for private rail companies strike, for example, this would normally be restricted to between, say, 6.00–6.30 in the morning, with workers giving information to any passengers who did turn up at that time about alternative routes they might try.
12 For a good overview of how this process was repeated over the past 100 years, see Gordon (1985).

12

CULTURAL DISEASE AND BRITISH INDUSTRIAL DECLINE

Weber in reverse

Mike Hickox

INTRODUCTION

In the post-Cold War era the concept of 'culture' as an explanatory device has been very much back on the theoretical agenda. Two recent examples of this trend have been Huntington's *The Clash of Civilisations and the Remaking of World Order* (1996) and Fukuyama's *Trust: The Social Virtues and the Creation of Prosperity* (1995). Huntington's thesis is an especially good example of this trend since he sees the post-Cold War era as one in which, after the bipolarity of the post-45 period, older deeper antagonisms between cultures will once again reassert themselves. This new stress on cultural identity, he argues, may be seen in part at least as a reaction against the increasing globalisation of the international economic order.

Typically, cultural explanations tend to stress deep-rooted endogenous social factors which are unique to the society or group of societies in question. The best known sociological explanation, couched in terms of cultural variables, is Max Weber's famous Protestant Ethic thesis (1958) in which he attempted to explain the unique development of modern rational capitalism in Western Europe in terms of cultural aspects of Reformation Protestantism. Modern rational capitalism, he argued, should be seen as an unintended consequence of certain aspects of the Protestant world view (this worldly asceticism for example) as this developed in the sixteenth and seventeenth centuries. More recently, similar types of explanation have been offered to account for a relationship between Confucianism and East Asian industrialisation (see Goodman, Chapter 11 above). Whereas prior to the 1960s Confucianism, with its respect for tradition, was depicted as a reactionary force inhibiting capitalist industrialisation, more recently emphasis has been placed on its ability to promote capitalism through such norms and values as frugality, discipline and self-sacrifice.

Generally, therefore, cultural explanations have been invoked to explain relative success in developing industrial capitalism. In this essay I shall examine one area in which they have been employed to account for de-industrialisation, Weber-in-reverse, as one might say. Thus the decline of the British economy over the past 150 years, in comparison with its major competitors, has often been explained in terms of a deep-seated, long-enduring and apparently irreversible 'cultural malaise' unique to Britain. The exact nature of this malaise has been endlessly debated, so that it threatens to become almost an industry in its own right. In general, I shall suggest, it has been defined in terms of the perceived 'complacent traditionalism' of British society in comparison with its major rivals. This traditionalism is seen to have found expression across many different areas – cultural, educational and economic – and has constantly reasserted itself against attempts to modernise British society. Indeed one might see some analogy with Third World societies where explanations of their failure to modernise have frequently been couched in terms of a contrast between traditional and modernising sectors, with the predominance of the former being invoked to explain the failure of these societies to industrialise.

THE DECLINE OF BRITAIN: CULTURAL-POLITICAL EXPLANATIONS

In many respects Britain has been an ideal candidate for such explanations, given the long-standing nature of its decline and the fact that this has occurred evenly across a number of different areas – political, cultural and economic. By contrast, the Fall of the Third Reich or the very rapid and unpredictable implosion of eastern Europe did not typically invoke explanations in terms of deep-seated cultural malaise. Indeed, an emphasis on the supposedly baleful effects of tradition on British society has been common to both the political Left and the political Right in Britain. Before the 1970s these critiques were mainly the preserve of the Left, for example, the Wilsonian attack on the political establishment and the imperial legacy in the 1960s. The same period also saw the development of a Marxist critique of British society, developed in journals like the *New Left Review*, which castigated Britain for its isolation from European culture and from continental intellectual traditions, both Marxist and non-Marxist alike. The political Left has typically tended to depict Britain as an 'Old Country' hopelessly enmeshed in pointless tradition, class prejudice, and the legacy of Empire, and isolated from the main currents of the contemporary world.

However, since the 1970s, these critiques have been increasingly adopted by the Thatcherite Right, which has defined the so-called 'British Disease' in a radically different fashion. This now became seen as clinging to a welfare-state dependency culture and to the illusion that Britain did not have to pay its way in the world. Again the New Right was inclined to see certain specifically British cultural traditions, albeit mainly of a different type from those identified by the Left, as at the root of the problem.

In fact the political Left and the New Right have tended to offer different role models of perceived economic and political success as a cure for the British disease. For the political Left, antagonistic to American imperialism during the Cold War period, the post-war examples of success have typically been drawn from Europe. In the 1960s and 1970s these pro-European attitudes tended to take a cultural form. Europe represented an intellectual alternative to a perceived British narrow-minded insular traditionalism and to the legacy of empire. Since that period, during the 1980s and early 1990s, the shift of the Left's critique, exemplified by Will Hutton's recent influential *The State We're In* (1995), has shifted to highlight Britain's post-war economic performance. Again, in this literature the role models of successful societies have tended to be European, although now the list has been extended to include Japan. These economies, with their corporatist high 'trust' cultures, long-term planning and social market capitalist contracts between capital and labour, are typically contrasted with British short-termism and antagonistic capital-labour relations (clearly such positions will have to be reconsidered in the light of the financial collapses in Japan during 1998).

For the Thatcherite New Right, on the other hand, the perceived successful role model has been the United States, with its traditions of minimal state intervention, entrepreneurial activity and flexible markets. Typically, advocates of Thatcherite neo-liberalism have been implacably opposed to two kinds of British tradition: on the one hand, what many of this persuasion saw as the post-war consensus to extend the role of the state and to buy off the threat of organised labour; secondly, to the aristocratic paternalistic one-nation aspects of Conservatism, embodied in a figure like Macmillan, which the Thatcherites saw as a major obstacle to their desire to return to the pure doctrines of mid-nineteenth century liberalism.

While, therefore, commentators have tended to disagree about the nature of this deep-seated British 'malaise', there has been a general agreement concerning the existence of a deep-seated and typically British cultural 'complacency'. This is often seen as reflecting a deeper insular chauvinism. This complacency itself can be seen to have a number of possible roots and two main sources are generally identified. For example, the Imperial Legacy is often seen as having left the British with an ill-founded feeling of superiority to other peoples. Equally, the fact that Britain survived the Second World War without suffering invasion or the horrors inflicted on many European countries is sometimes seen as having allowed an illusion of national greatness and a belief in perceived specifically 'British virtues' such as fair play to persist in the post-1945 period.

BRITAIN'S 'CULTURAL COMPLACENCY'

National complacency can itself then be used as a partial explanation for the survival of obsolete and outmoded institutional 'traditions', such as the lack of a written constitution and the monarchy, which may have held Britain back in comparison with other more dynamic societies. Lacking the external 'shock' imposed on other European societies, Britain, it has been argued, failed to respond adequately to the challenges of the post-war era. Thus it failed to revolutionise its industries and opted instead for the 'New Jerusalem' of the welfare state without first, unlike its European counterparts, having created an adequate economic base upon which to build this (Barnett 1986).

However, the thesis can be overstated. If we take, for example, the perceived effects of the imperial legacy, it is very clear that xenophobic and racist attitudes were common to all European societies of the late nineteenth and twentieth societies, irrespective of whether or not they possessed foreign empires. Germany is a notorious example of this. Without wishing to understate the effects of British imperialism in the past (both through direct colonialism and neo-colonialism) on Third World societies, it is still true that only a relatively limited fraction of the British population would have been directly exposed to the effects of empire. The average size of the British army garrisoning India during the twentieth century, for example, represents approximately the number of troops lost in one day on the Somme in 1916.

Moreover, the long-term political pressures in favour of imperial preference, which would have turned the empire into an economic bloc, as demanded by the Austen Chamberlain wing of the Conservative Party at the turn of the century, were successfully resisted until as late as the 1930s. Britain's move to imperial preference, far from reflecting a complacent traditionalism, represented a response to the same pressures, arising from the Great Depression, which were driving all other major economies towards autarchy and protectionism in the 1930s.

Indeed Britain was conspicuous perhaps for the speed with which it dismantled its empire in the immediate post-war era. In contrast, France fought long and bloody wars to defend its two major foreign colonies. Yet it was the more 'traditionalist' France, under the quasi-dictatorship of De Gaulle which glorified French culture and tradition, that enjoyed an economic miracle during the 1960s. By the same token, as Kyong-Dong has argued (1994), nationalist ideologies proved a powerful motor for much East Asian industrialisation in the post-war period.

Another argument linked with the complacency thesis relates to Britain's escape from invasion and conquest during the Second World War. This is sometimes seen as preserving a series of delusions and complacent attitudes across a wide social and political spectrum. Conversely, those economies which suffered the shock of external destruction (Olson 1982) and which were able to rebuild more successfully their economies and social structures from scratch, were more dynamic and successful during the post-war period. For example, one effect of the war and of the Soviet conquest was to eliminate the militaristic Junker aristocracy which had been one of the major sources of traditionalism in German society and an obstacle in the way of its modernisation.

One of the most influential advocates of this thesis has been the military historian Corelli Barnett, who, in two influential books (1986, 1987), has identified the immediate post-war era as the period in which British complacency was at its height and in which, in the temporary absence of competition from its major industrial rivals, Britain failed to build the foundations for a successful post-war economic revival. In the foreign policy arena it is similarly claimed that victory in the war, a victory in reality achieved mainly by American and Soviet power, left Britain in the post-1945 period with an illusion of national greatness. Hence the pursuit of nuclear status and the Atlanticist 'special relation' with the United States.

While there are undoubtedly some elements of truth in the complacency thesis, it has been wildly overstated. Indeed an opposite case can be made for the view that for much of the twentieth century British policy, in many key areas, has been dominated by a sober realism. British planners were long aware of the problems of what the historian Paul Kennedy (1988) has termed 'imperial over-stretch'. The policy of appeasement in the 1930s, for example, as some revisionist historians would suggest, reflected the realistic belief that a moderate-sized power like Britain could not afford both to maintain a world empire and to engage in a major European war. By the same token, in the post-war period, the special relationship with the United States can be seen as again a realistic response to Britain's new subordinate role in the world and to the overriding priority of maintaining American involvement in Europe (Macridis 1992). Indeed a recent revisionist historian of post-war British foreign policy (Charmley 1995) has argued that this was not assertive enough of Britain's specific national interests.

Moreover, the stress on discontinuity with respect to Europe may have been overdone. Simon Reich (1990), for example, has argued that the successful German post-war corporatist economy had its roots in the social and economic structures of the Nazi period. Notoriously, both in Germany and France, the post-war de-Nazification process was relatively superficial in view of the need to enlist these countries in the new anti-Soviet Cold War alliance. Indeed, New Right Eurosceptics would wish to turn the tables on the 'complacent traditionalism' thesis by arguing that it is precisely European countries which, with their less flexible labour markets and protected industries (Eurosclerosis as it is sometimes termed), have demonstrated this traditionalism most strongly.

BRITAIN'S 'CULTURAL CHAUVINISM'

To turn to the area of supposed cultural chauvinism, as I have suggested, part of the case against British traditionalism has involved portraying the British as chauvinists existing in an insular cultural and intellectual isolation. In the cultural areas, it is argued, the British intelligentsia has tended to remain provincial in its attitudes, making few genuine contributions to twentieth-century intellectual or cultural movements.

In fact a strong case could be made for the opposing point of view. Britain's long neo-liberal tradition, its position as a trading nation and, for a long period, as a World Empire, has exposed the British to a wide range of external cultural influences in a way that may not have been true for other European countries.

In most key cultural areas the British would appear to have been importers of culture over a long period. Britain has derived its high culture, for example, largely from Europe and its popular culture from the United States. The major exception to this, literature, arguably reflects the British incompetence at languages. But this should be seen as reflecting the position of English, itself reflecting post-war American hegemony more than British imperialism, as a world language rather than chauvinism as such. Indeed the most noticeable feature in most cultural areas, ranging from high culture to areas like the cinema, has been the inability of Britain to create and defend a version of its own 'national culture' *vis-à-vis* other countries of equivalent size. This itself can be accounted for by the weakness of a strong statist tradition in Britain and the British failure, as in manufacturing, to support and subsidise its own cultural industries.

One interesting example here is provided by the Pre-Raphaelites, who represent one of the few schools of painting that are truly autonomous and relatively untouched by continental influence, and as such might be claimed to represent a genuine native tradition in art. Yet as a group they have largely been disregarded by English art historians and intellectuals whose interests have remained firmly directed towards continental art traditions. Thus there is no national museum devoted to their work, which remains scattered among a number of different collections.

THATCHERITE DIAGNOSIS OF 'CULTURAL MALAISE'

As I have suggested, the Thatcherite New Right offered a diagnosis of cultural malaise which pointed the finger at the post-war welfare-state consensus. However, from the first, the Thatcherites, as we have seen, also tended to be hostile to the patrician aristocratic elements in the Conservative party, as classically typified by Macmillan, seeing them as representing a patrician 'one-nation Toryism' which had colluded in the building of the post-war consensus and which had abandoned the tradition of nineteenth-century *laissez-faire* liberalism. Thus there were clearly radical-right populist elements in Thatcherism antagonistic to the existing political establishment. These have been represented most strongly in the Murdoch press with papers like the *Sun* and the *Sunday Times*, whose former editor, Andrew Neil, has been bitterly hostile to what he has seen as an old-school-tie establishment controlling the civil service and the main political power. Ironically this attack on the 'establishment' echoes many of the themes of the Wilsonian Left of the 1960s. This in turn has been countered by the 'fogeyism' of papers like the *Telegraph*, as represented by columnists like Auberon Waugh, which deliberately makes a virtue of adopting a gentlemanly traditional lifestyle and scorns what it terms the 'New Brit' (i.e. uncouth) attitudes of the Murdoch press.

Typically the Thatcherites have appealed to the famous so-called Victorian values of thrift, hard work and entrepreneurial endeavour as advocated in the writings of Samuel Smiles, the great Victorian advocate of self-help, or in the novels of Arnold Bennett: values which, in the nineteenth century, were typically associated with the entrepreneurial middle class. Implicitly, the great enemy of the honest bourgeois in the rhetoric of the New Right is the leisurely, snobbish landed gentleman who disdains honest toil in favour of a leisurely *rentier* lifestyle. The Thatcherites saw this patrician wing of the Conservative Party as allied with the Civil Service values which had underpinned the post-war consensus and which, at the foreign policy level, had betrayed British interests.

So the question arises as to how the Golden Age of nineteenth-century liberalism, a 'virtuous tradition', had come to be replaced by a false tradition based on *rentier*, Civil Service values. One answer was supplied by a very influential work, published in 1981, which captured the spirit of the early Thatcher period, Martin Wiener's *English Culture and the Decline of the Industrial Spirit* (1981). His argument had the additional virtue, from a New Right viewpoint, of putting the main blame for the British Disease on the education system, since this already had low esteem in their eyes.

Wiener suggested that in the latter decades of the nineteenth century the entrepreneurial traditions of the middle class came to be usurped by the gentlemanly leisured values of the landed aristocrat. These values were transmitted through the public schools, and more latterly the grammar schools, which took in the children of businessmen and turned out gentlemanly civil servants hostile to the world of commerce. This animus against industry became reflected, Wiener suggests, in the school curriculum, which cultivated a taste for rural nostalgia and an aversion to the world of industry (Bernbaum and Mattheson 1991)

Ironically Wiener was, albeit from a more right-wing perspective, approaching a problem that had been much debated in Marxist sociology of the 1960s and 1970s, that of the political and economic role of the nineteenth-century English middle class. England seemed to represent a paradox. On the one hand, it was the first country to witness an industrial revolution and the English came to be seen in Europe as, in Napoleon's famous phrase, 'a nation of shopkeepers'. On the other hand, the landed aristocracy had managed to maintain its political hegemony well into the industrial period. In some deep cultural or political sense the English middle class would have appeared to have abdicated its hegemonic role, allowing the landed aristocracy to retain a controlling position in English society.

Wiener's thesis has been subject to considerable criticism from a variety of perspectives. Empirically, for example, doubt has been cast on the supposition that the public schools did in fact turn businessmen into civil servants (Rubinstein 1993). But perhaps some of the most trenchant criticisms have involved cross-cultural comparisons with other European countries. Britain was far from being the only European country to witness the continued political and cultural strength of the landed aristocracy persisting deep into the industrial period. Indeed the historian Arnot Mayer (1981), in a well-known thesis, has argued for the persistence of the landed aristocracy as a political force during the industrial period across Europe as a whole. By the same token, although this view has more recently been challenged (Blackbourn and Eley 1984), the peculiar path of German history from the late nineteenth century onwards has been frequently attributed to the political and cultural power of the Junker aristocracy and the relative weakness of the German middle class. Also, a nostalgia for the soil and for rural life, far from being unique to the British intelligentsia, was notoriously a marked characteristic of inter-war European fascism in most European countries.

A CONSENSUS OF OPINION

More recently a new consensus has emerged in the field (Cain and Hopkins 1993; Rubinstein 1993) which manages to provide a partial solution to some of the central problems raised in these debates. This centres around the concept of 'gentlemanly capitalism' and attempts to dissolve the traditional aristocrat–middle class contrast, which has dominated debates in this area for so long. Its proponents argue that, at the beginning of the eighteenth century, Britain witnessed the creation of a form of 'gentlemanly capitalism', involving a fusion between City-based finance capital and the landed interest. This pre-dated industrial capitalism, since Britain was a major trading nation and financial centre before it was an industrial power. For this reason, despite its short-lived pre-eminence in the middle decades of the nineteenth century, northern-based industrial capital was never successful in projecting its interests in the political arena over those of a southern-based gentlemanly capitalism which fused together civil service, financial and landed élites into one unified culture. In short, British industrial capital did not abdicate its power, since it was never a hegemonic class in the first instance.

From this perspective one might claim that there has been a peculiarly British tradition, stretching back to the beginning of the eighteenth century, which has seen the dominance of a specifically British type of 'gentlemanly capitalism', based on the social and political hegemony of finance capital. Characteristically, this form of capitalism has been associated with a neo-liberal market economy hostile to state intervention and the protection of national industry. Thus it was always in the interests of the City and the financial fractions of British capital to try to maintain world free trade and to reject moves towards protection, despite the fact that the latter was necessary to protect fractions of British industrial capital from the inroads of foreign competition.

Those who take this position might therefore claim that the Thatcher years, with their de-industrialisation and victory of financial over industrial capital, represent a successful extension of an earlier British tradition. From this perspective, Thatcherism, far from representing a deviation from British tradition or a form of cultural malaise, should essentially be seen as business as usual. Thus one might argue that these traditions represent a viable alternative to a European model based on essentially different principles and traditions in which, from its very inception, industrialisation was directed by a powerful central state. Taking the argument one step further, some might even suggest that these traditions now give Britain an advantage in moving towards a more advanced stage of capitalist development, giving it an extra edge in the switch towards a post-industrial economy based on services rather than manufacturing.

This approach is in conflict with a consensus that gained strength in the political Left in the late 1980s (Marquand 1988; Hirst 1989), which, as we have seen, argued that the post-war British model represented a deviation from a more successful European Social Market capitalism based on a compact between Capital and Labour. This form of capitalism was seen as associated with the 'virtuous cycle' of a high-wage economy producing high-quality products relatively immune from Third World competition. Consequently de-industrialisation was not inevitable and in Britain had reflected both policy errors of the Thatcher period and, more fundamentally, a failure of the State over many decades to protect and to further the interests of British industrial capital. These attacks have also typically been associated with attacks on the traditionalism of the British constitution as contrasted with the greater rationality of European systems reflected in written constitutions, charters of human rights etc.

One central problem for the proponents of the cultural malaise argument on the political Left was the rise of the Thatcherite New Right, since, as we have seen, the latter had appropriated this for its own purposes and saw itself as representing a discontinuity from the preceding post-war period. In reaction, opponents of Thatcherism tended to depict it as a blend of two types of so-called New Right. These, in turn, could be perceived as reflecting two malign British traditions. On the one hand they reflected a neo-liberal New Right espousing free market principles, the destruction of the social fabric and the atomisation of the individual; on the other, a neo-conservative chauvinist New Right advocating order, traditional family values and a strong central state (Gamble 1988) together with an Imperialist tradition which surfaced once again during the period of the Falklands/Malvinas war.

However, in the 1990s some of these elements of Thatcherism have started to separate, demonstrating that there is no necessary connection between them. This is illustrated, for example, in the conversion of the Labour party under Blair to the neo-liberal free market aspects of capitalism despite the Blairite adoption of a pro-European stance. In fact, as I have suggested, certain elements within Thatcherism – a constellation of disparate, mutually antagonistic elements rather than a single entity – have been inherently opposed to tradition. To take the single example of the monarchy, its treatment by the tabloid press during the past decade has effectively turned it into a branch of the entertainment industry with relatively few 'deferential' elements surviving from an earlier era. In some respects perhaps the Falklands/Malvinas War should be seen in a similar light. One might well wonder whether the undoubtedly xenophobic reactions this produced would have long survived a prolonged campaign or heavy troop losses. By the same token, many aspects of British traditionalism, so frequently berated by critics, should be seen as part of a highly successful and money-spinning 'heritage industry'.

In the economic context of the late 1980s and early 1990s, the Lawson inflationary boom, followed by the economic recession, seemed to weight the argument strongly on the pro-European side. However, much more recent events, such as the recovery of the American economy coupled with the economic problems of Continental Europe, have shifted the debate to some degree in favour of what has been termed the 'Anglo-Saxon' model of capitalism. The British exit from the ERM has been followed by a period of relatively steady growth and, by European standards, relatively low levels of unemployment. At the same time many of the features of the Continental model, praised by the opponents of British traditionalism, no longer seem so unequivocally beneficial. The flight of capital from the corporatist continental economies like those of Germany and Sweden has 'hollowed out' the industrial structures of these societies, leading to high unemployment and the development of Far-Right extremist movements in some of these societies.

CONCLUSION

I have tried to suggest that diagnoses of the 'British malaise' have varied widely in the post-war period and reflect both false comparisons between Britain and other societies and very frequently an over-idealised view of the latter. Much of this diagnosis, it might be argued, has been applied to what was a non-problem anyway. Britain's relative economic decline from the late nineteenth century onwards, as many have argued, was only to be expected, given what was in many respects a 'false position' in the middle decades of the nineteenth century. A relative decline *vis-à-vis* countries like Germany and the United States with much larger internal markets was inevitable. To some degree the relative decline of the 1960s and 1970s might be seen to confirm some aspects of the 'complacency' thesis. Yet even this was corrected in the 1980s, although without the accompanying economic miracle anticipated by New Right propagandists.

Increasingly, one might suggest, the image of a British traditionalism, when looked at closely, starts to turn into a mirage. Typically, the aspects of traditionalism stressed are either exaggerated, not unique to Britain itself or not necessarily as harmful in their effects on the British economy as sometimes supposed. As I have suggested, the case for seeing Britain as a uniquely chauvinist or inwardly directed society has been grotesquely overstated. Insofar as the development of nationalist ideologies involves the manufacture of tradition, this has been far from a uniquely British phenomenon. Many features of the British gentlemanly capitalist and neo-liberal traditions have arguably exposed it to outside cultural and intellectual influences to a greater extent than many other countries. In fact much of the anti-British rhetoric has tended to be confused and contradictory. On the one hand, Britain has been attacked for being culturally insular and, on the other hand, for not being nationalist enough in failing to protect her industrial base at crucial periods in the past.

But neither is it clear that British traditionalism, insofar as it has existed, has had the detrimental effects attributed to it. The example of East Asian industrialisation might well be used to demonstrate that perceived traditional values frequently contain a hidden potential with respect to modernisation. Just as some students of Japanese industrialisation have come to argue that Shintoism was after all functional for the industrial process, so many aspects of British tradition might easily be viewed in the same light. Much-derided public-school values, for example, might be seen as helping to build the 'trust culture' which writers like Fukayama have seen as the moral ingredient which will increasingly underpin successful capitalist development in the future. Indeed, one commentator (Honey 1987) has suggested that, in the past, public-school education functioned precisely to provide a 'trust culture', replacing the narrow social base of Quakerism, for the expanding financial service industries of the late nineteenth century. To broaden the argument, as I have suggested, the whole 'gentlemanly capitalism' thesis might be used to suggest that British cultural traditions give Britain a unique edge in today's globalised world economy, which is witnessing a general post-industrial move towards services and away from a high-cost industrial production menaced by cheap-labour Third World producers.

13

ETHNICITY

John Eade and Tim Allen

INTRODUCTION

'Ethnicity', 'ethnic group' and 'ethnic violence' are three terms which are increasingly used in everyday discussions of political and cultural developments around the world. During the last five years the most obvious illustrations of 'ethnic speak' have been provided by media reports of 'ethnic cleansing' in central Africa and the former Yugoslavia.

However, the frequent use of 'ethnicity' is not confined to these extreme conflagrations. One influential commentator has argued that the world has entered 'a period of ethnic conflict, following the relative stability of the Cold War' in which, as large formal structures break up and ideology loses its hold, people 'revert to more primal identities' (Moynihan 1993: v). Another has written with great concern about the ways in which ethnic war acts as a kind of metaphor for what he calls 'the modern conscience' (Ignatief 1998). It has become almost a truism to assert, as Manuell Castells puts it, that 'we have experienced, in the last quarter century, the widespread surge of powerful expressions of collective identity that challenge globalisation and cosmopolitanism ... .' (Castells 1997: 2).

This chapter investigates these ideas. It shows how social scientists, particularly anthropologists, have been deeply implicated in the growth of 'ethnic speak' during the last two decades and have also engaged with everyday assumptions concerning ethnicity. Their analyses of ethnicity as a general phenomenon and of specific ethnic groups has fed into the public arena through the use made by journalists, politicians and teachers – to name a few – of the statistical surveys, field reports, training programmes and briefings from research institutions and individual experts. However, this research has in turn been influenced by popular understandings, and sometimes reveals a similar confusion about what ethnicity actually refers to. Ethnicity is a concept which has escaped from scholarly discourses and now often shapes that discourse. The chapter tries to provide some conceptual order to the muddle. It discusses the development of theoretical understandings of ethnicity from three main perspectives, and in so doing it critiques simplistic and misleading assertions, and indicates ways in which some of the dangers of what might be termed 'cultural functionalism' can be avoided.

ETHNICITY AS AN ANALYTICAL CONSTRUCT

Definitions and uses of the term ethnicity have changed over time. The word 'ethnic' is derived from the Greek words *ethnos*, 'a people', and *ethnikos*, 'a heathen'. This early usage shaped subsequent interpretations across Western Europe so that from the fourteenth century 'ethnic' was employed in English to refer to pagan, i.e. non-Christian,

Plate 13.1 A refugee from Bosnia-Herzegovina, now in Croatia, holds a newspaper with the headline, 'Through Terror to Great Serbia'. She has lived through some of that terror.

Source: Photographer, A. Hollman, 1992. Reproduced with kind permission of UNHCR.

populations. During the mid-nineteenth century the focus on religion began to be replaced by a preoccupation with the relationship between ethnic and racial difference and, by the 1940s, the word 'ethnic' in the USA had become a polite way of referring to 'Jews, Italians, Irish and other people considered inferior to the dominant group of largely British descent' (Eriksen 1993: 4). As we shall see, it was later used to refer to aspects of collective identity in very different ways, but it has retained some of these earlier connotations.

Use of the term in English in the form of a noun is a much more recent introduction. 'Ethnicity' did not start appearing in most dictionaries until the turn of the 1970s – something which seems quite strange, given its current prevalence. In recent years it has become so widely used that it apparently requires little elaboration, and it is readily invoked both to describe and to explain behaviours and events. Its specific meaning, however, is often rather obscure. There is, for example, a common tendency both to distinguish between religion and ethnicity and link them together. Sometimes religious affiliation is presented as reinforcing ethnic divisions, even though it is often not clear what a particular ethnic identity is based on. Meanwhile, other commentators define religious communities as ethnic groups.

The ways in which ethnicity is interpreted also vary according to diverse national traditions. Within Britain, ethnicity is usually used in debates about minorities which have settled in that country since the end of the Second World War. In Africa, the discussion tends to focus on the changing character of post-colonial nations. Here ethnicity can be a euphemism for tribalism or may be used to describe a process whereby tribal divisions are being replaced by newer social solidarities, particularly in urban areas. Marxist ideological traditions within the former USSR have established another distinctive way of approaching ethnicity, linking it with an evolutionary theory of 'nationalities', while in North America ethnicity is often used to describe divisions among the majority white population. Sometimes, different meanings may be suggested by the place being referred to. The same analyst might use ethnicity to refer to minorities in England, religious groupings in Ireland, tribes in Kenya, nationalities in Eastern Europe, indigenous peoples in Latin America, and races in South Africa.

It is perhaps the muddled association between ethnicity and race which has become most important in recent years. It is no longer limited to particular places, and it has very disturbing implications in that it shares certain similarities with apartheid ideology, and helps disseminate racist ideas in a new form.

Since the 1960s a complicated conceptual relationship has developed between 'ethnicity', 'race' and 'culture', especially in the highly industrialised countries of the West. Considerable heat has been generated between anthropologists, sociologists, policy-makers and teachers over this tangle. School materials on the 'cultural traditions' of ethnic groups and the multicultural diversity of a particular country have been introduced which encourage students to regard racialised differences and conflicts between 'black' and 'white' citizens as unusual instances of generally benign ethnic group relations. In Britain, such multicultural policies played a key part in efforts to promote the integration of immigrant groups in the 1970s and 1980s. However, teachers, policy-makers and political analysts who have focused on the eradication of racial prejudice and discrimination have regarded this preoccupation with the cultural differences of different ethnic groups as a dangerous evasion of the institutionalised disadvantages and sufferings of racialised groups of citizens.

The emergence of ethnicity as a popular term was linked with these debates. In Britain, for example, its use was actively promoted during the late 1970s and early 1980s as part of multicultural policies, such as the setting up of ethnic minority units within local councils and the introduction of multicultural issues within the school curriculum. In effect, this meant that ethnicity became a term which could act as an alternative to the word 'race', and was often viewed as preferable because it was apparently less divisive and more neutral. However, with the growing popular use of the term, its connotations have shifted. Because there is a tendency in much popular writing on ethnicity (particularly in the press) to suggest that it relates to deep, fundamental and natural traits, ethnic identity is implicitly, or sometimes even explicitly, linked to ideas about biological distinctiveness. It has increasingly become an alternative word for 'race', not in the modern sense of a socially constructed categorisation of groups, but in the discredited, pseudo-scientific sense of differences between particular groups existing in nature.

Part of the problem is that discussion about ethnicity has tapped into a more general set of ideas which can be termed 'cultural functionalist' (Duffield 1996a: 174), and which have become very prevalent since the end of the Cold War. It is now widely assumed that culture and social identity coincide, and that they are discrete, each having particular sets of values and their own inner core. This is both a popular view, and one shared by many academics. Castells, for example, discusses 'surges of powerful expression' which are 'multiple, highly diversified, following the contours of each culture, and of historical sources of formation of each identity' (Castells 1997: 2). Multicultural policies might be viewed as a potentially benign expression of this cultural functionalist perspective in that they aim to promote respect between communities. But multiculturalism can easily be switched to something more pernicious, as racial discourse becomes imbued with the idea that cultural integrity must be preserved, largely by excluding other influences. As Barker pointed out as long ago as 1982, it has allowed racism to be modernised and made respectable (Barker 1982).

How may we map out what is clearly a complex and highly contentious area where academic and popular interpretations interweave? We have chosen here to establish some order by describing three theoretical approaches which have developed within anthropological studies of ethnicity. As we have already seen, these studies frequently involve discussions of 'race' as well, either explicitly or implicitly. However, for the sake of some notional sense of order and brevity, we will focus on *ethnicity* in the context of the following approaches: (a) essentialist, (b) instrumentalist and (c) relationalist.

ESSENTIALIST APPROACHES

These approaches come closest to the popular assumptions that ethnicity and conflict between ethnic groups is somehow natural and inevitable. Although many anthropologists have dismissed such assumptions on the grounds that social ties and differences are socially constructed, the anthropological focus on small localities or population groups has encouraged the implication that each community is characterised by a unique (social and cultural) essence. This implication can easily lead to a perception that the distinctiveness of particular communities is created by deep and ineradicable differences between 'us' and 'them'.

Such an essentialist view is often linked to ideas about family and kinship relationships, which are deemed to express 'people's close personal ties and early cultural experiences' (Calhoun 1997: 31). Geertz, for example, contends that ethnic solidarities are more primordial than national affiliations because they express more fully the attachments to family members and relatives (Geertz 1963: 110–1). However, most social and cultural anthropologists based in North America and Western Europe draw a distinction between the subjective experiences of their informants, which may well be highly essentialist, and their own analyses, which usually focus on the historical or social construction of collective identities. Overtly essentialist interpretations are avoided or actively disparaged. But this is less so amongst biological anthropologists and sociobiologists, and amongst anthropologists working in Eastern Europe who have been influenced by Soviet nationalities policies and by the 'ethnos' theories of Yulian Bromley and his colleagues. Moreover, while these essentialist interpretations have been rejected or ignored by the mainstream of anthropological writing on ethnicity in the West, they have remained a significant influence because they have contributed to understandings outside academic circles.

SOCIOBIOLOGY

Sociobiologists and biological anthropologists who have been interested in ethnicity have drawn on notions connected with the theory of evolution. They emphasise 'inclusive fitness': the idea that genes will spread if their carriers act to increase not only their own fitness or reproductive success but also that of other individuals carrying copies of the same genes (Reynolds, Falger and Vine 1987; Van der Dennen and Falger 1990). Some have argued that there is a genetic basis to group membership and the propensity for human warfare (e.g. Shaw and Wong 1989), but they also point out that this is not the same as asserting that ethnicity (or indeed any form of collective forms of violence) is inevitable. Genetic traits are not the same as unchangeable instincts. As Segerstale notes, the 'social message' of sociobiology 'could be that if we only adjust the environment accordingly, we would be able to avoid aggressive behaviour altogether' (Segerstale 1990: 277).

Despite this social message, popular interpretations of sociobiology have emphasised the determining role of human needs and drives in group dynamics (biological determinism). As a result many social and cultural anthropologists have fiercely criticised any biological interpretations of social action. In 1986, for example, the American Anthropological Association went so far as to pass a formal statement claiming, among other things, that it is 'scientifically incorrect to say that war or any other violent behaviour is genetically programmed into our human nature' (quoted in Tiger 1990: 100). Actually there does not appear to be adequate scientific evidence one way or the other, so the main aim of the statement appears to have been to discourage members discussing the issue.

ETHNOS THEORY

Eastern Europe and the former USSR have been more open to biological determinism and other overtly essentialist explanations of human behaviour. This openness was expressed through the development of 'ethnos theory' during the late 1960s by Yulian Bromley, the director of the Soviet Academy of Science's Institute of Ethnography in Moscow. Bromley contended that ethnicity possessed a stable core for which he used the Greek word 'ethnos' (Bromley 1974). While 'ethnos' was not eternal it still persisted across the evolutionary stages of human history, which, in conventional Marxist terms, moved from primitive communism to slave-ownership, feudalism, capitalism and finally, socialism. Ethnos was defined as a historically formed community of people characterised by common, relatively stable cultural features, certain distinctive cultural traits, and the consciousness of their unity as distinguished from other similar communities. Given this starting point, the anthropological problem was to determine how the fundamental characteristics of 'ethnos' manifested in such overt cultural forms as language, costumes and folklore, and integrated as a system.

Some of Bromley's colleagues took the theory a stage further by suggesting that 'ethnos' was, at least in part, not just a historical but also a biological phenomenon. Gumilev, for instance, argued during the 1980s that 'ethnos' was a 'biosocial organism' and that 'the rise to existence' of an 'ethnos' was due to the 'combined effect of cosmic energies and landscape' (Sokolovski and Tishkov 1996 : 191).

'Ethnos theory' gained a considerable following within the Soviet bloc largely because it appeared to resolve two key issues. It provided an explanation for the survival of cultural traditions in the face of rational socialist planning. It also fitted into official Leninist views concerning 'ethnogenesis', which defined the nation as the highest type of ethnic community, the most important social group upon which the state and its economic and cultural life rested.

Western awareness of 'ethnos theory' has recently widened and deepened with the opening up of Eastern Europe after the collapse of the Soviet bloc and the numerous upheavals across the region. Nationalist movements in the East have made use of Soviet ethnos theory to justify their ideological claims, which, in turn, have been reported by Western journalists and political commentators. The phrase 'ethnic cleansing', for example, has been employed by all parties in the Bosnian conflict and has rapidly changed from being a shocking new term requiring explanation for a Western audience to being seen as accepted fact (Banks 1996: 166–71). Indeed, academics in the former Yugoslavia have sometimes played important roles in providing conceptual sophistication and credibility to essentialist notions of violent struggle between ethno-nationalist communities.

Eastern European anthropologists now vigorously debate the ideological justifications which ethno-nationalists have tried to establish by drawing, in part, on 'ethnos theory'. There is deep scepticism about the validity and usefulness of universalistic and homogenising models of ethnicity for explaining what is going on in the former Soviet bloc. Nevertheless, the essentialist traces of 'ethnos theory' can still be seen in the analyses of contemporary Eastern European anthropologists, which place greater emphasis on the motivational force of corporate pyschologies and the use in Russia, for example, of concepts long discredited elsewhere about the importance of 'genetic pools' or the genetic history of given peoples to understand cultural behaviour in contemporary times (Dragadze 1995: 2).

INSTRUMENTALIST APPROACHES

Instrumentalism has developed in sharp contrast to essentialist interpretations. Instrumentalist analyses locate ethnicity within the conscious actions of individuals and groups who share the common interest of acquiring such scarce material resources as jobs, education and housing. The focus on these particular resources emerged from pioneering studies on ethnic groups within the USA by sociologists and political scientists rather than anthropologists. This approach also draws attention to the role of élites in shaping – manipulating even – the actions of others through political and ideological strategies (see Calhoun 1997: 31).

The most influential early study of ethnicity from an instrumentalist perspective was Glazer and Moynihan's book on New

York – *Beyond the Melting Pot: The Negroes, Puerto Ricans, Jews, Italians and Irish of New York City* – published in 1963. It controversially challenged the then popular belief that America was a melting pot where ethnic differences were transformed into an American national culture. Their study of the competition for scarce material resources in America's largest city revealed that its residents used their ethnic ties with families and friends to compete with other groups for these resources. Against essentialists who would see this ethnicity as 'a survival from the age of mass immigration' Glazer and Moynihan claimed that ethnic groups were a response to the new urban situation in which immigrants found themselves. Although residents could mobilise themselves as interest groups on the basis of class or race, they argued that ethnicity was and would continue to be a more popular source of solidarity.

While developments during the 1960s arguably confirmed the continuing significance of ethnicity across America, Glazer and Moynihan failed to anticipate the gulf which opened up between white ethnic Americans and their black compatriots. Civil rights campaigns and the emergence of black power groups across the USA emphasised the abiding racial rift between black Americans and all white Americans, irrespective of their ethnic differences. Indeed, when Glazer and Moynihan produced their revised, second edition in 1970, they acknowledged the importance of the racial divide within New York politics during the 1960s and the ways in which this fracture encouraged white ethnic groups to mobilise to defend their interests across ethnic boundaries.

Whatever the limitations of Glazer and Moynihan's book their approach to ethnic groups as new, emergent social forms engaged in competition for scarce material resources attracted anthropologists who were trying to explain social change in Africa during the 1960s. Research by Gluckman (1958, [1940]) and Mitchell (1956) revealed that this process of change had already begun in the 1940s and 1950s and it became clear that a new terminology was needed to analyse the social solidarities which were appearing. Migration from the countryside to the expanding towns and cities played a crucial role in this process of change, which entailed the transformation of tribal loyalties. The primordial model of African tribalism fostered by European colonial authorities was clearly becoming redundant.

Certain scholars began to turn towards ethnicity as a framework for analysis even if they did not necessarily accept Glazer and Moynihan's particular interpretation. Cohen, for example, in his book *Custom and Politics in Urban Africa* (1969) emphasised the practical functions of ethnicity in his analysis of Hausa traders in the Nigerian city of Ibadan. Ethnic groups were created out of informal political organisations which used cultural values in their competition for scarce material resources.

Instrumentalism has proved particularly useful for such analysis of social change. However, the approach is not without its problems. One of the key questions advanced by its critics is: why are certain ethnic solidarities possible and others impossible? Individuals may see certain utilitarian advantages in leaving their original ethnic group and joining another, especially where the new group is more powerful economically and politically. However, those individuals may not be accepted by the members of the group they wish to enter. The 'insiders' may reject newcomers because they do not speak their language or share their particular beliefs and values.

Of course, this example can be used by instrumentalists to demonstrate once again the importance of material interests. They can argue that it may not be in the interests of the insiders to accept outsiders. However, the example also shows that, in a clash of interests, it is the exercise of power by one group against another which decides the issue and makes some alliances possible and others impossible. The interaction does not involve the free play of equal groups and individuals rationally determining their strategies.

Glazer and Moynihan's study of New York politics actually confirms this unequal struggle and the limits placed upon the pursuit of material interests. They also showed that people may not decide to join a more powerful group even when it makes material sense to do so. They may prefer to stay with 'their own people' and their decision may be taken independently of the other group's beliefs about them. Many New York Puerto Ricans, Italians and Irish stayed within their ethnic group even though it made financial sense to marry into the more powerful and prosperous White Anglo-Saxon Protestant (WASP) community.

Some instrumentalists have responded to these points and produced more subtle analyses of ethnicity. Mahmood Mamdani's recent discussion of tribalism, for example, has incorporated discussion of power and resistance (Mamdani 1996: 183–217). However, the horrific events in Bosnia, Rwanda and the former USSR have demonstrated the extent to which groups will ignore their material interests in their deathly struggles with other ethnic groups. Many political analysts have responded by moving towards essentialism or producing an ill-conceived blend of essentialism and instrumentalism (a good example is Moynihan's 1993 book, *Pandaemonium*, which is quoted at the start of this chapter). Anthropologists have been more aware of this danger. They have responded by drawing on a less ambitious perspective which offers fewer explanations for ethnic violence, but is highly successful in describing and clarifying what is actually going on.

RELATIONALIST APPROACHES

Most anthropologists have never been totally convinced by instrumentalism. While they reject primordialist approaches because they overemphasise the fixity of social formations, they are sceptical of instrumentalist interpretations because they lay too much store by the rational, utilitarian dimensions of human behaviour.

Anthropologists have been concerned to explore the emotional, symbolic, unconscious and contingent aspects of human creativity – all of which tend to be set aside or treated as residual in instrumentalist interpretations of social action. They view these dimensions as expressed in the relations between human beings, and in the social boundaries which are constructed through those relations. This perspective has informed the work of such celebrated anthropologists as Evans-Pritchard, Gluckman, Leach and Douglas. However, it is particularly easy to understand in the context of ethnicity and ethnic groups, as Barth's classic 1969 compilation, *Ethnic Groups and Boundaries*, reveals. In his introduction to the volume Barth distinguishes between ethnic groups and cultural units and argues that the study of ethnicity should focus on the ways in which social (ethnic) boundaries are constructed and maintained. He criticises any attempt to link ethnicity to what may appear to be essential or primordial characteristics of particular groups because such an attempt 'allows us to assume that boundary maintenance is unproblematic' (Barth 1969: 11). People who claim a particular ethnicity may differ radically in terms of their overt behaviour, while those who share lifestyles may assert different ethnic identities.

Although the internal characteristics of a group must not be overlooked according to Barth, the crucial issue is the ethnic boundary which separates insiders from outsiders. People belong to group A because they distinguish themselves from those in group B. This distinction remains in force even when people and ideas flow across group boundaries. Ethnic groups, therefore, do not exist independently: they are created and sustained through their relations to one another. The apparent cultural uniqueness of many groups should, therefore, be analysed as emerging from a lengthy social process where this uniqueness is continually created in relation to the perceived qualities of other ethnic groups.

Barth's development of the relationalist approach tries to avoid essentialism but his arguments lead him perilously close to this perspective. He claims, for instance, that, while ethnic boundaries are not always important, they remain to be activated when the need arises. Social groups, therefore, appear to share some deep, psychological base which shapes the fluctuations of their social relations. Another problem involves the emphasis on social agency which, as we have seen in the section on instrumentalism, can ignore power relations and other constraints on human behaviour.

Despite these criticisms, most anthropologists see the relationalist wisdom of defining ethnicity as 'an aspect of a relationship' (Eriksen 1993: 12) and have rejected Glazer and Moynihan's claim that ethnicity is 'the character or quality of an ethnic group' (Glazer and Moynihan 1975: 1). They have focused on process and the dynamic, emergent qualities of relations between groups.

When anthropologists have examined violent social relations they have not sought explanations in terms of some fixed ethnic substance contained within particular groups. Turton, for example, puts the case very well:

> Ethnicity will not serve as a causal explanation of war because it is not a thing in itself, even though its power to influence behaviour is largely the result of being seen as a 'natural' property of a group. It is, rather, a relational concept: it refers to the way cultural differences are communicated and it is therefore created and maintained by contact, not by isolation.
>
> (Turton 1997: 78)

When civil war broke out in Yugoslavia during 1991, for example, the 'irreconcilable' differences between Serb and Croat were suddenly remembered. Although both shared many cultural characteristics such as language, were linked by high rates of intermarriage and had lived in harmony for over forty years, ethnic boundaries were reactivated. As one commentator noted: 'It is only when they *make a difference* in interaction that cultural differences are important in the creation of ethnic boundaries' (Eriksen 1993: 39).

While these recent discussions of ethnicity have confirmed Barth's earlier approach other anthropologists have extended and refined the relationalist perspective towards ethnicity. Worsley has examined the relationship between ethnicity and other social formations such as class, while numerous commentators have investigated the diverse ways and contexts in which ethnic identities are expressed. They have also shown how an individual's identity is bound up with the process of identification by others. The historical development of ethnic categories has also been investigated. Attention has been paid not only to the historical myths of a particular ethnic group but also to the ways in which ethnic identities and boundaries have changed over time.

If African tribal identities are seen as one type of ethnic formation, then it is now clear that many were largely created by colonial authorities, i.e. they are a modern creation. They have taken different forms in response to changing circumstances and have co-existed with other, non-tribal kinds of ethnicity (see Allen 1994; Campbell 1997). Similar developments can be detected in other parts of the world. The recent events in the Balkans have shown that ethnic identities were not frozen during communism and that the Balkan wars are not simply the result of an upsurge of primal identities after the end of the Cold War. Group identities changed during the communist period in response to urbanisation, industrialisation and secularisation. The communist political élite was well aware of these changes and adapted its policies accordingly.

Instrumentalists would agree with most of what we have discussed in this section. A parting of the ways emerges when relationalists assert the existence of unconscious as well as conscious motivations and the significance of people's perceptions of the primordial nature of ethnic identities. Although African 'tribal' categories may well be new, many people adopting them genuinely believe that their 'tribes' have a long and rich pre-colonial history. While Serb ethnicity has adapted to social, economic and political changes many Serbs believe that their 'nationality' enjoys a biological ancestry dating back to medieval times.

Ethnicity, therefore, takes on essentialist characteristics through people's perceptions and the systematic organisation of their beliefs. For relationalists, the horrors of Rwandan genocide and 'ethnic cleansing' in Bosnia cannot be understood without taking into proper consideration the power of subjective meanings and the ways in which these meanings interweave with other, objective processes. Clearly an analysis of objective political and economic developments has to be accompanied by careful attention to people's subjective beliefs and to the emotional power of symbols. Despite the rejection of earlier anthropological theories about the 'collective unconscious' and 'primitive mentality', there is a need to look more carefully at debates about crowd psychology or collective pathologies of mass slaughter.

RELATIONS BETWEEN LOCAL ETHNIC GROUPS AND GLOBAL FLOWS

We have so far outlined three different approaches to ethnicity. Although they differ in a number of respects, they have all focused on social groups within particular geographical locations. While some writers have assumed that ethnicity is a general property of human relations, a more limited approach is evident in the detailed studies of ethnic groups. Anthropologists have located these groups within local, bounded communities where social interaction takes place both within and across the boundaries of those communities. They have acknowledged the influence of developments taking place further afield, but their prime consideration has been the investigation of social and cultural developments within a particular local setting. In this more limited approach ethnicity is differentiated from the wider settings of the nation-state and nationalism.

It has become increasingly evident, however, that local events are deeply affected by developments within a rapidly shrinking world. Since the early twentieth century a number of anthropologists have acknowledged the wider context of colonialism and nation-state boundaries in their empirical studies of local communities, but recent developments have called for a more radical understanding. The shrinkage of both space and time entailed in contemporary debates concerning globalisation (see Harvey 1989; Giddens 1990) requires us to look beyond analyses confined within the nation-state. The compression of space and time forges new kinds of relationships between communities in very different parts of the world. In many contemporary situations, therefore, local boundaries must be understood not simply as the product of local social interaction but also of influences far beyond local and national space (see Hannerz 1990; Hall 1992; Held and McGrew 1993; Albrow 1996).

The global flows of people, information, images and capital across ethnic boundaries and beyond nation-state frontiers produce a much more complex and heterogeneous world, where the local and global interweave. Communities may no longer be based on local, face-to-face encounters. It is even possible now to talk about the emergence of 'virtual neighbourhoods' created by global communications. These are 'no longer bounded by territory, passports, taxes, elections and other conventional political diacritics' (Appadurai 1995: 219). Although these neighbourhoods are currently limited to transnational élites in highly industrialised countries, they can and do influence the way people perceive ethnicity in other parts of the world. The links between India and Indian migrants in the West (North America and Britain) provide an illustration of this global–local interaction. South Asian intellectuals in the West have played a significant role in the development of ethnic conflict across India. Global migration creates 'new patriotisms' through 'puzzling new forms of linkage between diasporic nationalisms, delocalised political communities and revitalised political commitments at both ends of the diasporic process' (Appadurai 1995: 220).

The effect of these contemporary developments is to make ethnicity a much more complex process, where local, national and global processes interweave and influence one another. Analysis has to move beyond the frameworks offered by the three perspectives outlined in this chapter. Furthermore, answers cannot be sought solely within one academic discipline. The search has to extend across the intellectual boundaries of anthropology, sociology, political science and geography.

ANTHROPOLOGY AND ETHNICITY IN INDUSTRIALISED COUNTRIES

Research into ethnicity across Western industrialised countries indicates a certain degree of erosion between academic disciplines and the growth of mutual understanding. A degree of overlap can be detected between anthropology and sociology (see, for example, Goddard, Llobera and Shore 1994). Moreover, one of the most influential texts for anthropological discussions of ethnicity in Western contexts has been Anderson's *Imagined Communities* (1991,[1983]), a book on nationalism written by a professor of international studies.

Yet, we must not overemphasise the disciplinary overlap in the discussion of ethnicity in industrialised countries. The development of an anthropological literature on ethnicity has inevitably been shaped by the three theoretical perspectives described in this chapter. Early studies of ethnicity in Britain, for example, concentrated on the construction of ethnic identities within rural communities (Cohen 1982, 1986) or ethnic minorities (Jeffrey 1976; Saifullah Khan 1979; Wallman, 1979; Bhachu 1985; Werbner 1990).

The decision to focus on ethnic minorities was not without controversy, however, since it suggested a departure from the basic anthropological notion that everyone, at least potentially, has ethnicity. We have already indicated this move in the preceding section, but, in the context of British ethnic minority research, the change seemed to lead back to an earlier view that ethnic groups were really non-white, i.e. racial, populations. Many anthropologists have tried to counter this implication either by studying white ethnicity or presenting white ethnicity as subsumed, theoretically or practically, in an overarching national identity (see Banks 1996: 156–60). Yet despite these academic manoeuvres, popular opinion within Britain still asserts that ethnicity applies only to non-white minorities.

A second problem has surfaced around the apparent influence of essentalism over anthropological studies of ethnic minorities. Because anthropologists have emphasised the social reality of essentialist ideas and values among ethnic minorities, there has been a widespread assumption that anthropologists' explanations of these cultural phenomena are themselves essentialist. Sociologists have been particularly prominent in pursuing this charge. The Centre for Contemporary Cultural Studies (CCCS) at Birmingham was an early advocate of this critique. In a CCCS-edited volume *The Empire Strikes Back* (1982) anthropological research on Asian ethnicity within Britain is dismissed for presenting 'culture' as 'an autonomous realm which merely "interacts" with other social processes (CCCS 1982: 113). The effect of such a presentation is to produce 'a static and idealised vision of Asian cultures which cannot help us to understand how or why those "cultures" have changed over time' (CCCS 1982: 113). Although the charge was based on a misunderstanding of the more nuanced position actually adopted within anthropology, some anthropologists have also expressed concern at what they believe to be essentialist influences over their colleagues' writing about British ethnic minorities (see the debate between Benson, Eade and Werbner in Ranger, Samad and Stuart 1996).

A third debate has emerged from the different ways in which anthropologists and sociologists have conceptualised 'race' and 'racism'. The controversy has brought fully into the open what has been implicit in some of the crucial conceptual issues discussed in this chapter. Sociologists have criticised anthropologists for ignoring the issues of structural inequality and power relations embedded in race and racism (CCCS 1982; Anthias 1992). Anthropologists have responded that many sociologists and, for good measure, political scientists and geographers ignore the salience of cultural processes and human agency (see Ballard 1992; Jenkins 1997).

Although we have described a considerable theoretical gulf between academic disciplines concerning ethnicity and race, attempts have been made to close the rift between them (see, for example, the volume edited by Werbner and Anwar 1991). Anthropologists are beginning to realise that part of the problem has been their old-fashioned interpretation of race as a biological phenomenon. Some are belatedly engaging with the modern sociological approach to race, which focuses on its social construction of human difference and hierarchy within the context of nation-states (see Shore 1997; Jenkins 1997). The anthropological tradition of eliding race and ethnicity is being challenged by anthropologists who realise that racism(s), as well as nationalisms, have particular qualities which make them different from what they might see as other ethnic phenomena. These specific qualities make racisms and nationalisms enormously important in contemporary nation-states and, perhaps, even more in the context of globalisation discussed earlier (see Sanjek 1996; Eade 1997).

Some anthropologists have gone further in accepting the charge levied by their sociological critics that racism has been institutionalised within anthropology. They have pointed to the support which some anthropologists have given to scientific racism (De Waal 1994). Others have also pointed to the vast under-representation of non-white people within full-time anthropological positions. In 1989, 93 per cent of US anthropologists in such jobs were white (Sanjek 1996).

CONCLUSION

It should have become clear during this chapter that anthropological approaches to ethnicity are neither homogeneous nor static. Yet despite the developments outlined in the previous section, certain distinctive anthropological themes run through the debates about ethnicity. Anthropologists have continued to remind the world that ethnic identities are commonly expressed as if they are natural, i.e. rooted in biological differences between people. They have shown how ethnic categories are sometimes perceived by people as having their own life. These perceptions are similar to many people's beliefs in the existence of a spirit world and the anthropological quest is to understand rather than judge such beliefs.

This process of understanding the subjective experiences of informants draws on the focus within the discipline upon *translation* – communicating knowledge across boundaries. Cultural creativity and human agency are placed within the context of social interaction embedded in particular contexts (historical, political and economic). The intention is not to present group identities as emerging from primordial motivations or some abiding, trans-historical essence, but to analyse the ways in which boundaries are constructed, maintained and changed.

We have seen in this chapter that despite this intention, anthropologists have, in practice, been drawn towards essentialist interpretations of ethnicity, partly because they are so prevalent amongst their informants. However, this tendency has been tempered by the influence of instrumentalist and relationalist perspectives on their analyses. Furthermore, despite a failure in the past to understand the influence of racism upon the intellectual development of their discipline, many anthropologists have recently radically revised their attitudes towards race and racism. This revision has enabled anthropologists to understand the impact of racist beliefs and practices on both their own ideas and on the groups which they have studied.

Significantly, anthropologists and other scholars are now challenging misleading use of the terms 'ethnic' and 'ethnicity' outside academic circles (e.g. Banks 1996: 165; Jenkins 1997; Allen and Eade 1999; Allen and Seaton 1999). It is recognised that the words are not only part of academic discourse, but have become embedded in popular debates, sometimes in dangerous or misleading ways. Some anthropologists may have, consciously or unconsciously, encouraged a conflation between ethnicity, race and culture, and this may have contributed to what are in effect racist applications of ethnicity as a concept. However, this chapter should have made clear the principled opposition of many contemporary anthropologists to these distortions. Through our uncovering of the limitations of anthropological and popular approaches to ethnicity we can not only contest dangerous and reductive simplifications, but also advance our alternative understandings.

V

CULTURE AND RESISTANCE

14

LOCAL FORMS OF RESISTANCE

Weapons of the weak

Hazel Johnson

INTRODUCTION

The title of this chapter alludes to the work of James C. Scott who wrote about the multiple ways in which poor rural people resist forms of exploitation, oppression or unwelcome changes to their lives. Scott used the terms 'weapons of the weak' (Scott 1985) and 'everyday forms of resistance' (Scott 1989) to describe covert action to lodge protest, bring retribution and try to undermine those with power. Such action includes: 'foot dragging, dissimulation, false compliance, feigned ignorance, desertion, pilfering, smuggling, poaching, arson, slander, sabotage, surreptitious assault and murder, anonymous threats' (Scott 1989: 5). As well as reacting against the power of individuals and groups, such behaviour may also take place in reaction to policies and processes of social and economic development intended to benefit the poor. Scott states: 'The problem with both the liberal-democratic and the radical view of development is ... that neither is sufficiently radical. What they miss is the nearly continuous, informal, undeclared, disguised forms of autonomous resistance by lower classes' (1989: 4).

Identifying such forms of protest enriches our understanding of the mechanisms by which people might protect themselves from perceived threats to livelihoods (understood as means of social as well as economic well-being) and to the person. They show that forms of protection and protest can be covert as well as overt, and individual as well as collective. However, it is not necessarily easy to interpret their role or significance in bringing about change (either for the individual or collectively) or to understand what they tell us about the relationship between culture and change.

The role of such forms of action and how analysing them informs an understanding of the relationship between culture and change are the central concerns of this chapter. The chapter first considers the concepts of weakness and power and their relationship to the concept of culture. The second section examines the links between powerlessness and human rights abuse, and asks whether weapons of the weak can protect human rights. The third section looks at whether weapons of the weak can bring about longer-term change. The chapter concludes by briefly reflecting on the relationship between culture and organisation in the everyday struggles of poor people.

WEAKNESS AND POWER

To talk of the weak presumes populations, or sectors of populations, who can exert little or no control over their everyday conditions of existence. By contrast to the weak, there are individuals and social groups who can influence their own and others' conditions of life by their command over economic resources, their control of institutions (such as the state, judiciary or religious organisations) and other sources of social control such as means of violence. Thus, in this sense, weakness is a relationship between individuals or groups of people: the weak experience inequality, subordination, and the inability to bring about change because of the power, control and influence of others, or of higher civil authorities such as the state and its administration. In reality, this weakness is often associated with poverty and economic inequality.

Power is also a relation between people, although there is debate about defining it and explaining its manifestations and dynamics. In his well-known essay, Lukes defined power in the following way: 'A exercises power over B when A affects B in a manner contrary to B's interests' (1974: 34). However, as well as 'power over', there is also 'power to'. This power can be seen as a positive and creative force, particularly for the relatively weak and powerless. Furthermore, ideas about power and how power is manifested will be mediated by variations in laws, customs, norms and values. Forms of behaviour linking the weak and the powerful comprise subtle and complex processes as well as those of evident physical or violent subordination and resistance. There are also many arenas of social life in which forms of power are evident. For example, Bardhan, an economist concerned with power as it is expressed in market behaviour, suggests that there are many processes, such as psychological ones, inherent in relations of power in the economic arena (Bardhan 1991). That power relations in the economic sphere involve complex and multi-dimensional processes is examined by Olsen, who investigated processes of economic exchange in villages in India. From this research, she concluded that people's beliefs and understandings of local social relations in the villages were as important in permitting different types of manipulation to occur as their material circumstances (Olsen 1993: 87–8).

As well as indicating that power relations (and people's behaviour) can take many subtle forms, these notions also suggest that the cultural context is a fundamental part of how they are expressed. A corollary is that understanding the nature and expression of weakness, and of the weapons that might be used by the weak to defend themselves, is also embedded in culture. An example of cultural embeddedness is provided by Naila Kabeer in a study of the personal survival strategies of Bangladeshi women in the changing context of *purdah*[1], which, she argues, is increasingly difficult to sustain in conditions of economic hardship (Kabeer 1990). The poor women she interviewed used many forms of creating personal income streams in a context where male members of the household usually controlled finances. For example, the women engaged in secret saving, borrowing and lending rice between them in the village; they used relatives to help provide cash-earning activities; they sold home-produced goods for relatively low prices to traders coming to the house rather than lose control of the proceeds by allowing male members of the family to take the goods to market, and so on. In addition, poverty and hardship were associated with increasing divorce and desertion, and women often resorted to activities, such as field or roadside labour, not normally included within the social boundaries of *purdah*. Kabeer states:

> The daily struggle for survival in these situations leads to an enforced rejection of the notions of femininity and modesty conventionally associated with respectable womanhood; poor women learn to develop sharp tongues and aggressive demeanours to deal with the hostility they encounter when they encroach on male space.
>
> (Kabeer 1990: 145)

This example not only shows how behaviours and attitudes are embedded in cultural practices but also how weapons of the weak can contest the practices of the past. However, Kabeer concludes that such individual actions are unlikely either to challenge gender relations in households in any fundamental way, or to change the conditions of poverty. Such processes, she argues, can only be realised by collective responses (1990: 146). I return to this point below.

WEAPONS OF THE WEAK AND HUMAN RIGHTS

The concept of weapons of the weak thus suggests that in practice people will resist or try to change conditions of powerlessness. Such resistance is often about everyday survival, although it may be to prevent the desecration of valued customs and beliefs (or to change them), or to counter specifics acts of abuse and physical violence. In broad terms, these actions could be seen as defending human rights in both the civil and economic arenas.

The abuse of human rights is usually associated with violence, whether arbitrary physical abuse against the person or limits on physical freedom such as arbitrary imprisonment. This sort of violence is generally seen as a transgression of internationally agreed codes of conduct such as those embodied in the UN Declaration of Human Rights. However, there are many cultural differences as to what are considered 'acceptable' forms of violence against the person, and which might change over time (for example, corporal punishment in schools, physical mutilation for certain crimes or capital punishment). Practices which are legal or acceptable in one society may be seen as an abuse of human rights in others (although practices may also be challenged from within by human rights activists). This relativism has been upheld by the governments of some countries (such as Singapore and Malaysia) to criticise the interference of the West (and the UN) in local practices. In this debate, economic development is given priority over liberal democracy, economic rights are more important than political and civil rights, and individual rights are subordinated to the presumed collective good.

For many people, the actual or potential threat of arbitrary physical violence is closely associated with poverty. For example, anthropologist Scheper-Hughes has documented the routine use of violence against poor people in north-eastern Brazil, in a context in which victims have little legal redress. However, it is not just that poor people often experience violent abuse routinely, but that their very poverty is a form of violence and an abuse of human rights. Scheper-Hughes cites the view given by a poor peasant from the Central American country of El Salvador to a North American visitor:

> You gringos are always worried about violence done with machine guns and machetes. But there is another kind of violence that you should be aware of, too. I used to work on a hacienda. My job was to take care of the dueño's [owner's] dogs. I gave them meat and bowls of milk, food that I couldn't give to my own family. When the dogs were sick, I took them to the veterinarian. When my children were sick, the dueño gave me his sympathy, but no medicine as they died.
>
> (1995: 444)

Human rights abuse among poor people can thus have many sites and manifestations. For example, Michael Watts, who appropriately named a study he made of famine in Nigeria 'Silent Violence', states in a later publication:

> If famine is the socially differentiated lack of command over food, it is naturally about power, politics and rights broadly understood, all of which are embedded in a multiplicity of arenas from the domestic ... to the national/state.
>
> (Watts 1991: 21)

Thus relations between the more and less powerful within and between households (where power may be based on gender, age or control over resources) can also be sites of abuse, as can the more evidently subordinate relations of the poor as a whole to those who hold economic and political power, including that held by the state.

As a means of protecting the human rights of oppressed or impoverished individuals and groups, weapons of the weak most frequently take the form of different mechanisms for survival in conditions of economic hardship. They usually include quite subtle forms of managing relations with the economically more powerful and privileged. Managing such relations may include forms of collaboration or co-operation with the more powerful to assist means of survival. They may also include changes in behaviour which result in overt ways of challenging these relations. For example, in a study of maize producers which

I carried out in the Central American country of Honduras in the 1980s, it was evident that some poor producers or peasants could use the patronage of landowners to help provide them with land and loans for growing food crops. In return, such peasant producers provided labour for the landowners on a regular basis. These exchanges occurred in a commoditised economy in which there were land, labour and product markets, but they had a personalised quality which involved loyalty and obligation, and implicit cycles of moral as well as cash indebtedness. However, such ties could potentially be undermined by personal disagreement, as well as by changes in market relations (for example, increased incentives for people to buy, sell and rent land on a commercial basis). There were also limited possibilities in the longer-term for such relations to be available (or acceptable) to many landless or near-landless peasant producers, particularly given increased pressures on the land and changes such as labour-reducing mechanisation. Thus while such individual responses and processes of negotiation with richer and more powerful farmers were a means of protecting livelihoods for some small producers, others organised and campaigned overtly for land redistribution and took widespread collective action in the countryside by occupying land. Although these more confrontational actions were often met with reprisals, they were also a source of pressure on governments for the reform of land distribution and tenure systems (Johnson 1995).

Coping with economic survival, changes in market conditions and ownership of assets can cause dissent and different protest strategies within households as well as between them (or between social groups differentiated by wealth and positions of power). Such an example is given by Mbilinyi from her research in Tanzania in the early 1980s (Mbilinyi 1990). Tanzanian agriculture combines peasant farming and large-scale crop production, drawing on family workers, permanent and casual wage labour and migrant workers. The increasing commoditisation of crops and labour and changes in male and female labour markets have affected the respective roles of men and women in agriculture. According to Mbilinyi, there is continuing use of casual labour (which is less costly to employers than permanent labour), and the proportion of women among this type of labour has increased. At the end of the 1970s, when she was carrying out her study, 63 per cent of the agricultural labour force (both farmers and labourers) was female. However, the relatively low incomes in farming, as well as resistance by women to being used as unpaid labour for male household heads, had encouraged both urban and rural women to take up non-agricultural occupations such as beer-brewing, food-processing, trading and prostitution (Mbilinyi 1990: 117). Women also protested against the general intensification of their labour, increasingly distributed between domestic work, their own farming, working on village farms, and working in women's co-operatives. They spoke out against sexual divisions of labour in the household and against the claims that their husbands could make on their labour. Some women refused to provide casual labour for large estates because of 'inadequate wages and oppressive work conditions' (1990: 118–21). At the time that this study was carried out, Mbilinyi argued that the prospects for positive outcomes for women were doubtful because the structural adjustment policies of the 1980s (involving the deregulation of trade, the removal of consumer subsidies and price controls and the introduction of user fees for services) limited opportunities for other, more independent forms of employment outside the customary relations of control over land and labour by male household heads in rural areas.

Weapons of the weak can take even more disguised forms than some of those described above, as well as being about many other issues than economic deprivation. Some of these have been analysed by J.C. Scott and are outlined in Table 14.1. Point 1 of Column 2 in the table gives examples of disguised protest against material conditions. The processes described by Mbilinyi might fall into this category. Some types of action may be to counteract the social devaluation of the person or group and may take symbolic forms, such as those in point 2 of Column 2. While resistance may take the form of hidden aggression, it can also be expressed in drama, dance and music, where the powerless or humiliated symbolically re-establish their social position or take revenge. In societies or contexts where this type of domination is prevalent, there is often a strong 'joke culture' in which the dominated make fun of those more powerful than themselves. This happened in Chile during the military government of General Pinochet (1973–89) (personal communications). During this time, women's groups also produced tapestries (*arpillerías*) telling stories of their experiences of the dictatorship (see Boyle 1993). In this instance, such symbolic actions were also related to underlying protest about political conditions, such as in point 3 of Column 2 (in the case of the tapestries, these acted as a means of testimony as well as solidarity between women and were a means of communication with the outside world as they were rarely on display in Chile at that time). Corcoran-Nantes refers to the practice of recounting oral histories among women organising in the *favelas* (shanty towns) of São Paulo in Brazil. The accounts of events, victories and confrontations helped to politicise the women as well as cement their solidarity (1993: 152–3). As an example of religious organisation acting as a means of political protest against domination, Allen (1991) has written about the Ugandan spirit medium, Alice Lakwena, who led an oppositional movement initially by challenging the authority of male holders of ritual knowledge. Through her capacity to communicate with spirits, she was able to gather a following of 6,000 people, including ex-soldiers, and to lead a peasant's revolt called the Holy Spirit Movement against the Ugandan government in the 1980s. In this instance, thousands died. However, the process of disguising what the protest 'agenda' is about can also enable people to preserve their anonymity and to protect themselves from reprisals.

WEAPONS OF THE WEAK AND LONGER-TERM CHANGE

Whether weapons of the weak can lead to longer-term change, as well as being means of protection and protest against the violation of human rights (broadly understood), depends in part on the relationship between covert and overt action, and individual and collective struggle. Can individual protest, and changes in individual consciousness about social abuse and inequality, be transformed into collective action which alters the conditions of the abused and subordinate?

Organisations of the poor and oppressed which can bring about longer-term change have typically been identified as trade unions, peasant organisations and political parties representing class interests, particularly within a Marxist perspective on social struggle. However, there has also been much debate about the role of 'new social movements' (peace movements, environmental movements, feminist movements and so on). They are called 'new' because they are seen as taking a different form and content of organisation, action and protest from other

Table 14.1 Domination and disguised resistance

FORM OF DOMINATION	FORMS OF DISGUISED RESISTANCE
Material domination, e.g. appropriation of food, taxes, labour etc.	Everyday forms of resistance, e.g. poaching, squatting, desertion; opposition by resisters in disguise, e.g. carnival.
Denial of status, e.g. humiliation, assaults on dignity.	Hidden transcripts or meanings of actions such as rituals of aggression, tales of revenge.
Ideological domination, e.g. justification by ruling groups for slavery, serfdom, caste, privilege.	Development of dissident subcultures, e.g. millenarian cults and other religious movements, myths about bandits and heroes.

Source: adapted from Scott (1989: 27)

organisations commonly considered as agents of change. For example, Australian writers, Jennet and Stewart, state that:

> Social movements aim to achieve change by asserting the moral superiority of their views and by changing people's hearts and minds ... In contrast with political parties and pressure groups which are involved in institutional politics, new social movements aim to *transform existing cultural patterns*. The movements have a loose and fluid structure; their membership is amorphous; they form spontaneously and can disappear quickly; they issue a clarion call for broad change; and they adopt a utopian goal of a new millenium.
>
> (Jennet and Stewart 1989: 1; original emphasis)

There is no clear consensus of how to analyse the genesis and role of such social movements, or even whether they are really 'new'. Jennet and Stewart list different approaches: those which try to identify what motivates the actors ('motivational view'); those which analyse social movements as deviating from institutional behaviour in existing social orders ('functionalist view'); and those which see social movements as challenges to the control of scarce resources ('resource mobilisation view') (1989: 2). Jennet and Stewart's own position is influenced by French theorist, Alain Touraine, and the idea that social movements are trying to change the cultural patterns of 'post-industrial' society.[2] Touraine's view has been criticised for concentrating on the ideologies of social movements (from which it is presumed that organisation will emerge) rather than issues of self-interest or the calculation of material ends which may act as mobilising forces (A. Scott 1990: 68–9). A. Scott's position is that social movements are generally struggles against different forms of exclusion from society, rather than attempts to put forward a complete alternative to the status quo (1990: 150). Thus while, within Touraine's framework, the disappearance of a movement might be equated with its failure, in Scott's view, it could be a sign of success because the goals of social integration have been achieved. Thus Scott is saying that social movements are not necessarily agents of societal transformation although they may achieve some changes. In addition, he holds that there are many continuities between so-called 'new' social movements and those embedded in existing social institutions such as workers' organisations (1990: 155).

The debate about new social movements is useful because it opens up the horizons for thinking about the possibly transformative character of weapons of the weak. Rather than conceptualising weapons of the weak simply as (often covert) individual or group action limited to self-defence, retribution or survival (and hence frequently to the perpetuation of weakness), there is a growing literature which shows the many ways in which poor people can organise collectively, and potentially create space for influencing longer-term change (see for example, Jennet and Stewart 1989; Wignaraja 1993a). How can one try to understand the nature of that action and its transformative potential? One approach has been to see the action of collective movements and organisations particularly among poor people in the Third World as a critique of Eurocentric or Western models of development, particularly to the extent that they have been extractive, exploitative and culturally insensitive. Just as social movements in so-called post-industrial societies are seen to be looking for new cultural patterns, it is suggested that social movements in poor countries (or among poor people) are seeking alternative models of development (Wignaraja 1993b; Verhelst 1990).

This has been one interpretation, for example, of the objectives of the Chipko movement in India. Initially a non-violent protest movement against commercial logging and the control of forest resources by the state and commercial interests rather than forest dwellers, the movement took on an environmental as well as a livelihood dimension as the links between commercial exploitation of forests and environmental degradation were made. Analyses of the Chipko movement tend to emphasise a number of aspects: first, the initial concern with livelihoods and control of forest resources by forest dwellers, and in particular how those resources were being undermined by the activities of logging companies endorsed by the state; second, the tensions between different actors among forest dwellers – the lack of livelihood opportunities for men, the implications of working as labourers for the logging companies, and the actions of women to protect access to forest resources by local people; third, the recognition over time of the relationship between commercial forest use and environmental degradation, including life-threatening erosion and landslips, as well as the threats to longer-term livelihoods (see Colchester 1994; Pathak 1994; Sethi 1993).[3] Who the central actors of Chipko have been and what have been their respective roles vary according to different accounts. For example, women played a fundamental role in the resistance to tree felling, but according to one account, 'the leaders of the Chipko movement are mainly educated men, whereas the women – most of whom are not trained – form the strength of the resistance' (Dankelman and Davidson 1988: 30). The direction and impact of Chipko has also been open to interpretation. Pathak states that 'in popular conception, Chipko is a celebration of peasants' environmental consciousness', and 'has also been read as the assertion of the unity of people with nature ... being ruptured by the industrial-urban pattern of development' (Pathak 1994: 42). His own analysis suggests that 'Chipko expressed itself as a conflict between the state, which was eroding the environment of the Himalaya by deforestation, which in turn was causing floods, and the people who were protecting the environment by resisting the state' (1994: 44). While he goes on to argue that this dynamic resulted in a change in local awareness of environmental issues among those involved in Chipko, he also suggests that the force of the impact arose from a broader changing awareness of ecological issues in central government which began to counteract the commercial interests supported by individual Indian states. Although this in turn provided legitimacy to the Chipko movement, 'Chipko was not accepted as a movement for peasants' rights but for the protection of the environment' (1994: 49). In documenting and analysing the twists and turns in legislation, and the dynamics of the relationships between central government, states, commercial interests and popular pressure from Chipko, Pathak concludes that while Chipko's main concern was control of forests by local people, the central government 'fragmented the message into two: the protection of forests for environmental reasons by the state and the raising of forests by the people ... The people's movement was conflated with people's participation in state-directed afforestation' (1994: 64–5).

Alternative models of development or changes in development priorities may be influenced by, if not entirely constructed from, norms and values of poor people resisting the forces of globalisation or the domination of particular economic or political interests. In this respect, Wignaraja states: 'to understand in depth the new people's movements ... a re-evaluation of some of the fundamental values in our own cultures and of the intellectual tools and resources at our disposal must be made' (1993b: 5). However, the cross-fertilisation of ideas as well as the role of dominant models of thought and action can present quite complex and interrelated processes involving the powerful as well as the weak, and those outside as well as inside the context of a given struggle. Different groups of people with different agendas may be drawn in, and the direction of social movements may change.

A further example is provided by case material (also on environmental issues) from Brazil. Vieira and Viola (1992) have analysed the processes and ideas behind action on the environment in Brazil. There have been many groups and currents at work on issues connected with the environment since the late 1980s, among which Vieira and Viola list: those involved in action on dams and reservoirs; the Amazonian rubber tappers; Amazonian indigenous organisations; rural worker organisations; women's movements; neighbourhood associations; the peace movement; consumer leagues; labour health movements; university organisations; and 'new age' responses. Vieira and Viola call this wider grouping of organisations an environmental social movement. However, Cleary (1992) challenges the idea that there is an environmental social movement in Brazil. He suggests that while there is a long intellectual tradition of Brazilian environmentalism, and while environmentalists have had a high profile and been influential in government circles, it is wrong to talk of a social movement of environmentalists which cuts across the different social groupings – including the relatively poor and powerless. Cleary sees groups such as the rubber tappers' union, indigenous groups or urban residents' associations as representing the particular and specific interests of their members rather than being part of a generalised environmentalism. He further argues that such groups have aligned themselves with the environmentalist cause because of world attention focused on the Amazon, which has put pressure on the Brazilian state and opened up space for different kinds of action. This in turn has also enabled the development of links between vulnerable groups and organisations in Brazil and other parts of the world, which has strengthened their cause and visibility (1992: 152–3).

EVERYDAY RESISTANCE, CULTURE AND ORGANISATION

Analysing the complex interaction of individual/collective, local/national/international, the material basis of people's existence, the power relations that affect them, and the different histories and agendas of organisations and movements, is thus all part of trying to understand the relationship between weapons of the weak, culture and change. This is a different approach from seeing culture simply as an obstacle to change, or as a vehicle for change (although at certain moments, it might be both). Forms of resistance may well challenge apparently accepted behaviours (as in the *purdah* example provided by Kabeer) and existing social relations, whether within the everyday life of households, or between social groups in communities and wider society. Whether this challenge results in wider or longer-term change depends on (i) whether the weak are able to organise on common ground for some collective purpose, (ii) whether they can form links or alliances with other organisations to bring attention to their cause, and (iii) whether there are parallel conjunctural changes in the behaviours and perspectives of other key actors, such as the state. However, the contribution of social movements of the poor to wider social change may in turn bring about changes to the social movement, including its reconstruction in relation to new agendas controlled by other actors.

Even for the poor and weak, everyday abuse or subordination does not usually go unchallenged. The myriad actions are part of an ongoing process, however gradual and imperceptible. When there are no or limited possibilities for wider organisation and change, these actions can take the form of self-protection and rationalisation (even if in behaviour which seems apparently irrational or mystifying to outsiders). When there are possibilities for organisation, change in the wider society may occur. The study of (new) social movements offers one avenue for investigating and interpreting the cultural foundations and dynamics of such action and its role in wider social change.

NOTES

1 *Purdah* is broadly the seclusion of women from men, particularly strangers. The *Shorter Oxford English Dictionary* states that the word derives from the Urdu or Persian '*pardah*' meaning curtain, and gives the contemporary meaning of *purdah* as twofold: a curtain 'serving to screen women from the sight of men or strangers' and 'as typical of the seclusion of Indian women of rank' (1965: 1622). The latter definition is put in historical context by Moreland and Chatterjee, who indicate that 'the usual view is that it originated under Moslem rule ... There are, however, some signs that, even in Hindu time, the ladies in a royal palace lived in apartments specially reserved for them' (1969: 29–30). However, the (Moslem) women referred to by Kabeer in this example from Bangladesh are not women of rank, even though some forms of seclusion might depend on economic resources such as the ability to set aside physical space for women, and for them not to need to work or carry out activities in the public arena or male domains. Seclusion can thus comprise many practices and operate relatively to social and economic circumstances. The implication also is that the conditions and operation of *purdah* are also changing over time. For example, an anthropological study of village health in India from the 1960s defines *purdah* in its glossary simply as a 'veil' (Hasan 1976: 219)

2 Giddens cites Bell (1973) as having the clearest view of post-industrial society, 'distinguished by the growth of service occupations at the expense of jobs that produce material goods' and a growth in the importance of white-collar and professional employees, particularly in the areas of information and knowledge (1997: 526–527). Schuurman states 'Post-industrial society is a "knowledge" society, in which a growing part of the labour force is used for the production of technical know-how' (1993: 24).

3 Many people have written about the Chipko movement and given it different interpretations. This should be borne in mind here. The sources quoted have in turn relied on other sources such as Bahaguna 1979; Bandyopadhyay and Shiva 1987; Das and Negi 1982; Guha 1989.

15

'THE PEOPLE'S RADIO' OF VILA NOSSA SENHORA APARECIDA

Alternative communication and cultures of resistance in Brazil

Vivian Schelling

INTRODUCTION

The aim of this chapter is to explore, through the study of a poor community of rural migrants from the northeast of Brazil living on the outskirts of São Paulo and the history of its own independent radio station, the much broader theme of the nature and experience of the modern in Brazil. This is a vast topic and the subject of considerable and fascinating debate. Here, however, I would like to illuminate only a particular facet as it relates to specific questions concerning the identity and integration of rural migrants in an urban context. In a society which is marked by uneven development, by the co-existence of ways of life relating to disparate epochs and economic structures, and in which a large percentage of the population has migrated to the cities in search of the fruits of 'development', the ways in which rural migrants negotiate their insertion in the urban context is a rich source of information on the ways in which modernity is experienced and constructed.

RADIO AND 'EMPOWERMENT'

Radio as a means of communication directed towards and used by members of the 'popular classes' of Latin America has a history that goes back to the 1950s with the creation by Bolivian tin miners of several independently controlled radio stations, and the deployment by Cuban guerrillas of *Radio Rebelde* (Rebel Radio) as a means of organising the insurrectionary forces leading to the Cuban Revolution in 1959 (Machado, Magri and Masagao 1987).

However, it is only in the 1970s that a distinctive theory and practice of 'popular' or 'alternative communication' emerged as a result of a variety of experiments taking place simultaneously in different parts of Latin America, using communications media to empower 'the popular classes'. These arose out of a confluence of ideas and social forces prevalent at the time: the election of a socialist government in Chile in 1970; the emergence of new social movements; the critique of modernisation theory and the concomitant belief that Latin America needed to follow an 'alternative' development path which satisfied the basic needs of its people and reflected its own history and experience. An essential precondition for achieving this consisted, according to proponents of 'alternative communication', in an overall democratisation of the media at national level and in the development of a 'popular hegemony' at the level of civil society (Kaplun 1987). This latter aim entailed transforming 'from below', through a variety of forms of popular organisation, the relations of economic, political and cultural domination which characterised the prevailing social order. In this process of transformation 'from below', the media would not only become an instrument of popular organisation, but also establish, through innovative experiments with radio and television, a 'new communicational order' based on participation and on dialogue between receivers and transmitters.

The history of 'The People's Radio', a community radio station set up in Vila Aparecida, a poor neighbourhood in São Miguel Paulista in the outlying areas of São Paulo, was one such experiment.[1] Until the 1930s, when new transport connections with the centre of São Paulo and chemical industries were established, São Miguel Paulista had been a quiet semi-rural district. Following this period, especially from the 1960s onwards, it has become primarily an area where migrants from the countryside, in particular the northeast of Brazil, seek housing or a piece of land (Caldeira 1984). In this process of gradual transformation São Miguel has become known as the *periferia*, an urban area like many others surrounding Brazilian cities, characterised by poor housing and a lack of adequate infrastructural facilities such as electricity, sewage and roads.

For rural labourers or peasants from the northeast of Brazil, migration entails a profound process of re-socialisation. Unskilled and shaped by extremely oppressive labour relations, they form an easily exploitable reserve army of labour employed in construction sites, domestic service, factories and in petty commerce. Integration in the urban context is frequently experienced as a form of cultural invalidation, entailing the loss of skills previously mastered without appropriation of the necessary material and intellectual means to participate in the urban context (Chaui 1987). While in the context of subsistence agriculture the cycle of production and consumption was concrete and transparent, in the city the rural migrant becomes an appendage of the machine and an abstract consumer engaged in the exchange of commodities for which his/her purchasing power is always insufficient. The majority of television programmes represent the industrialised south of Brazil and its 'Westernised' and comparatively affluent lifestyle as the 'true' image of Brazil. Other regions and their ways of life are either hardly represented or, when they are, it is in a stereotypical and derogatory manner. The overall effect of this is the cultural invalidation of the migrant. The force of this process of cultural invalidation, however, also depends on the extent to which 'cultures of resistance' – alternative representations of the migrant condition and positive strategies for constructing a new urban identity – are available.

'CULTURES OF RESISTANCE'

Cultures of resistance are ambiguous and amorphous phenomena, since what distinguishes them from more systematic and confrontational forms of struggle is that, as several writers on the topic have observed, they consist of more subtle and everyday practices of opposition to domination.[2] These can include, for example: 'the prosaic but constant struggle between the peasantry and those who seek to extract labour, food, taxes, rents and interest from them' (Scott 1985: xvi); the sabotage of machinery and work rhythms in off-shore transnational companies; inventive ways of circumventing anti-strike legislation; forms of transgressive popular celebration; as well as millenarian and religious movements in which the imminent overturning of the social order through divine intervention is announced. They may, as Gledhill (1994) notes with regard to neo-Inca Andean rebellions in Peru during the colonial period, leave 'a legacy in historical memory' which generates a 'counter-hegemonic indigenous historical consciousness which changed over centuries and could manifest itself in the form of participation in more conventional political and class-based organisations in recent times' (1994: 89). They may, however, also, by creating a space for 'contained' opposition, facilitate pragmatic adaptation to reality without fundamentally changing the balance of power and possibly even in this way reinforce domination. Movements of resistance are thus frequently shot through with elements of conformism. Consequently, it would seem that decisive in determining the direction in which resistance and the struggle for hegemony which takes place on this terrain move, is the existence of more structured and systematic forms of opposition. For, as Gledhill points out, it is the broader context of power relations which determines 'the precise structural implications of particular counter-hegemonic acts' (Gledhill 1994: 93).

SOCIAL MOVEMENTS IN SÃO MIGUEL PAULISTA

What distinguished São Miguel Paulista from other similar areas was the extent to which a 'culture of resistance' had developed over the years since it had become part of the periphery of São Paulo. Since 1945, its inhabitants have tended to vote for political parties representing the interests of the poor and the working class (Caldeira 1984). It has also been characterised by a variety of local associations involved in campaigning for infra-structural improvements as well as promoting recreational activities, many of which aim to recreate the cultural rituals and forms of sociability of the northeast. Since the early 1980s the inhabitants of São Miguel Paulista have also become organised within local and national social movements, in particular the *Movimento Sem Terra* (Landless Movement) and the *Movimento de Moradia* (Housing Movement), both involved in struggling for a more equal distribution of land. These and other social movements concerned with ethnic and gender issues emerged during the period of the right-wing military régime in Brazil (1964–85) in response to the growth in social inequality and the loss of civil liberties which characterised this period. After the return to civilian government these social movements played a significant role in redefining a political culture based on authoritarianism and clientelism. Mobilising around demands to share in the benefits of a growing modern economy and aiming to achieve this without relying on political parties and professional politicians, but rather on forms of independent popular organisation, the social movements have broadened the concept of democratic participation (Cardoso 1992).

Since the establishment of the military régime, the radical wing of the Catholic Church, tied to Liberation Theology and its vision of religious salvation as a liberation from all forms of economic, political and personal oppression, has been an active promoter of the social movements. Thus the marked presence of these 'Popular Movements' in São Miguel Paulista was partly due to the fact that this area of São Paulo was part of the Catholic Diocese of the Bishop Dom Angelico, a forceful advocate of Liberation Theology. In conjunction with the Labour Movement, the Catholic Church had set up Basic Communities throughout the region, including the neighbourhood of Vila Aparecida. The underlying purpose of the Basic Communities could broadly be defined in terms of two interlinked objectives: first, to carry out group readings of the Bible, developing the poor's own ideas and capacity for reflection through shared reading, discussion and personal interpretation; second, to 'awaken' the poor to the possibility of changing their condition by supporting forms of collective action – participation in co-operatives, local associations and unions which aim to liberate the poor from oppression.

In agreement with the broad aims of Liberation Theology, one of the key features of the grassroots movements in São Miguel Paulista was the adult literacy classes and forms of popular education informed by Paulo Freire's notion of a 'pedagogy of the oppressed' (Freire 1975). According to this pedagogy, any 'cultural action' which aims at liberation from oppression needs to reflect not only on the social and economic causes underlying the immediate and everyday experience of oppression but also on the self and the way it has psychologically internalised these forms of oppression in the form of feelings of, for example, fatalism, inferiority or displaced anger. Only this can ensure that the creation of a more just economic order will also lead to the construction of a democratic culture. As Freire notes, 'Culture as an interiorised product which in turn conditions men's [*sic*] subsequent acts must become the object of men's [*sic*] knowledge so that they can perceive its conditioning power' (Freire 1975: 35).

Informed by these ideas, the introductory section of the Popular Education Manual of

São Miguel Paulista states that 'Popular Education aims at strengthening the class consciousness of the popular sectors. Popular Education does not aim at knowing or contemplating reality from outside but at decoding its meaning while intervening in its transformation' *Catálogo Educação e Comunicação Popular,* undated: 7).

VILA APARECIDA AND 'THE PEOPLE'S RADIO'

The neighbourhood of Vila Aparecida, named after its patron virgin, *Nossa Senhora Aparecida* (Our Lady the Appeared) stretches across a gently sloping hill and valley at the end of a two-hour bus journey from the centre of São Paulo to the periphery of São Miguel Paulista.

In the 1980s, with the rise of Liberation Theology and the growing participation of São Miguel Paulista in local popular movements, Vila Aparecida was gradually transformed into a Basic Community. At the instigation of Catholic priests and lay Catholic workers, several local groups and institutions emerged: a catechism group and *grupos de rua* (street groups) to organise readings and interpretations of the Bible in the light of the Vila Aparecida inhabitants' experience; youth groups combining vocational and recreational tasks; and an urbanisation and adult literacy group connected to the broader Housing, Education and Landless Movements. Each group was represented in the local 'Council', whose task was to co-ordinate the work of all the groups and advance the process of popular mobilisation. With the help of funds raised locally, material was bought to build a Church and a community centre, and a shared journal recording the events and experiences of the local inhabitants was kept. Its purpose was to act as a memory of the *caminhada* – the way, the path of liberation taken by Vila Aparecida. Following a period of several years after piped water and electricity had been obtained as a result of the pressure exerted by local groups on the state authorities, the idea of setting up a local radio 'station' emerged.

Plate 15.1 A street in Vila Aparecida.

Source: Photographer, Vivian Schelling.

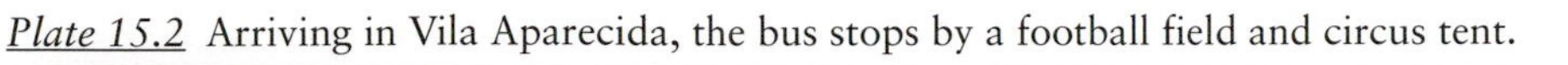

Plate 15.2 Arriving in Vila Aparecida, the bus stops by a football field and circus tent.

Source: Photographer, Vivian Schelling.

Plate 15.3 Photographs on the church wall record events and themes in the history of Vila Aparecida as a Basic Community. These photographs are images of the 'Day of the Migrant', when the church, 'committed to justice, the land and brotherhood', was inaugurated.

Source: Photographer, Vivian Schelling.

Plate 15.4 The church built by the inhabitants of Vila Aparecida with loudspeakers on the roof.

Source: Photographer, Vivian Schelling.

Inspired by the use of loudspeakers as 'popular radios' in Villa El Salvador, on the outskirts of Lima, and the subsequent creation of 'The Day of the People', a local 'Centre for Radiophonic Communication and Production', the adult literacy group of Vila Aparecida, set up a local radio 'station' using a tape recorder, a record player, an amplifier, microphones and loudspeakers wired up outside the local church tower, with a transmission radius of 3 km. It was called 'The People's Radio'. Given the importance of oral culture for rural migrants, their familiarity with the use of loudspeakers in village squares and the rate of illiteracy among the inhabitants of Aparecida, it was hoped that the use of an oral medium would be more effective as a vehicle for popular mobilisation. In the words of the local newspaper *O São Paulo*: 'In the shanty town writing doesn't work, nobody knows how to read, people have an aversion to paper' (*O São Paulo* 1986).

As a result of the success of this first experiment, a further nineteen radio stations of a similar kind, financed by the Catholic Church and European charities were set up in other settlements in the region. They were then transformed into an interconnected network by a

Plate 15.5 A cloth banner behind the stage where the Festival was being held: 'Blacks: A Cry for Justice'.

Source: Photographer, Vivian Schelling.

co-ordinating regional centre, *Proconel* (Project for Non-Written Communication of the Eastern Zone), made up of the representatives of each of the radio 'stations' in the region. These were in turn divided into the following three subgroups charged with developing the popular communication skills in each locality: a group for the collective production of programmes; a group for the training and formation of new leaders; and a group responsible for the exchange of the 'radio experiences' of different regions. With the growth in the number of radios, *Proconel* became a department of the broader Centre for Popular Education and Communication of São Miguel (CEMI). The task of CEMI was to run courses in popular communication, organise regular meetings with the local representatives of the community radios and promote the exchange of 'radio experiences' by providing the whole area with locally produced taped programmes, audio-visual materials and communications manuals.

Ideologically, CEMI adhered to the view that in contrast to the media of the culture industry, 'popular communication' was characterised by the fact that it was dialogic and participatory, enabling the receiver to become the producer and vice-versa. As such, 'popular communication' was an integral part of the life of the community, present at local activities, creating links between local issues and broader social and political processes and reflecting as well as elaborating in its programmes the needs and interests of the popular classes. In this way 'popular communication' would enable the poor and subaltern classes to 'discover their voice', become aware of their creativity and humanity and thus acquire the experiential precondition for claiming their rights as citizens.

How were the community radios thought to achieve these aims? First, the use of the radio itself by the poor was seen as empowering and as demystifying the medium; hearing their 'own voice', local inhabitants would realise that they were capable of manipulating the required technology and adapting it to their needs. Moreover, as transmitters–receivers speak freely, experiences are shared, giving rise to a language and symbols expressing hope and solidarity pre-figuring a new social order. In that sense, it could be argued, the ultimate purpose of 'popular communication' was to promote the development of a 'popular counter-hegemony' in the sphere of civil society.

In 'The People's Radio' locally distributed booklet the broader theoretical and ideological tenets of 'popular communication' were given concrete expression in terms of the following aims:

1. To recover the voice of the people, its history, religiosity, culture, tradition and legends.
2. To provide basic information, to support the organisation of community struggles and to communicate the hopes and problems of the people.
3. To promote the transformation of society through tasks undertaken in common.
4. To support popular artists as well as organise musical events and festivities.
5. To encourage participation at every level and engage in the formation of new community leaders.

Item 6 states: 'through following this path (*caminhada*) we began to perceive that the radio was no longer a dream and had become part of the history of our people. It is a difficult path but we resist because we believe in the power of the word of our people' (*Radio do Povo*, undated: 21)

The members of 'The People's Radio' were divided into four groups with different tasks designed to advance the process of popular communication: an 'interview group' was responsible for exploring, in conjunction with the inhabitants of Vila Aparecida, the ideas and issues which they would like to see included in the weekly programme as well as evaluating whether the radio was reaching them; a 'visiting group' was responsible for exchanging experiences with other communities; a 'programme dynamics group' was charged with involving the community in the work of the radio and popular struggles using other media such as slides, films and music; finally a 'news group' was responsible for reading daily newspapers and journals and reproducing them together with a critical commentary (*Ante Projeto do Grupo de Comunicação da Vila Nossa Senhora Aparecida* 1984).

The weekly programme prepared beforehand generally ran from two to four in the afternoon on a Saturday; during the rest of the week the radio was at the disposal of the community to transmit messages, for example, to call people together if someone needed help, to come to meetings or to announce events. The weekly programme usually began with a piece of Brazilian urban or rural music. To play Anglo-Saxon rock was frowned upon as 'anti-popular'. This was followed by a reading from the Bible interpreted in such a way as to make sense of an aspect of the local inhabitant's life. An important component of the weekly programme was the 'socio-drama', short pieces invented and recorded by the inhabitants of Aparecida or by other communities and distributed to Aparecida via the Regional Centre for Popular Education and Communication. These pieces dramatised typical aspects of their life which were considered 'conflictive but in favour of the people'. The purpose of the socio-dramas was to act as a mirror of their experience and by objectifying it to understand and invent new solutions to their problems. The manual 'How to Make a Socio-drama' states:

> By acting we are forced to imagine different solutions to the conflicts of our life. We are representing what we live but also preparing ourselves for what we are going to live. In the socio-drama we can *rehearse the future*: better ways of organising, new ways of demanding our rights, new forms of behaviour. We are rehearsing our liberation.
> (*Como fazer um socio-drama* 1987: 8)

Included in every programme were announcements relating to the organisation of the grassroots movements at local, regional and national level, announcements brought in by inhabitants of Aparecida as well as information on the price of food, employment opportunities, public services and available courses in literacy, carpentry and sewing. The programme ended with a piece of music. Two representative examples of 'The People's Radio' programmes are provided in Boxes 1 and 2 .

Despite the evidently didactic purpose of many of the programmes, such as the ones in the Boxes below, 'The People's Radio' did not, according to one of its members, campaign in support of the preferred Workers Party, defined as *o partido da casa* (the house party). It preferred to set out for listeners the aims of different parties and the current positions regarding single issues. When the new Constitution was being formulated in 1988, 'The People's Radio' walked the streets of Vila Aparecida discussing the items in the Constitution, and socio-dramas on the current anti-inflationary economic plan were enacted. Moreover, the radio changed over time as it was increasingly appropriated and managed by local inhabitants rather than being informed by the guidelines of the Church. As one inhabitant commented:

> In the beginning, the radio played only religious records and there were problems when we played other songs. Then we chose different songs to introduce a subject. If the subject was land then we would play something to do with the land and with news from the northeast. Now it is changing, the radio is no longer so serious – it is better, more fun. To make a programme talking only of problems is difficult, life itself is crushing, so we don't only talk about bad things, but also with more joy, to entertain and distract. Do you have problems? Millions.

'The People's Radio' functioned not only within the premises of Vila Aparecida but also in other locations in the form of a 'mobile radio'. It accompanied representatives from the grassroots movements on demonstrations and expeditions to lobby the state authorities, acting both as witness to and animator of the *caminhada* (the path). According to the testimony of Sonia, one of the main promoters of the radio:

> The radio is carried on a little car and the music and programmes animate the demonstrations while the state authorities make us wait the whole day long. In the (religious) processions and celebrations people carry the radio; the litter with the Saint goes in front and the litter with the radio behind. This year we had a pilgrimage in favour of peace very connected to the question of race and violence. And the radio was there.

Local attachment to the radio also became apparent on the occasion when the radio was stolen. Money was raised from the local

BOX 1. TYPICAL PROGRAMME SCRIPT

The programme begins with a song about the need for agrarian reform.

- Reading from the Gospel of St Luke: a parable about the 'offerings of a poor widow'. She, like the rich man, gives money to someone poorer than herself. The radio commentator highlights that the poor widow gave much more because she gave what she had and not what was left over and not needed, as in the case of the rich man. In that way, he concludes, 'although we ourselves are poor, we all have something to give.'
- A current popular 'hit' by the Brazilian band '*Paralamas do Sucesso*' (The Mudguards of Success).
- Ten Commandments on how to vote consciously (elections being held for the mayor of São Paulo were won unexpectedly by the candidate of the Worker's Party, Luisa Erundina).
- Socio-drama entitled 'which is the party on the side of the people?'
- Poems read by children of Aparecida on 'Children's Day', followed by a reflection on ecological issues and the question 'is nature being loved like the children?' This is accompanied by the jingle of an extremely popular and lucrative children's television programme presented by 'Xuxa', a blonde and blue-eyed young woman.
- Another song entitled: 'It is not necessary to have a degree, a youngster from the country is also of value'.
- The 'afoxé' song 'Oh Faráo' is played.[3]
- The programme ends with a request for help to repair a broken street pipe and the announcement of a radio course which is being run in order to ensure the continuity of 'The People's Radio'.

BOX 2. TYPICAL SOCIO-DRAMA

Frequently used was a taped socio-drama in which the central character was the construction worker 'Pedro'. Blending in with the song *Pedro Pedreiro* by Chico Buarque de Hollanda, a narrator recounts the process of Pedro's awakening to his predicament as worker:

> Pedro built houses where there was nothing but earth; like a bird without wings, he climbed up the houses which sprung forth from his hands ... but Pedro didn't know that the house he built was both his freedom and his enslavement; he didn't realise that the worker makes things and is in turn made by them.

However, one day Pedro notices that he, the humble worker, makes everything:

> the knife, the bench, the glass, the walls, the window, the house, the city, the nation ... he looked at his hand and in addition to his profession as worker he acquired a new dimension of poetry. And this event was noted – other workers listened and heard and thus the worker who always said 'yes' began to say 'no' ... noticing that his pot belonged to his boss, that the hardness of his day was his boss's night's rest, that his great fatigue was his boss's friend. The reasoning of a poor and forgotten man grew, the reasoning which made the construction worker into a worker in the process of construction.

At this point the music blends in with the first words of a very popular 'bossa nova' song of the 1960s *Opinião* (Opinion): 'You can arrest me, you can beat me, you can starve me, but my opinion I will not change'.

inhabitants to buy new equipment, which was then used to exhort the thieves to return the stolen equipment; after some persuasion they obliged.

To celebrate the creation of 'The People's Radio', a festival of locally produced music and poetry was organised annually on a specific theme. 1988 was the centenary of the abolition of slavery and the festival theme was: 'Black Brazilians – A Cry of Justice'.

The effect of the Festival was to consolidate the network of social relations and the sense of belonging to a *pedaço*, a particular neighbourhood. In contrast to the other spaces occupied by the inhabitants of Aparecida – the factory, the bus, the underground, the construction site and the domestic space of others – the *pedaço* embodied a shared identity and a history of struggle which enabled them to resist the experience of marginalisation and invalidation. This was reflected, for example, in one of the poems recited at the festival:

> Who is outside, wants to come in
> Who is in, doesn't want to leave
> Our shanty town is good
> I no longer want to leave
> We have a good church
> And a priest who prays
> We have piped water
> And a small school
> What we still need is a small hospital
> For we already have the school
> To teach us to read
> With the protection of the priest
> and the social worker
> We will struggle for land
> With our rights as voters.

In addition, the Festival itself gave rise to a local 'Black Roots' group and also fostered the development of small bands and duos who played throughout the year at birthdays, weddings and demonstrations.

'THE PEOPLE'S RADIO' AS RESISTANCE?

To return now to the problem raised earlier concerning the insertion of rural migrants in an urban context, how effective was 'The People's Radio' in developing a counter-hegemonic culture of resistance? And how did it contribute to articulating a new urban identity for the rural migrants? In order to address these questions it is necessary to discuss briefly the concepts of hegemony and counter-hegemony.

Hegemony theory is a refinement of the Marxist concept of ideology; it emphasises that, although acceptance by subordinate classes of belief systems which bind them to the existing power structure is essential to the maintenance of the social order, this does not occur through a process of simple imposition of an articulate system of ideas and values. Rather, it refers to a situation in which, as a result of the unequal relations of cultural power, subordinate groups actively subscribe to values, aspirations, meanings and life styles which secure their adherence to the prevailing social order but which are not in their interest. The contribution of hegemony theory in this sense is to have pointed out that a social order is maintained when particular values and meanings become, to use Raymond Williams' phrase, 'practical consciousness', that is, when they not only give shape to consciously held ideas but 'saturate the whole process of living'. These values and meanings thus come to 'constitute a sense of reality for most people in the society, a sense of absolute because experienced reality beyond which it is very difficult for most members of the society to move, in most areas of their lives' (Williams, undated: 110).

This is achieved significantly through containing the public discussion of issues and the circulation of meanings relating directly or indirectly to the distribution of power and wealth, within a framework which seems to respond to and accommodate the interests of subordinate groups, without, however, altering the prevailing power structure in any significant way. Conflict is covered over or partially resolved and the power of alternative meanings is mitigated through partial incorporation in the hegemonic discourse. In this way hegemonic discourses retain their claim to being the only way of thinking, rendering the prevailing social order 'natural'. Moreover, language is used in such a way that the values implicit in the dominant or preferred discourses are invested with a positive emotional charge, thus promoting subjective identification with them and transforming them into 'practical consciousness'.

By contrast, counter-hegemonic forms either openly or implicitly aim to foster 'structures of feeling' and ways of thinking outside the parameters of the hegemonic configuration, revealing sources of social conflict so that meanings and values contained in the hegemonic discourse are no longer seen as universally valid but as directly or indirectly benefiting a partial interest, that of the dominant classes, rather than society as a whole. This in turn has the effect of 'denaturalising' the social order while opening up a space for a counter-hegemonic interpretation and experience of social reality.

The responses of the inhabitants of Vila Aparecida to the work of 'The People's Radio' varied depending on his/her status within the community and how closely involved he/she was with the radio. From the testimonies of those actively involved in the radio's popular communication groups, it was apparent that the public discussion of issues such as racism, the right to land, the Constitution and the right to the product of one's own labour – as in the socio-drama on the construction worker Pedro – had played an important part in building links between their problems, the broader socio-economic structure and the possibility of imagining and bringing about a different future. In addition, working on the weekly programmes in conjunction with participation in the grassroots movements had fostered an awareness of their identity and rights as citizens, while working with a simple form of technology adapted to local needs had demonstrated to them that it was possible to break the communications monopoly of the culture industry, in however limited a form. It thus seemed that they felt they had been given the means to address the contradiction or duality of their predicament in that the 'cultural action' promoted by 'The People's Radio' emphasised the value of rural culture while also providing tools through which to gain access to modernity as creators and knowers. This was reflected in the frequent use of the word 'transformation' and 'path' or 'way' to refer to the changes experienced in living in Vila Aparecida.

In the case of those inhabitants who were within the three kilometre radius of the loudspeakers and who participated intermittently in it, the radio was seen very much as a public utility. In that sense 'The People's Radio' functioned more instrumentally as a vehicle for improving their life chances in the urban environment through the information it provided on the price of food, employment opportunities and the state of negotiations with the authorities over land. Nevertheless, despite its instrumental value in securing some form of social mobility and comments that the radio was noisy and interfered with listening to 'international' music on the official radio stations, the inhabitants' responses also revealed that, since the emergence of the radio and local involvement with the church, Aparecida had been transformed. While it was seen as a place previously characterised predominantly by mutual suspicion, violence, despair and social disintegration, it was now a community with more resources, with an identity which the inhabitants were quite proud of and with a shared memory of struggle and celebration expressed in the community journal, in the photographs of the community's activities which covered the walls of the church, and in a relative degree of political awareness. This view was reflected in the frequent reiteration that the quality of life had improved:

> The atmosphere has changed in the Vila; there is less hostility between the children. There used to be much violence. The community resists and confronts the violence – it has occupied a space and it continues to celebrate [*fazer a festa*], the groups meet and the joy of living encourages people; the community is a small light perhaps.

Nevertheless, in comparison with those consistently involved in the radio and the grassroots movements, this group was far less ideologically coherent, sometimes expressing on the one hand pleasure in the positive changes in Aparecida and on the other dislike of Blacks – this being a fairly marked feature of northeastern rural culture. In fact, the contradiction between a manifestly anti-racist stance and the extent to which Brazil's black inheritance is still devalued and denied is also apparent in 'The People's Radio's' programmes. These included discussions of questions of race and the fate of Zumbi, the leader of a major slave rebellion in the seventeenth century, as well as an uncritical acceptance of the popularity of the children's television presenter 'Xuxa'. Young, white, blonde and blue-eyed, she is the embodiment of the stereotypical North American ideal of female attractiveness and a glaring example of the hegemonic aspiration to 'whiten' Brazil, despite the widespread claim that Brazil constitutes a 'racial democracy'.

In the testimonies of the women, who, in contrast to the men, spent a greater amount of time in Aparecida, the radio had to a greater extent, through naming and symbolising their experience, fostered what could be defined as a 'counter-hegemonic sensibility':

> The Radio plays a different kind of music; it tells the story of unemployment; it talks of bread, flour and the rise in prices and it is all so true – I think it is beautiful – people like it. When the light or water bill arrives they call you; they invite people to come and help when someone is ill, when there are disasters; when someone needs a document they help; they call people for meetings. All that helps a lot. And it has gotten better – every year there is a festivity – I think its beautiful, the children dancing in a circle. I wash and listen to the music they play, which is about what we go through: one day you go to the supermarket and the milk has gone up; it talks about bread, flour, salaries; I think that is important.

There was also tension and conflict surrounding the work of the radio: some of the core members of the radio were seen as having acquired élite status in the community and within this group itself criticisms were voiced that 'the people have to come to the radio not the radio to the people'. In that sense 'The People's Radio', it could be argued, had predominantly transformed a small number of the inhabitants of Aparecida into transmitters–receivers. Although there was no censorship, there were differences in the selection criteria of material to be used in the weekly programmes between church workers and local inhabitants, as exemplified in the above quotation on how the radio had become less serious. This difference also became apparent when comparing the radio programmes with the repertoire of the festivals celebrating the anniversary of the radio, which were more profane, melodramatic and influenced by the language of the culture industry. However, these were differences and tensions which the church workers, in accordance with their radical democratic ideological position, appeared to want to deal with through dialogue rather than the foreclosing of difference.

Notwithstanding these limitations, in the context of Brazilian society, where the view of the popular classes as unfit for self-government and incapable of thinking is a central element in the hegemonic discourse, cultural practices such as 'The People's Radio' constitute, despite their shortcomings, a counter-hegemonic force. For a subordinate social group to validate its needs and interests, to experience relationships and create symbols and meanings not characterised by domination, and to relate this with whatever degree of clarity to how the social order is constructed is the precondition for the development of a counter-hegemonic cultural force. In that sense it could be argued that 'The People's Radio' contributed towards the development of an alternative modernity privileging a form of citizenship in which the moral, political and cultural resources of the poor form a central part of the way modernity is defined.

ACKNOWLEDGEMENTS

Research on the culture of northeastern rural migrants in São Paulo was made possible by funding from the *Conselho Nacional de Pesquisa* (National Research Council) in Brazil. I am most grateful for their assistance. My thanks also to Jorge De Carvalho and Rita and Luis Alberto Peluso Segato for their support and collaboration.

NOTES

1 The research on 'The People's Radio' in Vila Nossa Senhora Aparecida was carried out between 1988–9 and funded by the Brazilian National Research Council (Conselho Nacional de Pequisa). It was part of a broader research project which included, in addition to the study of 'The People's Radio', investigating

other aspects of northeastern culture in São Paulo and the way it is transformed in the urban context, for example, oral poetry and northeastern dance halls.

2 See in this respect Scott 1985; Gledhill 1994; Chaui, 1987.

3 'Afoxé' groups are groups of musicians, dancers and percussionists connected to the Afro-Brazilian religion of *Candomblé* in Salvador, Bahia who perform and enact aspects of their religious ritual during the Carnival period. The song 'Oh Faraó' celebrates black culture.

16

THE NEW MIGRANTS

'Flexible workers' in a global economy

Chris J. Martin

INTRODUCTION

It has been estimated recently that there are currently over 100 million international migrants in the world today (Castles and Miller 1993: 4). This means that nearly two per cent of the world's people do not live in their country of origin. Many of these are refugees – at least 19 million, according to the UNDP (1994, Table 3: 135). Migration across frontiers is nothing new. One of the most well-known is documented in the Old Testament of the Bible, where the whole Jewish nation took to its feet to escape servitude under the Egyptians, crossing land and sea to create a new nation. Many polities were built on migrations from kingdoms of Africa to the modern republics of the USA and South Africa. Yet the volume of national and international migration since World War II has increased greatly, continuing up to the present.

Why has there been this increase, or to put it another way: why is it that, now, nearly twice the number of people that make up the population of France are living as 'foreigners' in the world, mostly in the advanced industrial economies? What are the characteristics of this migration, who are the migrants and where do they come from and go to? Why do these people uproot themselves to live in an alien, often hostile environment? How do they make the move and how do they 'make out' when they reach their destination?

The focus will be on the most recent migrations and their implications for receiving and sender countries, and on what is normally termed 'voluntary migration'. Although the 'forced migration' of refugees constitutes an increasing proportion of contemporary migration, in the space of this article it is impossible to do justice to their special circumstances. Recent migration is also characterised by large numbers of undocumented workers and by migrants of all kinds from countries with little or no previous links with the destination country. In other words, in contrast with older movements, which were organised or which sprang from agreements, the new migrants typically pre-date, pre-empt or ignore official policies and agreements. Migrants are forging their own destinies outside any formal framework. This fact is all the more remarkable when one considers that, typically, modern-day international migrants are people with limited economic resources and without the advantages of formal education and training. The extent to which migrants respond creatively to given circumstances, and how they transform latent conditions into real opportunities, is something which I will emphasise in this article. The approach I am taking here treats migration as a series of interactive relations between migrants, kith and kin, employers, as well as with broader circumstances, thereby qualifying conventional views which have generally treated the phenomenon as the response of the downtrodden to forces beyond their control.

INTERNATIONAL MIGRATION IN THE POST-WORLD WAR II BOOM

Before focusing on the most recent migrations, I will sketch the main features of post-1945 international migration and the explanations it elicited in social science.

The growth of international migration was notable after World War II principally from the Less Developed Countries (LDCs) to the Developed Market Economies (DMEs) of Europe, North America and Australia.[1] Workers arrived in response to active recruitment policies by advanced industrial countries, whose labour demands outstripped local supply during the post-war period of economic reconstruction and growth. Workers were recruited either from colonies and ex-colonies or through other labour agreements, such as the famous *Bracero* programme between the USA and Mexico in the 1940s. Some measure of the growth of immigration in Western Europe is indicated in the growth of the minority populations of the period. Between 1950 and 1975 the ethnic minority population tripled from 5 to 15 million in Western Europe (the UK, Belgium, France, West Germany, the Netherlands, Sweden and Switzerland combined) (Castles and Miller 1993: 87–8).

Underlying the movements of labour is not just the active recruitment for labour by the DMEs, but also the increasingly precarious living conditions in Third World agriculture. The flight from the countryside is a long-term historical trend. Several factors underpin it, especially the following. The growth in markets for land and labour releases workers from personalised obligations to former landholders but also dispossess these workers from previous rights to land and its use. Also, as industrialisation and trade increase, terms of trade for agricultural products decline, impelling the mechanisation of agriculture, which in turn reduces labour requirements. Increasing people–land ratios caused by land subdivision accentuates the flight from the land among the smallest landholders.[2] The exodus of the rural dispossessed or the under-employed provides industrialists with abundant supplies of cheap labour, if not in the home country, then across national borders. As Wolf explains, from the migrant's point of view, a sharp distinction between national and international migration may be irrelevant to the overriding need to make a living (1982: 361).

The explicit relation between demands for labour from industrialised nations and its acquisition through labour recruitment from LDCs in the period 1945–75 appeared to corroborate the view among scholars of migration that movement was a direct response by the poor in the LDCs to economic opportunities in the north. The basis of this argument is that the disparity between the LDCs and the DMEs is the pivot of migration flows. These movements are produced by a combination of 'push' factors such as rural dispossession and urban unemployment in the LDCs, and the 'pull' factors, particularly economic opportunities in the DMEs (see, for example, Jackson 1969). Some studies within this framework concentrate particularly on income differentials to assess the strength of migration flows (Lewis 1954; Fei and Ranis 1964). Individual migrants are thus seen as rational decision-makers responding to the forces of supply and demand. Todaro (1989) refines this model by adding in the subjective component of perceptions. Thus migration flows represent *expected* opportunities, which may diverge from actual ones.

Radical political economic positions criticised such neo-classical abstractions for being a-historical, for not addressing the causes of the disparities provoking migration and for reducing migration decisions to an abstraction of the individual, optimising '*homo economicus*'. Yet the polarity between north and south continued to be accepted as the chief motive for migration. However, instead of being accepted as a manifestation of differential stages of modernisation, it was treated as a proof of neo-colonialism and 'underdevelopment'.

Two typical examples of post-war migration are Caribbean and South Asians to the UK. Both groups arrived in response to both deteriorating rural livelihoods and in response to explicit labour demands. However, particularly in the case of the South Asians, migration flows proved unresponsive to shrinkages in employment. In the case of South Asian immigration documented by Robinson (1986), labour demand and immigrant supply were negatively related. In summarising a number of studies on these and other migrations of the period, two broad reasons help explain these anomalies (Portes and Borocz 1989; Castles and Miller 1993; Robinson 1986 and Sassen 1988). The first is that aggregate demand and potential supply from LDCs are poor predictors of international migration because they obscure changes in the specific types of occupations taken by immigrants. For example, in the South Asian case just mentioned, this group entered low-paid strata of the declining manufacturing industry, replacing outwardly mobile indigenous workers. Secondly, supply and demand factors assume a notional interrelation between donor and receiver economies which simply may not exist because local and subjective economic and non-economic factors influence decisions. As Scott has made clear in other contexts, the rational economic individual of classical economics, is not always the 'optimiser'. A businessperson may play the stockmarket with a designated amount of funds; peasants do not gamble with their livelihoods; safety first is the rule (Scott 1976). Furthermore, migration decisions are rarely taken by individuals alone: they are more likely to be co-ordinated household decisions, in which the support of wider kith and kin networks is expected.[3] For this reason, migrants may be particularly able to withstand the risks of migration and unconducive working conditions. Thus South Asians, relying on flexible domestic relationships among friends and family, were able to take on night shifts in British factories; their need to do so in the first place was their desire to avoid racism in the workplace (Anwar 1979). The very flexibility of workers opens up possibilities that indigenous workers are unwilling or unable to accept. Yet by the same token, this very flexibility makes them especially vulnerable to abuse in the workplace (Burawoy 1976; Legassik and Wolpe 1976).

The non-economic factors which interfere with the smooth operation of 'push and pull' include immigration control, geographical proximity (important in the Mexican case to be discussed below), traditions of migration and historical ties. The two latter factors are cited by both Foner (1977: 122) and Thomas-Hope (1986: 31) as significant in having facilitated the Afro-Caribbean gravitation to the 'mother-country'. One final non-economic influence on migration is the mismatch between the rhythms of migration and of the labour market. The build-up of migration experience spans large parts of individual lives and even generations, increasing the facility for finding work, as does educational attainment.[4] Migration flows therefore transcend the more immediate staccato rhythms of supply and demand in the wider economy and sustain themselves for longer time periods through fluctuations in supply and demand.[5] They can also help redirect that movement into new areas of demand obscured by aggregate trends, but discernible to those whose livelihood depends on sensitivity to new opportunities. It is as if labour market signals are picked up by means of self-made receivers built from the accumulated and up-dated practical knowledge, relationships and resources of the migrant community. This counterpoint of experience, social relations and market forces is what guides action, creating migration flows. Informal knowledge and ties have special significance in the most recent migrations. Thus the economic dimension of migration is embedded in the accumulation and mobilisation of cultural resources; 'cultures of migration' give the specificity to the population movements just outlined. Some of these will be examined in more detail below.

INTERNATIONAL MIGRATION IN THE AGE OF UNCERTAINTY

According to the predominant views of migration mentioned above, following the 1973 oil crisis and the subsequent world recession, what should have happened is for migration from the less to the more developed countries to tail off as a result of shrinking labour demand in the DMEs and increased immigration controls in many countries. In fact, flows to Western Europe remained fairly steady and to the USA they actually increased (Harris 1995: 13). This mismatch between demand and supply has turned out to have been be more than temporary. Migration movements to the advanced industrial countries have ridden roughshod over the recessions of the early 1980s and early 1990s.

Is the failure of falling demand in the DMEs to discourage northward migration explained by economic stagnation in the LDCs? Not so. In spite of the 1973 oil shock, overall growth in the LDCs was high – higher, in fact, than that of the DMEs. In the 1970s, when recession and unemployment were especially high in DMEs, the main immigrant-sending countries were growing at between 5 per cent and 9 per cent (Sassen 1988: 94) For example, between 1965 and 1980 Mexico grew at an annual average of approximately 6.5 per cent, South Korea at 9.5 per cent and Pakistan at 5.1 per cent. Between 1980 and 1991 (a period including the austerity years) LDCs grew at 4.6 per cent compared with DMEs' 3 per cent. In the austerity years between 1980 and 1991, LDCs grew at 4.6 per cent compared with DMEs' 3 per cent (World Bank 1993, Table 1: 238–9). The relatively more favourable growth and employment creation in the LDCs actually apply to the post-1945 period as a whole. Thus labour has been moving continuously from areas of highest and towards those of slowest growth – the opposite of what would be predicted by both push-pull and radical political economic theories.

The unique features of post-1973 migration make this anomaly particularly difficult to explain. In the pre-1973 period, the shedding of rural labour as a consequence of agricultural modernisation helps explain south–north migration. But more recently, industrialisation has been the main engine of growth, and industry tends to create jobs rather than shed them. Yet not only does labour emigration from the industrialising countries continue, it is women as well as men (with urban employment experience and with some education) who are migrating and not just displaced non-literate rural workers. Also, these migrants originate in the more rapidly, industrialising countries of the Far East and Latin America, rather than from just the ex-colonies or countries with fewest economic opportunities. Furthermore, they are so keen to penetrate the recession-torn DMEs, that increasing numbers enter illegally. Why and how are relatively well-endowed workers with reasonable resources abandoning areas with opportunities for the ever more hostile and restrictive conditions in the advanced industrial countries?

Sassen (1987, 1988, 1989) explains the seemingly perverse migration flows by arguing that it is, on the one hand, the kind of industrialisation occurring in the LDCs and, on the other, the kind of economic restructuring occurring in the DMEs which accounts for current labour flows. The particular nature of export manufacturing is such that it creates sudden, large demands for labour, much of which is for women. The classic cases are the Export Processing Zones (EPZs). The loss of labour to the already fast-changing rural areas dislocates what remains of the household economy – particularly the social reproduction of the family. Cultural and social bonds with the rural economy are lost. Given the 'flexible' labour requirements, pay and conditions typical of export manufacturing, there is a high turnover of the workforce. Departing workers do not necessarily regard a return to their regions of origin as desirable or even possible. High turnover not only means departure from jobs, but entry into new, hopefully, better ones.

This restlessness leads to mobility beyond national borders, into the DMEs. Here, synonymous with the dislocation of workers in the south, is the rise of new opportunities for migrant workers in the north. Whereas in the 1970s, immigrants tended to replace indigenous workers in the lowest levels of employment in declining industries, in the 1980s and 1990s, migrants have begun to enter new sectors, or even generate them themselves. The new types of employment are noticeable in the personal services, especially those related to the rising élite of the burgeoning financial and global management sectors. Typical are catering, domestic service and cleaning. In addition, migrants have found their way into new industries, either in capital intensive industries where task fragmentation has occurred, such as in high-tech, or in industries where labour-intensive 'putting out', or sweatshop production is possible, such as in garment making. Changes in Mexican–US migration in recent years exemplify the above features.

MEXICAN–US MIGRATION

Recent trends in Mexican–US migration are consistent with Sassen's findings. The bulk of such migrants, until the early 1980s, were young males, migrating seasonally from Mexico to work as agricultural labourers, principally in the southwest USA. Their wives back in Mexico supplied the (mainly unpaid) domestic labour to sustain the household in the worker's absence. From the mid-1980s this pattern began to diversify, with increasing volumes of workers from urban backgrounds, many of them women, travelling to the USA with or without male companions (Bustamante 1993). An explanation for these changing trends is suggested by Gabayet and Lailson's study of the place of females in migration from Guadalajara to the USA, which corroborates and gives body to Sassen's somewhat unsubstantiated restlessness notion. Gabayet and Lailson (1992) note both the diversification of women's roles in migration, including active, unaccompanied labour migration to the US and the growing trend for many of these women to have already worked in export manufacturing before migrating.[6] Given the relatively privileged position of in-bond workers (workers in tariff-free enterprises, typically in export-processing zones – EPZs), earning, as they do, pay above the local average, the authors ask why they should want to move on. The answer given is that the high turnover and insecurity created by this situation leads women to move on. On the other hand, 'since these positions are thought to be privileged, companies demand a higher educational level for work that in the end turns out to be repetitive and uncreative'. This, combined with the lack of incentives and promotions, 'produces discontent among workers, particularly the more educated and skilled – precisely those most likely to migrate to find better jobs in the United States' (Gabayet and Lailson 1992: 198). Here a connection between Sassen's paradoxical relationship between employment creation in LDC export manufacturing and out-migration is made explicit.

If responses to structural economic changes in the sender country help explain emigration, this is no guarantee of success in the destination country. What happens there? In the Mexican–US case what has been widely recognised is a shift in occupational destinations from agriculture into services and sweated industrial labour. Mexican employment in Los Angeles exemplifies these features. Los Angeles is not only one of the chief urban destinations for Mexican migrants, boasting also a well-established Mexican-American community, it is also one of the major industrial centres of the USA, and one of the largest in the world. While the smokestack industries on which the northeastern towns depended declined, as they also did in Los Angeles, high technology in the Los Angeles region grew. Immigrant labour suited some of the new labour demands of the region, particularly in the high-tech assembly plants, where female workers were considered particularly able (Soja 1987: 192). Apart from the attributed dexterity and patience of women in performing fiddly but repetitive tasks, they were not unionised, and adapted more easily to the labour-saving, part-time and casual 'flexible' requirements of the high-tech companies (Storper and Scott 1990: 587).

Contrasting with low-paid labour in the capital-intensive high-tech industry, the other main industrial sector employing Mexican labour, is the 'down-graded', labour-intensive jobs in garment and toy factories, either in sweatshops or in the home. Apart from Los Angeles, New York, a recent destination choice for Mexican workers, is notable for sweatshop employment growth in the 1980s (Sassen 1989). This is particularly remarkable in the light of the much vaunted structural unemployment crisis in New York, (Fitch 1994). The third area of immigrant employment in which Mexican workers have been conspicuous in northeastern towns and in the southwest, is in the expanding service sector. This sector of course encompasses a wide variety of employment, and a particularly large polarisation in job statuses and income levels. Mexican workers are found especially in the low-wage levels of catering, domestic service and cleaning. Many of these jobs, though they are tied to the formal growth of the service sector are informal in that they are unregulated or undocumented (Fernandez-Kelly and Garciá 1989).

Mexican undocumented workers find themselves in these jobs, (a) because English language skills are not essential, (b) because the undocumented or temporary status of the majority of migrants enables employers to pay little and to hire and fire according to fluctuating requirements, and (c) because the support networks established among the Mexicans, in the USA and Mexico, permit them to adapt to the unconducive working conditions just described. To put it briefly, Mexicans, in common with many other immigrant groups (though to a greater extent because of the history and geography of Mexican migration) are much more adept at taking advantage of the increasingly deregulated working conditions of contemporary economies than are most indigenous workers.

Mexican workers do not only respond to external demand, they also help form it. One of the characteristics of the rise of the high-paid jobs in business services – finance, insurance and real estate (FIRE) – is that they have spawned a demand for customised personal services. Fernandez-Kelly and Garciá (1989) associate this with a change in living styles of the 'new rich'. What the supermarket, the freezer and the washing machine were to the post-war middle-class aspirants, the delicatessen, the ethnic restaurant and Hispanic maids are to today's FIRE élite. These services are being provided, often informally, by immigrants, the goods and services of whom have become increasingly attractive and status-conveying among the upwardly mobile. In addition, ethnic undocumented workers are unable to barter over pay and terms; yet they can rely on some degree of social support from family and friends at home and abroad. What this means is that in the context of an increased potential demand for customised services, Mexicans and other minorities are not simply replacing existing labour, much less displacing it, but are creating new 'niches' both for the white upwardly mobile and for their own ethnic groups (Fernandez-Kelly and Garciá 1989; see also Portes and Zhou 1992).

If economic restructuring explains *what* the preconditions are for changing migration trends, it does not explain *how* those preconditions are translated into action by the migrants themselves. Why is it that Mexicans and Colombians migrate much more than, say, Bolivians, or people from Chad, whose material conditions are generally worse? Two factors are crucial. The first is the existence of strong ties between sender and destination areas (Portes and Borocz 1989: 612). The second is that migration is not an individual but a collective decision, in which kith and kin relations play a central part in facilitating migration and subsequent insertion into the destination economy. The Mexican case is well documented in both respects. Research on sender Mexican communities reveals that future migrants have a detailed knowledge, a mental map, of their future destination, communicated to them through friends and relations. Mass media also familiarise would-be migrants (Macías 1991). Macías reveals how modern Mexican songs are particularly important in conveying and expressing the emotional cross-currents about migration, thereby preparing potential migrants for the situation they may enter. Thus migration has both generated, and has been generated by popular culture.

Another study of a small town in west Mexico, reveals how the culture of migration is planted in the deepest layers of personal relationships in family and friendship groups (Ortíz 1995). Family ties are the most important source of information and support for migrants, but friendship groups are the foundation for other basic networks. This is how they form and operate. Initial neighbourhood groups of friends eventually get pared down to tiny 'trust groups' of two or three lifelong friends. These groups in turn broaden into generational groups and, though increasingly associated with school, are recognised as having historical antecedents in traditional age-sets. These groups were known for carrying out daring exploits in times of war and peace, thereby acting as rites of passage for the young. Today they help prepare the young for the contemporary endeavour of adulthood – international migration. This is achieved through the intra-generational bonds formed among the young people, in concert with the inter-generational ties with adults and older generational groups. Informally in family and friendship circles and in larger social events (typically in town festivals and in the *plaza*), knowledge and contacts build up which facilitate movement. But if inter-generational ties facilitate out-migration, it is the peer pressure and bonds within the generational group which spur the northward venture: once one or two have migrated, the rest do not wish to remain behind, for this (together with marriage and departure from Tizapán) would result in losing one's generation group. Generational groups for girls are particularly important as they are legitimate means of chaperoning the girls in their interrelations with boys, and most recently, of facilitating migration to the USA.

Individuals waiting to migrate may preempt invitations by contacts already working in the US, arriving near the border and asking for help. They may even ask to be sent a *coyote* (a guide for illegal border crossing). These requests cannot be denied, but the rules of reciprocity are made clear from the beginning. It is imperative that jobs are found for newcomers, since, in this way, maintenance costs (rent, food etc.) are reduced. The contacts built up by such groups are so successful that no one fails to get a job in the USA, where the current unemployment rate is around 8 per cent! Jobs gained are in the sectors already mentioned, and are low-paid and insecure; yet the returning migrants are conspicuous by their relative wealth, which they display in the town square. Their distinction materially and culturally from the locals is sufficient partially to displace original affiliations with peers from the same generational group who remained at home, in favour of the ties built up from the experience of migration. Once established, these newer migrant groups serve as hotbeds of information exchange, further facilitating the migration flow. Generation groups eventually break up, since their chief purpose, that of emancipating the young, has been accomplished.

This kind of study enables us to see how, in times of recession in Mexico and of restructuring and high unemployment in the USA, migrants not only find jobs, but actually establish conditions to reproduce the Mexican job market in the USA. Muller and Espenshade (1985) and Waldinger (1989) have argued that Mexicans have helped revitalise the once depressed economy of Los Angeles. The striking thing about migrant communities is the social embeddeness of their economic activities. This has led some authors to speak of 'social capital' (Bonacich and Modell 1980; Coleman 1989). What migrants lack in formal education and training (human capital), and material and financial resources (economic capital), they make up for in social capital, which is the 'know-how' necessary to make a living and possibly even a business success. Portes and Zhou (1992) identify two main relational characteristics of migrant 'social capital', explaining small business success among Dominicans, Cubans and Chinese in the USA. The first is 'bonded solidarity' – the sense of common nationhood and cultural identity, which helps focus the groups resources. The second is 'enforceable trust' which controls the mutual assistance supplied and demanded, permitting a higher degree of resource sharing than would be conceivable through more informal channels.

Immigrant workers are unlikely to compete with the better-placed, if unemployed, domestic labour force. However, ethnic groups may compete with each other, one group eventually replacing another. Waldinger (1994) distinguishes three phases of this process. The first phase occurs when pioneers enter a niche previously held by another group. At this stage, success tends to combine luck and a change of employment practices or circumstances. Thus the success of Waldinger's Egyptians entering public service coincided with the downgrading of public employment, the upward mobility of previous incumbents and their being in the right place at the right time. Similarly, the presence in Baytown, Houston, of Mexicans from Tizapán with good contacts with carpenters and builders, at a time of the mid-1980s office building boom, paved the way for Mexican establishment there (Ortíz 1995). Once established, hiring through the ethnic network saves employers time and money, and assures them of a known type of worker through bonds of mutual, if unequal, interest.

CO-OPERATION AND COMPETITION AMONG MIGRANTS: LATIN AMERICANS IN THE UK

Within minority groups, as seen above, bonds and collective identity capitalise on the limited resources of immigrant groups. Yet all is not co-operation and solidarity in immigrant groups. Clearly tensions between individual or family interests may pit sub-group against sub-group, in the struggle for survival and mobility. Werbner's (1987) study of the Pakistani rag trade in Manchester demonstrates how this business has consolidated itself through horizontal multiplication of firms setting up beneficial 'clusters' of information and resources – 'entrepreneurial chains'. At the same time, the same competition pushes the more successful business upstream to the supply and, eventually, the production end of the business, while the least successful get shaken out. The net result is an economic sector which is competitive with 'outside' enterprises, one with strong interconnections, yet one which is highly internally co-operative. The dialectic of competition and co-operation has underpinned these trends.

Werbner's research on the Pakistani business community is a useful starting point for examining the much newer Latin American population in London, though there are divergences. If the Pakistani business community is characterised by a 'critical mass' of participants (bound by co-operative and competitive relations), some immigrant groups, such as the Hispanics in London have been able to establish themselves by making maximum use of limited networks of social relations. Although they are recent arrivals, they have become one of the largest new minorities in London. Estimates of their numbers range from 50,000–70,000, most of them, Colombians. Nevertheless, their presence is practically unrecognised. They have rapidly developed employment niches in catering, cleaning and ethnic music and dance, so that it is rare for a Latin American to lack some kind of work. How have the Latin Americans managed to earn a living in the UK in such unpropitious conditions, and why did they arrive in the UK? Most of the first migrants arrived in the early 1970s, a period of socio-political unrest, of a rapid rural-urban shift of population and of the beginnings of labour demand in export agriculture and industries as well as in services. As in the Mexican case, many future migrants had worked in such sectors.[7] From this time onwards, the demand in the UK for domestic and catering staff was known through a variety of channels, from adverts in language schools to word-of-mouth contacts with British expatriates in Colombia.

Given these labour market contacts, the relative ease of securing work permits up till the late 1970s and the perceived strength of British currency, the UK was seen as an attractive destination by Colombians. But at the end of the 1970s, stricter application of the 1971 Immigration Act and severe reduction of work permits, rather than deterring Colombian migration, simply increased the proportion of undocumented workers. In addition, the number of family reunions and asylum seekers fleeing the increasing violence in Colombia grew as a proportion of in-flows in the 1980s.

Family, friendship and formal and informal ethnic associations have been the informal 'employment agencies' for the Colombians. As discussed above (Portes and Zhou 1992), the ties of personal loyalty built up by early employment among the Colombians have been the foundation for mutually beneficial ways for employers and employees to meet. From the employer's point of view, relying on a trusted employee, a broker, for supplies of workers cuts down recruitment costs, is self-reproducing, because of the interpersonal links between employees of the same origin and, in conditions of illegality, ensures that workers are compliant and cheap. From the broker's angle, the chance to develop a power base is offered, through which relationships of dependence are built. Employees, for their part, though not well paid, at least can be assured a source of income. In this way the cleaning niche has been established among London Colombians (though they are complemented by the West Africans, in the same line of work). Contract cleaning is particularly well suited to employment brokerage of this kind, since, by definition, it favours a discontinuous hierarchy whereby the foreman is responsible for the scheduling and management of a working group. Neither the company managers nor those for whom they clean want to know much about their cleaning workers, since many highly prestigious public buildings are cleaned cheaply by illegal workers. It is not surprising then that most cleaners have never even seen the company managers. The growth of subcontracting in the public sector in the UK is well known to have been actively promoted by the Conservative governments from 1979 onwards as part of the larger project of privatising public services. In the private sector, the casualised employment practices associated with contract-cleaning, and domestic and other personal services, should be considered as examples of labour flexibility employment strategies, also promo-ted by government economic policies in the UK and the USA in the 1980s, and they particularly affect women (Walby 1989; Jenson 1989; Storper and Scott 1990). The particular implications for immigrant women are examined in Morokvasic (1993).

Restaurant work has also emerged as a Colombian employment enclave, but for different reasons. In the late 1970s and early 1980s, when continental, particularly Italian food was gaining popularity in London, Colombians became equivalents to the domestic labour which has predominated in much of the catering industry. The uncertain employment status of Colombians meant that wages could be kept low; their visual similarity to Italians helped make Colombians substitute Italian waiters. Like cleaning, language presented no problems, in fact worked to reinforce their attractiveness as 'Latin' waiters. Like cleaning too, high labour turnover was not an impediment to either employers or employees in this unskilled work, though unlike cleaning, in restaurants, employees have to be closely supervised to maintain acceptable levels of service. The catering business has been one of the few obvious niches of another new minority in the UK, the Vietnamese. Though as programme refugees, they have the right to government social benefits, which many Colombians lack, they have been impeded by their limited, mostly rural work experience and their arrival in a country which many did not choose, having few prior ties here. In addition, they had to overcome the enforced fragmentation of their family and friendship networks, which would otherwise have assisted their resettlement. Such fragmentation was mostly the result of persecution and flight, but was also furthered by the misguided government policy of dispersing them throughout the UK. The Vietnamese have steadily regrouped themselves in the large UK cities, where they have begun to reconstruct their cultural and economic life. Among other economic opportunities, employment in the established Chinese food business in the UK has proved to be one area of employment (Lam and Martin 1994, 1997). Their cultural closeness to the Chinese community helped especially the ethnic Chinese Vietnamese in this respect.

In neither cleaning nor catering have many Colombians made the transition from employee to employer, though there are a few notable exceptions. Small enterprises have emerged in response to the burgeoning interest in Latin American dancing and music. The relatively low capital demands in this area of entertainment, together with vital contacts with gatekeepers of public venues, has favoured Latin American performing arts enterprise. Though initially the music and entertainment enterprise was a response to the social demands of the Latin American population, it has become chiefly a Latin American 'export' to other sections of British society.

CONCLUSION

The picture of Mexican workers discovering, opening up and eventually organising an informal but international employment agency, of Pakistani entrepreneurs establishing themselves in the British garment industry and of Colombians establishing employment niches in recession-torn Britain, with few antecedents, casts doubt on two modes of thinking about immigrants. The first is that migrants' movements are a response to existing forces of labour demand. This is false in three senses. In the first place there is no obvious demand drawing many of today's migrants to the commonly oversaturated labour markets which are their destinations. In the second, even if there were a demand, this in itself does not guarantee any potential supply would be able to satisfy it; all manner of obstacles may exist, which, to be overcome, require considerable effort and ingenuity in the case of long-distance migration. In the third place, if demand is *not* obvious, then very special ingenuity is required to discern it, then to satisfy it and then to sustain it.

The second misconception fails to acknowledge such capacities. Migrants, because they are from poor backgrounds, are considered hapless, lacking in vitality and entrepreneurialism. This view, though challenged by many academics, lingers in the popular mind; hence the recent tabloid scaremongering against illegal immigration in defending the 'deserving' potential immigrants from Hong Kong (see *Daily Mail*, 26 September 1995) As seen, however, it is increasingly obvious that most migrants, *especially* illegal entrants, represent the more, not the less, dynamic sector of workers in their countries of origin. It is curious that success in the business world is not treated as the outcome of simply responding to market forces in the business world. On the contrary, the subjective characteristics of pro-active entrepreneurialism are considered indispensable for 'creating opportunities'. Yet international migrants, whose collective impact is nothing short of historic, have most commonly been seen as merely reacting to external circumstances.

In this chapter it has been suggested that migrants do indeed react to circumstances, but creatively and interactively rather than according to the pre-set terms of a supposedly given market. Through planting the roots of labour mobility deep in the heart of the communities of origin, while cultivating a wide variety of social relations, they create a dynamic of migration which over-reaches temporal and spatial fluctuations in labour markets. In this manner they been able to compensate for their disadvantages in education and bargaining power. In their flexibility, they are the archetype of the 'flexible' labourer of the late twentieth century. They are in the economy but not of it; they define themselves not by their working lives, but according to the world beyond, in the migrants' case, in the traditions which their family and friendship networks perpetuate. For this, migrants are often said to live in a fantasy world, neither here nor there. The word 'fantasy' was suggested to me by a Latin American community leader in London who used it in much the same way that Anwar coined the phrase about the Pakistani 'myth' of return to the home country. But perhaps these fantasies or myths enable them to cope with the insecurities of the age better than indigenous workers who are still trying to rationalise the uncertainties of modern employment.

NOTES

1 The terms 'less developed' or 'developing' denote countries with low per capita incomes as compared with the 'developed market economies' or 'advanced industrial countries', a difference which is mainly due to the lower levels of industrial productivity and output.
2 Mechanisation does not necessarily expel small producers. Modernised smallholder commodity production based on family labour has been resilient in some areas. Not all mechanisation means large farms. The penetration of the market in the countryside arises from the more general growth of industrial capitalism.
3 The key factors to be taken into account are the following: (1) the dependency ratio in the household; (2) the gender balance with respect to the availability of male or female employment; (3) the terms of trade between domestic and income earning work; and (4) a variety of opportunity cost calculations. Above and beyond the household are wider sets of relationships along which resources flow. These networks share information, resources and the burden of work or income forgone during a migrant's absence. In so doing, resources are pooled and the load is spread. The wider the networks and the resources pooled, the greater the economy of scale supporting migration, and benefiting from its returns.
4 Godfrey (1976) shows how education also loosens ties to the locality. Even when, as education becomes more universalised, the edge on employability by the educated is lost, the diffusion of education's tendency towards mobility offsets this loss. Thus the non-economic unsettlement syndrome outweighs the economic diploma disease factor.
5 The long-standing scholars of migration, Portes and Borjas (1992), have reached the conclusion that the fluctuations in labour demand and subsequent income differentials are much more capable of predicting labour migration flows than sociological indicators – particularly the extensiveness and strength of networks.
6 Women are increasingly (1) supplementing the male wage through local wages; (2) living and working in cities in the *maquiladoras* (literally, assembly plants, but also understood to be foreign-owned factories, often on the Mexican–US border), having cut their immediate domestic links with the rural areas; (3) crossing to the US without male companions, upon invitations from relatives, and finding work there, and (4) accompanying their husbands to work in the USA.
7 Service employment nearly doubled from 34 per cent of all jobs in 1965 to 66 per cent in 1990, and export processing, notably in the EPZs, where female employment predominates, grew in the 1970s.

VI

CULTURE AND HUMAN RIGHTS

17

THE WEST, ITS OTHER AND HUMAN RIGHTS

Rolando Gaete

THE WEST AND THE OTHER

What lies beyond the West? Who is the West's Other? In the age of 'discoveries',[1] American natives were the paradigmatic non-Western Other: fallen, pagan, maybe soulless. The West claimed to have 'discovered' the Other, and the question of the nature of this exotic otherness was raised: could these people, who were called 'Indians' by Western conquistadors, be rescued from the world of Satan? This question has continued to underlie the attitude of the West towards the Third World.

The Pope settled the issue in 1537, proclaiming in his papal bull *Sublimis Deus* that these people were 'true men' and capable of receiving the True Faith of Christianity. This project of massive conversion continued to drive Empire-builders for centuries, but it was gradually secularised under the influence of the Enlightenment. During this second stage, the imperial masters educated their colonial servants without necessarily converting them: In Defoe's novel, Friday, the island native, is Robinson Crusoe's Other, presented as an ignorant and superstitious creature who will be enlightened by Western knowledge.

Finally, during the last stage of colonialism, the colonial Other was seen as a human being 'like us' who has not reached 'our' level of development. The task, especially after the colonies were granted independence, was no longer the revolutionary one of religious conversion or scientific transformation but the reformist one of gradual evolution along a straight evolutionary line. Subsequent stages of development were theorised along this progressive line leading to the promised land of full development as illustrated by the West. Sometimes, bringing development to the colonies was understood as a divinely inspired task. For example, Charles Grant, chairman of the East India Company's Court of Directors, wrote that God gave Britain its Indian colonies

> not merely that we might draw an annual profit from them, but that we might diffuse among their inhabitants, long sunk in darkness, vice and misery, the light and benign influence of the truth, the blessings of well-regulated society, the improvements and comforts of active industry ...
>
> (cited in MacGrane 1989: 51)

At the same time, and in contradiction to this evolutionary concept of historical time, the concept of culture appeared. Increasingly, anthropologists discovered that the language, myths and forms of organisation of 'primitive' people were not primitive at all and could not be allocated to a precise point along the evolutionary process of 'development'. Relativism seemed to be inevitable: the Other was another culture, and could not be measured using the criteria of 'our own' culture.

Current debates on human rights are determined to a great extent by these two contradictory pressures: on the one hand, the

Western pressure on the contemporary Other – 'the Third World' – to adopt human rights standards as they have been developed during the last three centuries in the West; and, as a response, the claim by some Third World leaders (whether from newly industrialised Asian countries or from African countries) that the concept of human rights and of the individualism underpinning them is alien to their culture. This debate raises a question that touches upon the very relevance of human rights: can the 'cultural Other' be criticised? Or is any criticism an abusive imposition of the critic's own values on the Other?

As I will show, the criticism of human rights records is a method by which the critic achieves an ethical identity, that of 'the West'. The critical West is, by implication and in contrast to the object of its criticism, the land where human rights are realised, the righteous Defender of the Faith, to borrow a religious expression.

WHAT IS THE WEST?[2]

The modern West, the West that has made of human rights the source of its legitimacy, was arguably born as a reaction against cultural relativism. Cultural relativism was common among the sixteenth-century humanists (Erasmus, Vico, Montaigne), who developed sceptical views before the rise of modern science brought intellectuals back to the search for absolute certainty and truth. In fact, the late Renaissance can be seen as a brief interval between the absolute truths of the Middle Ages and the absolute truths of modernity.[3]

Originally, the ideology of human rights belonged to this emerging field of unquestionable truths. It emerged after the West had reinvented itself, developing a new set of absolute truths, whether in scientific and philosophical words such as Galileo's Dialogue concerning the two great systems of the world, Hobbes' *Leviathan* and Descartes' *The Discourse on Method*, or in the writings on moral theory by the Cambridge Platonists who had made of ethical questions an abstract discipline of general principles. From then on, Science would provide the model for modern rationality, which would take the form of abstract and general axioms and principles and theories.

It is against that intellectual background that the talk on 'the natural rights of Man' (and they did only really mean men) emerges, first in the rationalist political philosophies of Hobbes and Locke, transcribed eventually to political proclamations like the American Declaration of Independence, which defined human rights as modern rational truths without need of proof:

> We hold these truths to be self-evident, that all men are created equal, that they are endowed by their Creator with certain unalienable Rights, that among these are Life, Liberty and the pursuit of Happiness ...
> (The Unanimous Declaration of the Thirteen United States of America of 4 July, 1776)

This was not a manifesto or political programme. It was not understood as a cultural product, an expression of the American Spirit. Although the Bill of Rights would become in time the official self-image of the United States, and by extension of 'the West' ('what the West stands for'), it was seen originally as a scientist's discovery of the general laws that govern social and political reality. Similarly, in the Declaration of the Rights of Man in 1789 and in the Universal Declaration of Human Rights in 1948, human rights were not enacted (as laws are); they were 'declared' to the world at large in the same way that Kepler declared the laws of planetary motion and Newton declared the law of gravity.

In other words, human rights were not only a feature of the American system of government but were thought to be some general and enduring features of human beings everywhere and in all times. They were universal and timeless rights, independent of culture and community. Their historical emergence in the West was presented as the discovery of the final and universal truth of legitimate authority against the untruth of medieval and monarchical justifications of authority.

The idea of universal rights was consistent with an abstract idea of justice, a justice that in general claimed not to be concerned with values or consequences but only with general principles. The concept of 'nature' replaced the concept of 'culture'. As a consequence, the cultural and historical nature of human rights was denied. The Classical political theorists of Modernity (Hobbes, Locke, Rousseau, Paine and others) called them 'natural rights'. They stood for something more primordial than culture, so that it was possible to argue that 'people are fundamentally the same everywhere and that "cultural differences" are merely something like different mental images of the same basic reality ... ' (MacGrane 1989: 118).

However, today, a confident West no longer needs to root human rights in Nature. The dominant claim now is that human rights are a cultural artefact but that this artefact is grounded on a universal consensus. The ideology of human rights is today arguably global because there is no alternative ideology of justification of power. At its most optimistic, this globalist thesis claims that we have come to the end of History, which will blur local particularities, and that all countries of the world have become Western-liberal or will become so in the relentless march of world history towards planetary enlightenment and freedom on the Western model. This historicist view does not make particular claims actually to know the nature of human beings but offers, instead, a view of History as a totalising planetary progression towards a community of free individuals. Two recent highlights in this relentless historical movement which can be used to illustrate this movement (if not to prove it) are the fall of the Berlin Wall, which represents the end of the ideologies of total state organisation, and the end of apartheid in South Africa.[4] The triumph of the liberal West has consolidated the concept of 'the West' as the leader of an increasingly global civilisation.

WHAT HUMAN RIGHTS STAND FOR

Inherent in the concept of human rights is the old Judaeo-Christian idea of a transcendent God beyond 'the world' and of a law beyond the law of the State – a law which is 'more legitimate than any law, more just than any justification ... the measure of all law and, no doubt, of its ethics' (Levinas 1993: 116); an authority which is more primordial than will and reason: 'God's original coming to the mind of man [sic]' (1993: 117).

In the secular modern world, this law beyond or above the law of the State is the law of human rights. It is enacted as a higher form of law against which the validity of ordinary law is measured. While US society is generally religious, the Bill of Rights guarantees freedom of and from religion and the secular character of the State, formalising what has been called 'the culture of disbelief'. Thus, the American Bill of Rights has become an unsurpassable obstacle for those who would like to enact a law establishing prayers at schools. Similarly, France is one of the most secular countries of the world, constitutionally speaking.

Rhetorically, if not in reality, 'the West' stands for common secular values in pluralist societies. This secularism – which prevails over both State law and Church law – results in a double separation which is characteristically Western: the separation between Church and State and the separation between the private and the public (religion is a private matter). This separation creates a space for the development of civil society, 'a system of fortresses and earth works', in Gramsci's words, which occupies a large gap between State and individuals.

Commenting on the fact that the Communist revolution happened in the East, beyond the Urals, Gramsci contrasts Russia and the West, writing:

> In Russia, the state was everything, civil society was primordial and gelatinous; in the West, there was a proper relationship between state and civil society, and when the state trembled a sturdy structure of civil society was at once revealed. The state was only an outer ditch, behind which there stood a powerful system of fortresses and earthworks.
>
> (Gramsci 1971: 238)

In Gramsci's formulation, the State is usually a threat to civil society. It is not an accident that the State is the very target of human rights. And yet human rights are also the official ethos of the Western State, at least since the outset of the Cold War. During this period, the US, followed with some embarrassment by the rest of the West, transformed the discourse of human rights into a discourse at the service of the domestic and foreign policy objectives of the State. Human rights became the morality of the Western State, ideological weapons of propaganda and of reinforcement of the allegiance of citizens to the State (I analyse this trend at length in Gaete 1993). From being a Cold War weapon, the discourse of human rights has now become a weapon against commercial competitors. Thus, the crusade against the use of prisoners for the production of export goods in China and the use of child labour in other countries has emerged at a time of grave concern in First World countries about cheap imports from the Third World. A moralism based on commercial interests is nothing new in Western history but it undermines the credibility of moral claims.

ARE HUMAN RIGHTS RELATIVE VALUES?

From being a discourse of protest against the State, human rights became, during the Cold War, a State ideology that was used selectively against what the West saw as its external enemies. The West presented itself as a post-imperialist, ethical force, encapsulated, in the 'Free World' formula, as a homogeneous and coherent ethical entity.

As recently as 1993, at the World Conference on Human Rights in Vienna, the United States could present the Western position as an uncompromising ethical commitment to the integrity and indivisibility of human rights against the views of the Chinese and other Asian countries that human rights should be interpreted in the light of prevailing cultural values.

That is the rhetoric. But the practice is different. In the United States, the practice of human rights revolves mainly around the Supreme Court, which is the political agency with the authority to determine the meaning of human rights concepts and principles. The Supreme Court is the interpreter of the Bill of Rights. While the Bill of Rights states that everybody has the right to be free from torture and cruel punishment, it is the Supreme Court that decides what cruelty is and what is cruel. Is the death penalty cruel? The answer to this question will vary according to the political climate, and to the balance of liberals and conservatives within the United States Supreme Court. Thus, a Court dominated by a liberal majority held in *Furnam* v. *Georgia* (408 U.S. 238, 1972) that capital punishment was a breach of the Constitution, but a new majority held the opposite four years later (*Greg* v. *Georgia*, 428 US 153, 1976). More recently, the Court held that the execution of somebody who was sixteen years old at the time of the crime was 'cruel and unusual punishment' (*Thompson* v. *Oklahoma*, 487 U.S. 815, 1988); yet one year later two Reagan appointees formed a new conservative majority which held that the execution of sixteen and seventeen-year-old murderers and mentally retarded murderers was not 'unusual and cruel punishment'.

Similarly, the decisions of the European Court of Human Rights are strongly influenced by culturally conditioned conceptions of the good life and even by contingent political views. In some cases, the Court has even investigated 'evolutionary trends' in European culture before giving an answer (cases illustrating this judicial policy can be found in Gaete 1993: Part 4).

In other words, Western human rights courts practice a healthy relativism that Western governments denounce in China and South Asian countries. Neither is 'the West' and its version of human rights the kind of homogeneous entity offered in images often projected by Foreign Offices or some academics. From within the West, two increasingly important movements have raised questions about the metaphysical[5] individualism underlying the discourse of human rights:

First, feminists have developed a powerful critique of the Western metaphysics of the subject, which is at the basis of human rights. This metaphysics is rooted in the Christian doctrine of the individual soul, which is taken to signify the quality of humanity, anterior to and separate from less 'relevant' aspects such as gender, ethnos and culture or wealth. A Catholic priest, Father Las Casas, explained how human rights were equivalent to the Christian doctrine of the soul:

> The … rights of men [*sic*] are common to all nations, Christian and gentile and whatever their sect, law state, colour and condition, without any difference … All the Indians to be found here are to be held as free; for in truth they are so, by the same right as I myself am free.
>
> (quoted in MacGrane 1989: 15)

This metaphysical subjectivism, with its emphasis on the autonomy of the self, is characteristic of Western culture, especially on the northern fringe of Europe, but it has become of the essence of modernity:

> That period we call modern … is defined by the fact that man [*sic*] becomes the centre and measure of all beings. Man is the *subjectum*, that which lies at the bottom of all beings, that is, in modern terms, at the bottom of all objectification and representation.
>
> (Heidegger, quoted in Gaete 1993: 106).

Feminist authors have contrasted the masculine ethic of universal rights, which emphasises individualism and separation, with the 'ethic of care', which is an 'ethic of sympathy and effective attention to particular needs' (I. M. Young 1990: 306).[6] Underlying an ethic, there is always a dream of the good society. The dream underlying the ethic of rights, that of a society where everybody asserts his or her rights (keeping lawyers busy) – in fact, the American dream – is seen by many in the West as the model that development strategies should emulate in the Third World, especially in China. But many feminists are sceptical about it: the dream may turn out to be a destructive nightmare. 'The dream of an infinitely striving self, unfolding its powers, in the process of conquering externality, is one from which we have awakened' (Benhabib 1990: 107).

Second, a new interest in culture and community is also sceptical about the Western emphasis on human rights (see for example Glendon 1992). This new interest is to a great extent the result of an extension of anthropology to Western cultures and subcultures. Disillusionment with a culture of neo-Darwinian individualism, free markets and media-driven electoral politics has generated a new interest in more communitarian forms of life at the end of the twentieth century. While many liberals think that the nostalgia and anxieties for community are growth pains in the irresistible march towards the liberal end of history, when a single global culture along Western lines will exist on the planet, many in the West, on the contrary, have 'gone relativistic', arguing that no culture has the right to impose its metaphysical conceptions and values on other cultures.[7] In fact, the best arguments supporting cultural relativism – the belief that all values are culture-bound and that there cannot be such a thing as universal human rights – are put forward by Western anthropologists and scholars, not by Third World despots trying to invalidate any criticism of their practices.

THE VOICE OF THE OTHER AND THE UNIVERSAL RIGHTS OF THE RELATIVE

The question of universal validity of human rights has split Third World cultures as well. As Ann-Belinda S. Preis has argued, these cultures are complex fields of overlapping cultural traditions under permanent negotiation and debate (Preis 1996). *Dow* v. *State of Botswana* is one of the examples she gives. In this case, a woman sued the government of Botswana for the first time in history, challenging the validity of a law that discriminated against women on grounds that such a law was a breach of human rights. When she won the case, government ministers, in Preis' words, 'vehemently attacked the decision as an unacceptable affront to Tswana culture' and the Attorney General expressed the view that a Constitutional amendment allowing gender discrimination would be enacted. Yet this did not happen. An essentialist view of Tswana culture may have regarded gender equality as alien to it, but the facts seem to indicate that such an ideological stance was only one of the components of a complex culture under permanent evolution.

In fact, one could argue that the rejection of extreme forms of brutality is a component of all cultures. Genocide, massive practice of torture, wars and forms of State Terror producing millions of refugees, religious persecution, residual forms of slavery, all these practices can be found in the world today, but we (we, human beings) all condemn them, whether we take part in them or not. Humanity did not wait until human rights appeared on the world stage to proscribe some of these practices. England abolished interrogation by torture in 1640 and authoritarian Prussia abolished it in 1740, a long time before the first Declaration of the Rights of Man. We do not need to master the language of human rights or a modern ethics of general principles to condemn genocide.

Whatever intellectuals may argue about the universalistic claims of rights protecting bodily integrity against extreme forms of brutality, a universal consensus on their undesirability has been achieved.[8] The international community considers these rights as part of *ius cogens*. This Latin expression refers to that part of International Law that cannot be derogated or changed by Treaties. For example, a Treaty regulating the slave trade between two nations would be null and void, whatever the governments of those nations say, because the prohibition of slavery is *ius cogens* today. Clearly, any relativism in this area is hard to justify.

Cultural relativism is more persuasive in relation to the area of human rights that paradoxically affirms relativity. I am talking of the sphere of privacy and of freedom of thought and religion (including the right to freedom from religion, that is, from the institutions that regulate life in the name of Religion). This area includes the principle of free speech insofar as this principle is based on a combination of private autonomy and secularism.[9] It is a sphere of rights that has had an important development in the West and that, more than any other value, constitutes perhaps the very ethical identity of the West.

I would call these rights the universal rights of the relative. They assume a concern for the Other, a readiness to listen to the Other as another, a respect for the Other's private life and for the Other's speech, which is only possible if the Other can have a voice, if one's own relativity in relation to the Other is not denied. The absolutist, the person who denies this relativity in the name of absolutist beliefs, cannot listen to the Other.

The most recent developments in the area of human rights – the right to non-discrimination on grounds of race, religion or gender and the respect for minorities – are examples of this respect for relativity, which is another name for the affirmation of difference and otherness. These rights are calls for an openness to the Other beyond the individualistic egolatry, the exclusive concern for the self's protection, that was thought to be the ground of human rights in Classical writings (particularly in Hobbes' writings, but also in Locke's), and beyond the tribal egolatry that many communitarians and nationalists support.[10] Tribal egolatry is universalistic in the same way that the modern State is universalistic (ethnic genocide is an extreme form of tribal egolatry in action).

The cultural identity of a community, especially when it is exploited and reinforced by nationalistic leaders, is similar to universalism in that both claim to be the law of the law, the meta-law, the ultimate ground of all justification, whether it is 'We, the People' or the Nazi myth of blood and soil. One should not forget that the great revolutions inspired by the Rights of Man (as understood at the time), the American and the French Revolutions, were exercises in unitary nation-building. The American Revolution did not change the treatment of Native Americans or African Americans. The French Revolution denied particularity in the name of the general will. Still now, the French Constitution states optimistically that there are no minorities in France.[11] Against this old State-centred control of human rights, the respect for otherness and difference calls for a prescriptive relativism, based on an ethos of respect for difference, pluralism and diversity (Turner 1991: 23).

THE OUTSIDER'S DILEMMA

These 'rights of the relative' are universal in the sense that they are extra-territorial: they are calls to justice coming from beyond the State and from outside the (often empowering, sometimes stifling and oppressive) boundaries of culture. Maybe despotism can be defined as the direct exposure to the gaze of the despot (Turner 1991: 23). The despot is often an arbitrary ruler, but, more often, despotism lies in the institution and our love for the institution, what Legendre has called 'the love of the censor' (Legendre 1974). The censor is nowhere and everywhere; faced with some particularly brutal forms of women's subordination and rituals of initiation and passage, an observer comments:

> I looked for villains in this conundrum and I found none. I found instead men and women entrapped in an antiquated ritual, dating God only knows how far back into history, unable to free themselves from its centuries old enmeshment, all of them prisoners.
>
> (Legendre 1974)

Often, the observer is aware of her own relativity while she painfully observes the universalistic claims of some rites. For example, in this testimony by an anthropologist observing practices of genital mutilation in a village in Sudan:

> the relativism which imbued the way in which I carried out some of my work also became a means by which to avoid addressing a difficult and conflict-ridden area ... retreating into a world of relativism, (I) managed to avoid discussions about female circumcision with women... [But not completely,] just as I struggle to understand why it is necessary to remove a young girl's genitalia, so they appear mystified and astonished that the operation is not performed on girls in England.
>
> (Parker 1995: 512, 510).

There is an asymmetry here between the Other's absolutism and the anthropologist's relativism. For the Other – the object of anthropological inquiry – rites that seem to have something to do with the subordinate role of women are irrefutable, and this absolutist position is often reinforced by the arguably wrong belief that this kind of practice is an essential part of Islam, the Will of God defining the nature and destiny of womanhood. For the Western anthropologist, relativism requires a paradoxically absolutist professional respect for the Other's absolutism. This asymmetry is a perfect reversal of the imperialist relationship.

Ritual practices of this kind as much as repressive governmental practices raise questions that cannot be solved by appeal to absolute principles – neither to the general principle of international (i.e. Western) intervention 'to put things right', reassuring one kind of liberal consciousness, nor to the absolute principle of cultural relativism, which advocates non-intervention out of respect for alien cultures, to reassure another type of liberal. Either choice addresses the agonies of the liberal mind, not the problem of how effectively to help the Other. Justice requires the examination of the particular case, not the blind application of absolute rules. It requires the careful scrutiny of any claim by foreigners of a right to intervene. There have been far too many crusades, 'wars for the hearts and minds' and humanitarian interventions serving the security and commercial interests of the West.

Yet globalisation is having an effect in all cultures. No culture is today an island, a world closed to technological and cultural influences. The victims of political and cultural practices of oppression are supported by outside forces, be it letters from Amnesty International members or diplomatic pressure from foreign governments. During the painful cultural changes that the process of globalisation is producing, no government has the right to deny its people the right to challenge sometimes not only their rulers but also their ancestors.

NOTES

1 On the various constitutions of the Other by cultural imperialism, I loosely follow MacGrane (1989). I use 'discoveries' with caution because, of course, the lands the Europeans 'found' were already occupied and known.

2 As imaginary constructs with no clear referent, 'the West' and 'the Third World' should be surrounded by quotation marks. However, this procedure is too cumbersome, but the reader should recall that they are constructs and certainly problematic.

3 Montaigne, for example, wrote :

> What am I to make of a virtue that ... becomes a crime on the other side of the river? What of a truth that is bounded by these mountains and is falsehood to the world that lives beyond? [Yet] those who have feelings very contrary to ours are not, for that alone, either barbarians or savages, but ... use reason as much or more than we do.
>
> (quoted in Guignon 1983: 22)

4 The merits of this argument, developed by Fukuyama when Communist regimes were coming to an end and apartheid was entering its last stage, are less important than the dramatic impact that this author's thesis had. In theatrical terms, the 'end of History' thesis was a hit among Western audiences, especially in the United States, who understood it as the end of ideological conflict and the final triumph of unfettered free markets and human rights. Human rights were seen as a science of the limits of power, not as an ideology.

5 Our metaphysical beliefs are those beliefs than cannot possibly be verified empirically. For example, the belief in a free will or, alternatively, in destiny, or the belief or lack of belief in God, etc. When McIntyre writes of human rights that 'belief in them is one with belief in witches and in unicorns' (McIntyre 1985: 69), he simply means that they are a metaphysical concept.

6 Naomi Sheman traces the masculinity of the Rights of Man to the patriarchal structure of the family (Sheman 1983).

7 Thus, commenting on Foucault's work and perspective, Dreyfus and Rabinow write that an act of justice 'cannot be grounded in universal, ahistorical theories of the individual subject and that, in fact, such attempts promote what all parties agree is most troubling in our current situation' (Dreyfus and Rabinow 1986: 118).

8 Rights recognised by the Convention on the Prevention and Punishment of the Crime of Genocide (1948), the four Geneva Conventions of 1949 on conduct during armed conflict, the Convention on the Abolition of Slavery, the Slave Trade, and Institutions and Practices Similar to Slavery (1956), the International Convention on the Elimination of All Forms of Racial Discrimination (1965), the International Convention on the Suppression and Punishment of the Crime of Apartheid (1973), and the Convention Against Torture, and Other Cruel, Inhuman and Degrading Treatment (1987).

9 Ironically, in the United States, in particular, the appeal to the First Amendment (which guarantees free speech) resounds like a holy invocation. For a good account of the formation of a public space of free discussion as a phenomenon that has happened only in the West, see Habermas 1978.

10 I borrow the term 'individualistic egolatry' from Levinas. It means: ego's idolatry, that is, the transformation of the ego into an idol. For example, Hobbes, who is reputedly the 'inventor' of the concept of natural rights, portrays human beings as lonely, sovereign selves whose only concern is to protect themselves and who regard all other human beings as cannibalistic predators.

11 In *M. K.* v. *France* (Doc. A/45/40 Apx.) and in *T. K.* v. *France* (Doc. A/45/40, Apx.), the Human Rights Committee stated that a declaration made by France upon accession to the Covenant stating that there are no minorities in France according to the Constitution was equivalent to a reservation to the Covenant. Therefore, it had the legal effect of exempting France from the application of its Art. 27 that provides for the protection of minorities.

18

FEMALE CIRCUMCISION AND CULTURES OF SEXUALITY

Melissa Parker

INTRODUCTION

Female circumcision is a subject which arouses great interest and concern among academics in the biological and social sciences as well as human rights activists, politicians and journalists. Intense emotions underlie this interest and concern and, more often than not, this has detracted from the quality of research and writing in the academic and popular arena.

For some, even the use of the term 'female circumcision' is infuriating. One of the most widely used guidebooks on Kenya, for example, asserts that it is 'a classic bit of male anthropologese', and indicates that the only politically correct term is the more graphic and derogatory 'female genital mutilation' (Trillo 1993: 63). It is likely that some of my readers will find me guilty of this 'male anthropologese', if only because I am not writing about the issue only to assert abhorrence (and in spite of the fact that I am a woman while the author of the guidebook on Kenya is a man).

The purpose of this chapter is to highlight the different ways in which the emotions aroused by investigations of female circumcision (and encapsulated by Trillo's statement) have influenced the collection and presentation of data, and to demonstrate the incapacity of many Western researchers to write about the practice in a 'scientific', 'neutral' or 'relativistic' way. Indeed, it is suggested that female circumcision acts as a metaphor for conflicting cultures of sexuality, and most of the academic and popular writing about the practice thus tells us more about sexual subjectivities in Western countries than about the phenomenon itself.

The paper is divided into three parts to illustrate these points. The first part describes one of the circumcision ceremonies that I attended while undertaking anthropological fieldwork in northern Sudan, as well as some of the discussions that I had with friends and colleagues once the fieldwork had been completed. The second demonstrates the resonance and wider repercussions of these conversations in the biomedical and anthropological literatures. The third and final part of the paper shows some of the ways in which emotions associated with sex, sexuality and the self have influenced the type of research undertaken by Western investigators on female circumcision.

FEMALE CIRCUMCISION IN OMDURMAN AJ JADIDA, SUDAN

Between April 1985 and May 1986 I lived in Omdurman aj Jadida, a village in the Gezira/Managil irrigation scheme, Sudan. The village is located approximately 240 kms south of Khartoum between the Blue and White Nile. It has a population of 1,115 and, by any criteria, is economically very poor. There is no electricity, no piped water (so water for domestic use has to be carried by hand from the canal), no sanitation facilities, and, inevitably, a large number of infectious diseases are endemic. These include malaria, schistosomiasis, diarrhoeal diseases and acute respiratory infections. All the villagers are Muslims, speak colloquial Sudanese Arabic and their social and economic activities are characterised by a marked degree of segregation between the sexes.

Pre-pubertal female circumcision is a feature of life among the majority of Muslim populations speaking colloquial Sudanese Arabic in northern Sudan (El Dareer 1982), and all girls are circumcised in Omdurman aj Jadida. Throughout my fieldwork, I was primarily concerned with investigating the effects of a parasitic infection, schistosomiasis, on the health of women and children. However, I attended a number of circumcision ceremonies and this reflects the fact that, over time, I became involved in all aspects of daily life. Indeed, it was assumed that, as a woman, I would want to participate in these ceremonies. The following accounts describe some of the circumcision ceremonies that I attended. The first extract was written a few hours after I had attended my first circumcision ceremony, and the second extract was written a few weeks later.

THINKING BACK TO MY REACTIONS TO FEMALE CIRCUMCISION IN THE 1980S

There was very little time for reflection during my fieldwork and it was not until I returned to England in 1986 that I fully appreciated how divided I had felt by many of the issues surrounding female circumcision. There were, of course, several signs indicating distress and discomfort. Shortly after attending Nijat's circumcision, for example, I walked across the village to give some Dettol antiseptic lotion to her mother. I knew that I should find my assistant or some other woman to accompany me but there was no one around and I was far too upset to wait. To my mind, a blunt razor blade carried a multitude of risks and I was convinced that her life was in danger. I simply did not care what type of reaction I provoked, as I felt I had to get the Dettol to Ziyarra, whatever the cost. This was the first time that I consciously and defiantly challenged expected codes of behaviour in Omdurman aj Jadida.

As time progressed, this distress was compounded by confusion. On the one hand, I wondered whether I should have witnessed

JOURNAL EXTRACT ONE: NIJAT'S[1] CIRCUMCISION, NOVEMBER 1985

'The knife is above her, the knife is above her – come quickly! come!' Selwa and I were drinking tea one morning when a young girl placed her head over the wall and shouted these words. We upped and left – and walked at a quick pace to her brother's house (Abass Mustaffa), where his seven-year-old daughter was about to be circumcised.

Together we entered a crowded mud-built room and greeted some fourteen women gathered there. A little later we were joined by an old woman, who, taking an axe head, soon began to dig a small hole – 6 × 5 inches – in the middle of the floor. Next to this she placed a cushion. And then Ziyarra picked up her daughter, placed her on this cushion, and with her arms around her waist held her tight. Two other women held her legs – straddled open above the hole. The blade was tested for its sharpness, and while we all looked on, the old woman began to cut. Slowly but surely, she took the girl's clitoris and all other loose remaining flesh.

The blood flowed and the girl screamed. And as she screamed, she tried to kick herself free. But the women held her tight – though anxious at any further damage she might do, they called on others to hold her to the ground.

The job was not yet done. More blood flowed. The woman cut, and cut. And as she cut, young children – largely girls, but some boys – beat drums, sang and danced outside. Inside, the women looked anxiously on. Had she removed enough? Was there anything left? And then it was announced: 'aiewa, maa fii eeyi shi' ... 'Yes, there's nothing left.'

They flushed the wound with [hot] water, bound and strapped her legs with cotton cloth and lifted her from the ground to a low-lying '*angarib*' (a low-lying wooden bed strung with cattle hide). Her mother held her head, comforting her as best she could: 'Don't cry' 'don't cry,' she said. But the girl wept and wept, for an hour or more. The flesh had since been buried and dampness on the ground was the only sign of the blood which had flowed so freely.

Tea and, later, coffee were brought for the gathered women. We drank – partly in celebration and partly to seal the event. And the girl, Nijat, was given a glass of sugared water – a treat for the harshness of the pain endured.

The conversation turned my way ... 'Keef el Arab?' As if to say: 'How is the Arab life with you?' 'Harr', I said. But I could not look them in the face and mumbled the word to the ground. 'Harr' carries a multitude of meanings but for the moment it seemed sufficient: 'hard', 'hot', 'severe' in every sense. They all agreed: 'Yes, it is hard, very hard.'

At this point, the old woman relaxed and lay outstretched on an '*angarib*' opposite. She smoked a (manufactured) cigarette and the conversation wended its way to a discussion of past circumcisions. Everyone agreed that bit Ali (Ali's daughter) had been done well. 'Samha khalaas' ... but there were problems with Sakeina's ... the midwife had come from another village and she had not taken enough. Perhaps it should be done again. But Allaweeya, the midwife, insisted that she would charge an additional LS 5.00 (£1.00) if she did it again.

When a girl is pharaonically circumcised in a village such as this, there is no anaesthetic, no sterilisation of the knife and no use of antiseptic to help heal the wound. Simply, the cut is made and the girl's legs are strapped together so that she can do no more than lie or sit with her legs outstretched for 15 days – waiting and hoping for the wound to heal. Many, of course, run the risk of contracting tetanus and other infections. In fact Ziyarra knowing full well the dangers at hand, came up to me a little later to ask for Dettol. This I gave her, and she was glad – greeting me with the word *mushkoora* (kind) when I brought a small bottle to her house later that day. For Dettol, like other disinfectants, is not available in the markets and is well beyond their means.

Several hours later, with time to sit and think about all that happened, I am struck by how little was actually said. One woman, who came a little late, greeted us all and congratulated Ziyarra with the words '*mabrouk 'alek*' – meaning 'congratulations to you' – no more, no less. Others simply greeted the gathered women as if it were any other occasion ... *salaam 'alekum* ('peace be upon you') etc, drank coffee and sat down to chat. So other than the moment when the old woman cut, and the gathered women grimaced at the pain, there was a sort of calmness about it all. I can't put my finger on it, but it somehow didn't seem wrong.

People clearly sympathised with the pain – for they themselves have all been pharaonically circumcised – and everyone was gentle with her. In fact, Ziyarra never left her daughter for the two hours or so that I was there. 'Mat guum, mat guum' ... 'Don't leave, don't leave', the girl kept saying, hanging her arms around her mother's neck. And the mother held her tight, gently assuring her that she would not go. Yes, she seemed calm and at ease with all that was going on.

If there was panic, fear and anxiety, then it was vivid and clear among those children who had sung outside the house and drowned the girl's cries. Many of these boys and girls (aged between 3–5 years) continued to cry intermittently for several hours afterwards. Incidentally, no men were in sight. They were all out working in the fields and when they returned they did not come to the house.

A little later I was pressed further about my thoughts on circumcision. But all that I could say was that it was not something familiar to my own culture and that I was not sure that I had understood what had happened. But just as I struggle to understand why it is necessary to remove a young girl's genitalia, so they appear mystified and astonished that the operation is not performed on girls in England ... How can a girl find a husband and make the transition to womanhood if she has not been circumcised?

How these women look on me now, I don't know. I hope it hasn't made much difference and that what matters is the things I do and the way I behave. And I ... do I see them differently? I'm not sure. I think I'm in too much of a state of shock to really think straight and I'm struggling to adjust to the normalness of events.

For just now I'm in Mohamed At tyib's house. It's 3.00 pm and nearly lunch time. A group of men sit outside the house discussing the price of grain in the market. Amna is washing clothes and Khadiga – still in her state of seclusion after giving birth to her second child – is cleaning some pots and pans. And me? I'm sitting in Zeinab's shop scribbling away. Yes – everything is as normal, but I don't feel at all 'normal'. And I don't feel able to discuss this morning's events with the women around me. That said, I am sure of one thing: there is an awful lot more to say about female circumcision than to state that it involves physical mutilation.

JOURNAL EXTRACT TWO: DECEMBER 1985

A month has passed and I have attended several other circumcision ceremonies in the village. A couple of times I did not respond in the expected way, as I did not congratulate the mother and I remained fairly quiet; and I was struck by how astonished the women seemed to be. Bakhritta, sensing that my interpretation of events was not hers, could not believe it: 'Don't you believe it's a good thing to do, Melissa?' And she said this with amazement rather than aggression. Similarly, Khadiga said: 'Don't you think it's fine and lovely?' – genuinely surprised that there could be any other interpretation.

Hannan, describing how a knife had been used to tear open her vagina before she had given birth (i.e. to deinfibulate her) also used the word 'good' to describe the practice. In common with other women she had been reinfibulated after she had given birth. While she did not deny the pain or any of the other problems which are frequently associated with circumcision, there was no question of it being anything other than *kwaiys* (good), *tahir* (pure), *nazif* (clean) and smooth (*na'im*). And I felt humbled. My questions were so useless, so utterly irrelevant to that which was seen as being important. Of course women do not circumcise their daughters to create problems for them later on. They do so to protect them. An uncircumcised girl is unmarriageable and would bring undying shame to her and her family. People would call her '*kaaba*' (bad), '*waskhan*' (dirty) and '*nigsa*' (unclean). Her life would be intolerable as she would be taunted by friends and relatives wherever she went. In brief, the practice of circumcision is bound to beliefs of honour, shame, purity and cleanliness. And it is these beliefs which need to be examined and interrogated if any headway is to be made in bringing an end to such a custom. It seems almost comical that Western and Sudanese feminists have spent so much time tackling it simply at a level of female oppression when it is rooted in so much else as far as those women who experience it are concerned.

these circumcision ceremonies and whether my presence gave tacit approval to something I found disturbing and abhorrent. On the other hand, I felt flattered, even honoured, to be able to attend such important events. I had no intention of betraying trust and confidences by deriding the practice of female circumcision. The fact that women portrayed female circumcision in a positive light merely drew my attention to how much I had to learn about women's lives in Omdurman aj Jadida. I thus continued to attend circumcision ceremonies but rarely took the opportunity to explore the issue in more depth. In this sense the relativism which imbued the way in which I carried out some of my work also became a means with which to avoid addressing a difficult and conflict-ridden area. Indeed, by the time I returned to England, I found it difficult to be critical or objective about social relationships in Omdurman aj Jadida and I felt much happier talking about life in the village in a relativistic way.

Not surprisingly, therefore, I was taken aback when I realised that the abhorrence of female circumcision and the reification of sexual enjoyment were widespread among my English friends and colleagues. In fact, I was shocked by the number of English anthropologists who felt able to describe female circumcision as a practice which was 'disgusting', 'revolting', 'abusive' and 'inhumane' without enquiring about the meanings attributed to this practice. One colleague even read the description of Nijat's circumcision and commented, 'General Gordon should have murdered the lot of them'. It was difficult not to resent the fact that very few people appreciated the importance of thinking about the issue of circumcision in terms other than physical mutilation and the denial of sexual pleasure. Their views became increasingly offensive and the confidence with which they espoused them was, it seemed to me, little short of racist. There were times when they appeared to amount to the following: circumcision is a barbaric practice. It is carried out by simple and uncivilised people. If they were sophisticated and educated like 'us' in the West, they would realise that there were new, different and better ways of behaving. The solution is simple. They should behave like 'us'.

The issue never died. In fact, the reaction of friends and colleagues to the issue of female circumcision and the desire to know more about the topic led me to the biomedical and anthropological literature. Here again I was struck by the ferocity of feeling expressed in a variety of academic journals and the narrow range of questions which research workers sought to address. The following section focuses on the biomedical and anthropological literature, with particular reference to research undertaken in Sudan, to illustrate these points.

BIOMEDICAL AND ANTHROPOLOGICAL RESEARCH ON FEMALE CIRCUMCISION

BIOMEDICAL RESEARCH ON FEMALE CIRCUMCISION

A wide range of terms are used in English to describe operations involving the removal of female genitalia. These include female circumcision as well as the more emotionally charged and judgmental terms 'female genital mutilation' and 'female genital torture'. In addition, researchers often employ a variety of different terms to convey the fact that the severity of the operation varies within and between populations according to ethnicity, geographical region and so on. For the purposes of this paper, the term female circumcision is used to describe all operations involving the removal of the female genitalia. Three different types of circumcision are referred to: sunna, intermediate and pharaonic. These distinctions were identified by Shandall (1967) and they are employed in this paper as they are increasingly referred to by researchers writing about the practice of female circumcision in northern Sudan.

The sunna type of circumcision involves the removal of the prepuce or head of the clitoris and the intermediate type involves the removal of the prepuce and glans of the clitoris with all or part of the labia minora. Pharaonic circumcision (which is also known as infibulation) involves the removal of the clitoris, the whole of the labia minora and most of the labia majora. The two sides of the vulva are then brought together with acacia thorns and held in place with catgut or sewing thread. Alternatively, the vulva are scraped raw but, either way, the girl's legs are tied together for 15–40 days until the wound heals and there is only a small hole (usually the size of a matchstick) to allow for the passing of urine and menstrual blood.

Whatever the terminology, it is striking that the presentation and interpretation of clinical and epidemiological data documenting the physical responses to different types of circumcision is greatly influenced by the intense emotions aroused by the subject. These emotions are sometimes made explicit by researchers in their academic articles. Allan Worsley, for example, wrote an article in the British Journal of Obstetrics and Gynaecology (1964) documenting, among other things, the consequences of female circumcision for the health of an unspecified number of women in northern Sudan. Much of the article wanders freely from the standard biomedical protocol and clearly demonstrates the depth of feelings aroused by the subject. At one point in his article, for example, he comments on the fact that men are never present at female circumcision ceremonies but then goes on to describe what happens at one of these ceremonies. To quote:

> The naked girl is laid across a bed, being securely held by the arms and ankles, while the midwife, with a deft sweep of the razor, removes the anterior two-thirds of one of the labia, together with the clitoris. The unfortunate girl's shrieks are drowned by loud shouts of 'That's nothing to make a fuss about!' – and the midwife proceeds to remove the other labium in the same way. There is always a sadistic smile of delight upon the face of the operator, and the whole business is thoroughly enjoyed by the privileged spectators.
>
> (1964: 687)

Worsley goes on to say 'One hopes that, with the passing of the older generation, this evil may cease to be the curse of a splendid and lovable race' (1964: 690).

It is, of course, difficult to gauge the extent to which the horror and outrage expressed by this particular gynaecologist continue to be shared by other gynaecologists and obstetricians. It is likely, however, that little has changed and Worsley's views are widely shared by biomedical practitioners and researchers from the Western world. Indications of the unchanging emotionality associated with the topic include the following: first, a general statement made by the Royal College of Obstetricians in 1982 referred to female circumcision as 'barbaric, futile and illogical' (Newsweek 1982: 55 quoted in Kouba and Muasher 1985: 101). Second, there is a widespread tendency among researchers to accept uncritically biomedical data documenting the detrimental effects of female circumcision for female health and well-being. This can be usefully illustrated with reference to some epidemiological research undertaken in northern Sudan by El Dareer (1982).

El Dareer's research documented the immediate and longer-term consequences of different types of female circumcision on the physical health of girls and women in northern Sudan. It was undertaken between 1977 and 1981 and a total of 3,210 women and 1,545 men were interviewed from five provinces in northern Sudan. Some of the most important findings include the following: first, over 98 per cent of women participating in the study had been circumcised (2.5 per cent with the sunna procedure; 12.17 per cent intermediate;

83.13 per cent pharaonic); second, 75 per cent of pharaonically circumcised girls had parents who had not received any school education (and the sunna and intermediate types were more likely to be undertaken in households where the women had received at least some education); third, more than 90 per cent of operations were performed by midwives who had not received any biomedical training; and, fourth, few women related the complications of circumcision to the operation, since it was generally believed to be harmless. El Dareer points out, however, that only twelve women agreed to have a physical examination and it was not possible to corroborate a lot of the information elicited from the interviews.

El Dareer's research is a substantial and useful piece of work, but it would be fair to say that there has been a blanket acceptance of her findings and these have been quoted and re-quoted in a variety of biomedical and anthropological books and journals (e.g. Sami 1986; Gordon 1991; Gruenbaum 1988). This is inappropriate as there are a number of reasons to be cautious about some of her data. Some of the most striking limitations include the fact that a considerable part of her research examines the relationship between the severity of the operation and the subsequent experience of infection and disease. Three types of circumcision are identified (sunna, intermediate and pharaonic), but it is not clear whether the interviewers explained their understanding of the differences between these types of circumcision to the study participants. It is possible that they did not and, if this is the case, it is most unfortunate, as the term 'intermediate circumcision' is not an indigenous category throughout northern Sudan. In those parts of Sudan where 'intermediate circumcision' is performed, it is possible that women had different ideas as to what this involved compared with the researchers.

It is also likely that many females do not fall neatly into these three categories. A gynaecologist working in Nigeria, for example, recently observed considerable variation in the severity of the operation among women from the same region and ethnic group. That is, some women had been excised while others had been partially excised but still had their clitorises intact (Murray Last, personal communication). It is reasonable to suppose that many females in El Dareer's study transgress the three types of circumcision identified in her study as a substantial number of operations were performed, without an anaesthetic, by midwives who had never received any formal biomedical training. It is also worth noting that it is difficult to cut with surgical precision when a girl is kicking and screaming and blood is flowing freely.

The difficulties of assessing El Dareer's research are compounded by the fact that many of the questions asked by field staff were rooted in biomedical conceptions of infection, illness and disease. Unfortunately, she does not tell us how, if at all, she set about translating biomedical terms into colloquial Sudanese Arabic and she does not convey any information about how participants may have interpreted these questions. This is unfortunate as a substantial number of the participants (43 per cent) had never been to school, let alone attended medical school, and it is not clear how they would have interpreted some of her questions. The following question is particularly awesome:

> Have you ever suffered from any of the following: keloid at site of circumcision; vulvar abscesses; inclusion cysts; recurrent urinary tract infection; chronic pelvic infection; difficult or impossible sexual penetration; pain during intercourse; difficulty in passing menses; infertility; vaginal deposits/stones; nervous troubles; none of these; others, specify; don't know.
>
> (El Dareer 1982: 113)

In sum, El Dareer's research has drawn attention to the large numbers of females in northern Sudan who have been circumcised and the circumstances under which the operation is performed. There are, however, a number of reasons to be cautious about some of her data, particularly those examining the associations between different types of circumcision, infections and diseases. The difficulties of assessing her findings are exacerbated by the fact that there is a dearth of other biomedical data enabling a detailed understanding of the effects of female circumcision on overall morbidity and mortality to be achieved. The fact that research workers have suspended their critical faculties and uncritically accepted el Dareer's findings suggests a desire to confirm their own perceptions of the 'true horror' of the practice at the expense of generating a detailed and rigorous understanding of the practice.

It should also be mentioned that biomedically trained researchers are by no means the only group of researchers to allow the strength of their emotions to distort the presentation and interpretation of biomedical data. Hayes, for example, wrote an interesting anthropological article entitled: 'Female genital mutilation, fertility control, women's roles and the patrilineage in modern Sudan' (1975). Here, again, the emotions aroused by the subject for this American anthropologist have undoubtedly impaired her ability to interpret data exploring links between infertility and female circumcision. Indeed, she is so convinced that the practice of female circumcision impairs fertility that she goes on to suggest that attempts to prohibit female circumcision would alter the rate of population growth. To quote:

> ... an analysis of the available literature on the subject, combined with the data I collected in 1970, shows that infibulation has functioned as one of several factors slowing population growth in Sudan. Its sharp curtailment could have serious demographic consequences unless other practices are introduced to suppress fertility.
> (1975: 619)

But it is not clear what 'available literature' Hayes is referring to. In 1996, let alone 1975, there were several references to a possible link between circumcision and infertility in a number of biomedical journals, but none of the authors refers to case material and an association between female circumcision, pelvic inflammatory infections and infertility is simply assumed.

ANTHROPOLOGICAL APPROACHES TO THE STUDY OF FEMALE CIRCUMCISION

The anthropological literature on the practice of female circumcision in different parts of the world is diverse and substantial. It is also limited and this can be usefully illustrated with reference to research undertaken in Sudan by anthropologists from the Western world such as Barclay (1964), Hayes (1975), Constantinides (1985), Gruenbaum (1982, 1988), Boddy (1982, 1989) and Kenyon (1991). In fact, this section suggests that the intense emotions aroused by the subject and illustrated in the preceding section reflect the influence of social movements in Britain, the United States and Canada. Unfortunately, these movements have affected anthropological writing to the detriment of generating a detailed understanding of the diverse meanings given to the practice.

In the 1970s, for example, the women's movement had a profound impact on anthropological writing. Feminists frequently emphasised the fact that female circumcision denied a woman's right to a full and satisfying sexual life and several popular and academic writers drew attention to the asymmetrical power relations between men and women (Daly 1978; Hosken 1982). Daly's article encapsulates attitudes which had a currency among radical feminists in the USA and Europe and the following passage is particularly illuminating:

> I have chosen to name these practices for what they are: barbaric rituals/atrocities. Critics from Western countries are constantly being intimidated by accusations of 'racism', to the point of misnaming, non-naming, and not seeing these sado-rituals. The accusations of 'racism' may come from ignorance, but they serve only the interests of males, not of women. This kind of accusation and intimidation constitutes an astounding and damaging reversal, for it is clearly in the interest of Black women that feminists of all races should speak out. Moreover, it is in the interest of women of all races to see African genital mutilation in the context of planetary patriarchy, of which it is but one manifestation.
> (1978: 154)

Anthropologists, while not necessarily aligning themselves with the women's movement, let alone the more radical element of the women's movement (such as 'SCUM' – the Society for Cutting Up Men), also emphasised these asymmetrical power relationships; and several writers felt no compunction about viewing female circumcision as a manifestation of oppression by men (Hayes 1975; Constantinides 1985).

In the late 1980s and 1990s, however, increasing publicity was given to the need to protect children from abuse and particularly sexual abuse in the Western world. A variety of papers began to discuss whether female circumcision was a form of child abuse and this is beginning to be reflected in anthropological as well as sociological and legal writing on circumcision (Le Vine and Le Vine 1981; Slack 1988; FORWARD 1989; van der Kwaak 1992).

The nature and depth of anthropological interpretations of female circumcision in Sudan vary a great deal, but, in common with many other anthropologists writing about female circumcision, these authors share one thing in common: sex is rarely, if ever, mentioned. In fact, anthropologists have tended, until recently, to avoid discussions about sexual behaviour and, therefore, the relationship between sexual behaviour and sexuality (Ortner and Whitehead 1981; Lindenbaum 1991). This reticence can be attributed to a variety of reasons, including unconscious anxieties and prudery on the part of the ethnographer (Devereux 1967); as well as practical, methodological and ethical difficulties of directly observing or obtaining reliable information on sexual behaviour.

With reference to circumcision, Lyons (1981) has also pointed out that the publication of the Rites of Passage (van Gennep 1960 [1909]) encouraged anthropologists to 'look beyond the 'genital' in genital mutilations and to see them in relation to other social and cultural forms' (Lyons 1981: 507). Whatever the reasons, a great deal of the anthropological literature on female circumcision in Sudan and other parts of Africa and the Middle East misleadingly suggests that circumcision is ' a physiologically trivial but socially important procedure mainly concerned with establishing clan membership and adult status' (Lyons 1981: 508). It thus appears that many anthropologists are reluctant to admit that ' the genital is not a nose ' (Vizedom 1976: 23).

The exception which proves the rule is Janice Boddy (1982; 1989). Her perceptive and insightful writing about female circumcision in a small village north of Khartoum suggests that women are not so much preventing their own sexual pleasure by removing their external genitalia (though this is presumably an effect) as enhancing their femininity. It is an assertive and symbolic act, controlled by women, which emphasises 'the essence of femininity: morally appropriate fertility, the potential to reproduce the lineage or to found a lineage section' (1982: 696).

The apparent dissociation between femininity and sexual pleasure among women in northern Sudan is challenging and provocative. Nevertheless, it is not at all clear how Boddy's interpretations relate to the historical spread of female circumcision in Sudan and to more recent observations about the tendency among some ethnic groups inhabiting the southern borders of northern Sudan to circumcise their daughters. These groups appear to be adopting the practice of pharaonic circumcision as a way of affirming their identity with the dominant Arab and Muslim populations of the north. In short, anthropological writing about female circumcision in Sudan is curiously uninquisitive about exploring many of the ethnic, religious and social dimensions of the practice which, in all probability, are enormously significant to the people concerned.

PRESENT REFLECTIONS

More than ten years have passed since attending Nijat's circumcision ceremony. Where has the reading of the literature and reflections of past events taken me? If ambiguities in my own position remain, it nevertheless seems clear to me that intense emotions often underlie popular and scholarly discussions about female circumcision. Among Western researchers, this often prohibits rigorous and detailed discussions of this topic. The question which flows from this observation is: why does female circumcision generate such powerful and emotional responses? There are no simple answers to this question, as researchers are influenced by different ideas, outlooks and social movements over time. Nevertheless, several trends can be discerned among research workers from the Western world. First, it is possible, if not probable, that some of my reactions to witnessing the removal of female genitalia in Omdurman aj Jadida convey a great deal about European and North American conceptions of sexuality, including a tendency to define the self in sexual terms.

Readers will recall, for instance, that a few hours after attending Nijat's circumcision, I walked across the village, without a female chaperon, to give some Dettol to the young girl's mother. She had been cut with a blunt and contaminated razor blade and I 'knew' the consequences could be fatal. Several years later, however, it dawned on me that the bluntness and contamination had been a fiction of my mind. I had only seen the razor blade at a distance and I could not, therefore, have had any way of knowing whether it was in fact rusty and blunt. This simple 'error' draws attention to the powerful associations I had made between pharaonic circumcision, the death of the young girl's sexual life and the denial of pleasure integral to her future well-being. Indeed, the distress and anger aroused by the event had been so powerful that I had, unwittingly, exaggerated the risk to her life.

It is likely that frequent, but undocumented, references to mortality from circumcision by Western journalists and academics also convey an association between circumcision, sexual death and the denial of self, rather than death itself. A recent article about female circumcision in the *Guardian* (25 April 1994), for example, says: 'Death from blood loss or shock is not uncommon'. The association is, however, assumed, as Mohamud (1991) is the only investigator to have examined the impact of female circumcision on mortality and he was unable to confirm this relationship.

The association between female circumcision and the death of a female's sexual life is not surprising. Several academics (Duffy 1963; Sheehan 1981) have shown that American and European surgeons in the nineteenth century sought to cure physical, mental and moral disturbances by practising clitoridectomy. Masturbation, for example, was considered to be a physical as well as a moral evil, and its practice was perceived to cause specific illnesses. Clitoridectomy was thus advocated as a 'cure' by many surgeons, as it was unthinkable that decent women should derive pleasure from sex.

Most of the research referred to in this paper was undertaken between 1960 and 1990 and there is no doubt that the social changes which have taken place during that time have profoundly influenced the responses of Western (and probably many non-Western) researchers. In particular, the 'sexual revolution' which took place in the 1960s implied a separation of sex from reproduction. It involved an increasing willingness to talk about sex publicly as well as the widespread availability and use of contraceptives such as the pill and condoms. These changes contributed to the emerging tendency to define the self in sexual terms.

Subsequent political and social changes in the 1970s and 1980s reinforced this tendency. The 'sexual revolution' was increasingly seen to be ' ... a revolt of young men. It was about the affirmation of young men's masculinity and promiscuity; it was indiscriminate, and their sexual object was indeterminate (so long as she was a woman)' (Campbell 1987: 21). The women's liberation movement (and a variety of other political movements) emerged, in part, as a critique of the sexual revolution. They focused on the quality of the sexual act and hence, perhaps, a shift from the vagina to the clitoris, in the representation of women's sexuality. Indeed, if Hite's investigation of female sexuality in North America is anything to go by, female orgasms by clitoral stimulation became a pre-requisite of 'good' sexual intercourse and, even, 'good' sexual relationships. To quote: 'There is a social pressure [in North America] that says a woman who has an orgasm is more of a woman, a "real" woman' (Hite 1992: 131). Similarly, Germaine Greer, drawing attention to the oppressive pressures placed on women, has written: 'The state of being inorgastic is sometimes described as being out of touch with oneself ... or not into one's body ... women in this contemptible state feel as much guilt as once they felt for experiencing spontaneous sexual desire' (1984: 201). Against this background it is understandable that the removal of female genitalia, an integral part of a woman's being, is such an emotional issue.

It also reflects a broader tendency which

affects men as much as women. In modern Western society, one's sexual orientation is a very important part of one's identity. It has been suggested that this behaviour may be essentially because 'gender here is not so much ascribed on the basis of physiological sex as of *achieved*. People are encouraged to see themselves in terms of their sexuality, which is interpreted as the core of the self' (Caplan 1987: 2; Brake 1982). It is partly for this reason that homosexual relations may be so vociferously stigmatised. They are not just to do with individual choice, but threaten heterosexual norms which link notions of a proper person with the culturally regulated expression of 'natural' sex drives. This perhaps also helps to explain the abhorrence of female circumcision. Many men, especially highly educated and sensitive 'new men', define their sexuality partly in terms of producing orgasm in others.

Thus female circumcision generates fears and anxieties about castration, not just for the women whose genitalia have been removed. A man is not a 'real' man unless he can sexually satisfy a woman and the chances of a man sexually satisfying a woman (that is, enabling her to become a 'real' woman) are much diminished. These fears may well have been enhanced by the declining birth rate in Europe, North America and Canada and the fact that masculinity is no longer so closely tied to reproductive prowess. In other words, female orgasm by clitoral stimulation has taken the place of procreation as the manifestation of male virility. Given the male dominance in biomedical science and practice, it is perhaps hardly surprising that this emasculation affects interpretations of female circumcision – arguably even by women who, while they may purport to be presenting a feminist case, may, in this crucial respect, be caught up in the cultural hegemony of male dominance they so resent.

NOTE

1 All names in the text have been altered to maintain confidentiality.

19

STREET LIVES AND FAMILY LIVES IN BRAZIL

Tom Hewitt and Ines Smyth

INTRODUCTION

Perceptions of street children are mostly based on notions of 'abandonment'. It is the absence of a home and a family which defines such children. But interestingly enough, while notions of childhood have been explored in trying to understand the causes and consequences of this social phenomenon, those of the family have not, though it is presumably by the family that such children have been abandoned.

In this paper we try to show how most reactions to street children are tinged by two, often coexisting, emotions: pity and fear. The paper will also indicate how the dominance of these two sentiments affects the actions which are undertaken to ameliorate the situations under which so called street children live. The paper claims that the dominance of these two emotions can be traced back to culturally constructed notions of the family.

Where possible, reference will be made to the Brazilian situation, since it has attracted considerable academic and media attention.

ABANDONED CHILDREN AND THE FAMILY

There are at least two categories of children who are normally referred to as 'street children' (see Figure 19.1). One is working children, who may be more visible than others because of their engagement in productive activities and thus could be called children 'on the street'. It has been estimated that up to one third of 5–15 year-olds in Latin America are economically active (Ennew and Milne 1989). Of these, only some are street children and even fewer could be described as abandoned children.

Children on the street are mostly visible in cities. For example, since around 75 per cent of Brazil's population is urban, it is not surprising that the majority of the country's working children live and work in cities. This is not to say that child labour does not exist in the countryside. On the contrary, rural child labour, often in slave-like conditions, has been well documented (Whittaker 1987; Lee-Wright 1990). Children work on plantations, in logging and mining. Official statistics show that four and a half million Brazilian children worked in agriculture in 1975, 68 per cent of whom worked more than a forty hour week, and there is no reason to suspect a reduction in such figures. But it is in urban agglomerates that there is a high incidence of child labour. The evidence seems to indicate that this has increased during the recession years of the 1980s (Myers 1988).

However, it is important that we do not fall into the trap of assuming that children who work are separated from or abandoned by their families. On the contrary, much of the work children perform is in their own homes and for their own families. White (1994), for

Figure 19.1 Diagram to show the complexities of children's lives on and off the street in Brazil.

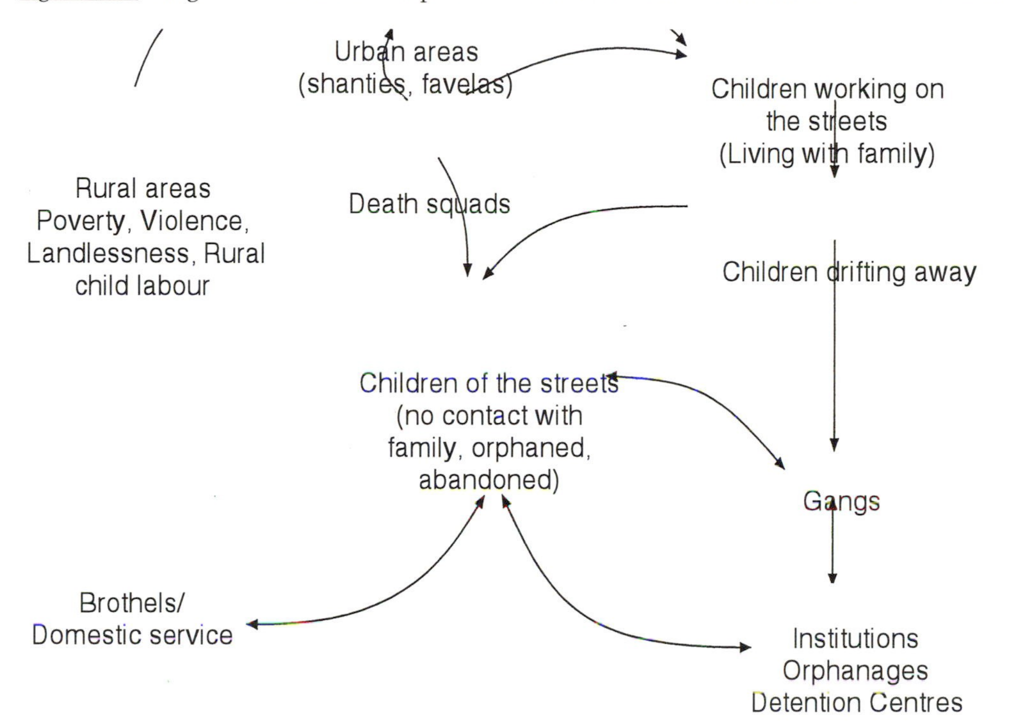

Source: Based on Richman 1994.

example, argues that the work children perform in the domain of the family can be much more exploitative than other kinds of work, and that it is often unpaid and instead of, rather than in addition to, school.

The second, smaller, group are children 'of the street' who have no contact with families and relatives and have no shelter. Perhaps this is the only category which should rightly be termed street children. The journalist Jan Rocha describes the conditions of street children in Brazil:

> Only a minority of Brazil's 7 or 8 million street children are totally on their own – orphaned, abandoned or without any contact with their parents. Most maintain some contact, however tenuous, with their family. On the streets, home is a shop doorway, a bench in a square, a hot air duct outside a restaurant, a bonfire on the beach, the steps of a railway station. Bed is a piece of cardboard, an old blanket, newspapers. Some sleep alone, others huddle together for warmth or protection. They never know when they might be woken up by a policeman's boot, a jet of cold water from a street cleaning truck, or even a bullet from a vigilante group or gun-happy officer of the law. ... During the day, the street children's main concern is survival – food. To get it they beg, pick pockets, steal from shops, mug tourists, look after parked cars, shine shoes or search litter bins. Frequently glue takes the place of food. They sniff it from paper bags and for a few glorious moments they forget who or where they are.
>
> (Rocha 1991: 2)

But while this description applies to far too many children and young people in Brazil and in other cities across the world, it would be greatly mistaken to see it as the reality of all the children who either work or spend some or even most of the day in the streets and squares of cities.

The distinction made above between children of the street and children on the street is an important one. However, in the popular perception, these distinctions are blurred in the uniform category of street children.

This is linked to a broader notion of

childhood. With notable exceptions, a persistent feature of both international initiatives for, and media coverage of, street children has been that they are led by western[1] perceptions of childhood. That is, for example, in most countries of the west, adults expect their children to be educated in schools at least until their mid-teens, they do not expect children to need to seek employment (beyond the paper round type) before this age and they expect children to have a 'home' to go to whether in a family or an institution. Such notions of childhood, which persist despite the growing incidence of young homelessness in Europe, make it difficult to accept other realities as anything but wrong.

However, the notion of 'childhood' itself is not unproblematic, as Ben White warns:

> 'Childhood' itself is not a universal or absolute category; its definition varies from one society and from one time to another, and also according to both class and gender. Research must therefore deconstruct the category of childhood. ... by identifying the social, economic and political factors contributing to its changing definition and to the activities defined as suitable for children.
>
> (White 1982: 468)

While the notion of childhood has attracted attention, it is less so for that of the family, at least in this context.[2] In many western countries, sociological thought and the shared culture of people often coincide in perceiving the family as a 'haven from a heartless world'. Functionalist notions of the family have been challenged by feminists of various persuasions, mainly because they ignore the fact that families are highly hierarchical, ruled by power relations, and often violent (Firestone 1974; Counts *et al.* 1992; Davies 1994). Furthermore, anthropologists have emphasised that the nuclear family typical of western countries is neither universal nor unchanging (Moore 1988). Finally, and more controversially, it has been argued that 'selective neglect accompanied by maternal detachment' towards certain infants (frequently leading to their abandonment or death) is widespread in some poor communities (Scheper-Hughes 1987, 1995). Despite these challenges, those who perceive street children as abandoned rely on the cultural construct of the family as a safe and nurturing place, where children are socialised into the dominant norms of society (Allen and Barker 1975).

The insistence on perceptions which see street children as having been abandoned by their families ignores the fact that:

> families may be other than safe for children who belong to them
>
> children are able to take their own decisions, including that of forsaking their family
>
> the socialisation of children may take place through other institutions.

THE FAMILY AS A PLACE OF SAFETY

As said earlier, feminists have been at the forefront of those contesting the cultural image of nurturing and protective families. World-wide, statistics on the incidence of physical, psychological and sexual abuse of women, children and the elderly are showing that families can be and are extremely dangerous places (UN 1989). Furthermore, the evidence presented by Scheper-Hughes from her work in the Brazilian Northeast suggests that 'the violence of everyday life' manifests itself in the selective neglect of children by mothers making the 'circulation of babies through informal adoption or abandonment' a commonplace (Scheper-Hughes 1995). Such families, it is argued, live:

> in an environment in which loss is anticipated and bets are hedged. 'Mother love', with its attendant emotions of holding, keeping and preserving, is replaced by an estranged and guarded 'watchful waiting'. What makes this possible is a cultural conception of the child as human, but significantly less human than the grown child or adult.
>
> (Scheper-Hughes 1995: 204)

CHILDREN CAN MAKE DECISIONS

Children in families are often the least possessed of social power. When this is combined with poverty and destitution, the options opened to them are few. Amongst these options, children may choose to leave their families.

So for some 'abandonment' may be a better, or the only, option. This is not necessarily abandonment of children by parents, although this certainly happens. Children make the rational decision to leave. Adults unable to provide for children have little control over the situation. It may not be a one-off decision, but something which develops over time. In the process, children will build up multiple sets of allegiances (e.g. commonly to gangs) or will be controlled by relations of exploitation (e.g. as apprentices, factory workers or domestic servants). In short, children may decide that they will be better off if they go it alone or make an allegiance to a gang on the streets than if they stay with their family or relations. The street can offer freedom and escape from abuse, extreme poverty and neglect, loneliness and seclusion.

THE FAMILY AS SOCIALISER

Families, we have said, are also perceived as socialising units. The family introduces children to the larger world, teaching them dominant norms and rules of behaviour. This perception is particularly strong in urban areas of western and northern countries, where the physical separation from the street has emerged with changes in patterns of productive activities and with urbanisation, and often has been made necessary by hazards of traffic, lack of safe open spaces and, in some circumstances, the fear of abduction. This is so even if many of the socialising roles the family used to perform, primarily that of education, have been taken over by the state (Nieuwenhuys 1993: 4).

In many developing countries, the boundaries of the family are much more fluid, and children belong to, and hence are socialised by, kin and non-kin drawn from a much broader pool of people. Furthermore, in rural areas, the street remains the place where they are socialised into the norms of their community. But in dominant perceptions, the family and the home are still the best institutions for the nurturing and socialisation of children. Children who have been 'abandoned' do not enjoy these benefits and are, as a consequence, to be pitied and feared: the needy poor are to be pitied; the delinquents have to be feared (De Swaan 1988).

PITY

Pity for children runs strong in western societies. Charities use this to great effect. For many years, fund-raising by charities has been based on images of children in distress. At the extreme, images of children suffering famine are used to engender pity for those in need. They have been relatively successful in raising funds for famine relief, but have done little to give a fuller picture of the lives of the children they portray.

Street children attract similar, and more complex emotions. Films of street children such as *Pixote* and *Salaam Bombay*[3] to some extent reflect the conditions of life of street children in Brazil and India, but their appeal to European audiences has been to engender pity for those in need (of a childhood) rather than recognition of the more immediate needs (of food, shelter, friendship and hope). Mixed in with this is a sense that street children enjoy a freedom (from adult power) that is not fitting for people of this age. An additional complication in the response to street children which engenders further pity is the idea of lost innocence. Once outside the nurturing family, children are forced to grow up unnaturally quickly.

Whether charitable and voluntary organisations such as non-governmental organisations (NGOs) intend it or not, the way street children are portrayed is that they are victims of circumstances and deserve help. This view is accurate but limited on two counts. First, it is caught in notions of what childhood and the family should be, rather than what they are for many of the world's children. In many western countries, popular culture reflects a shared imagery of innocent childhood. In nursery rhymes and stories, children often fall prey to dangers from which they escape unharmed exactly because of their 'innocence' (so, for example, Little Red Riding Hood, Hansel and Gretel). This is reflected in contemporary English language where, for example, the adjective childlike is said to mean 'having the good quality of a child, simple and innocent' (*Oxford English Dictionary* 1987).

Second, there is an unspoken assumption that being on the street is the worst of all options available to children. How children got to be on the streets in the first place and the options available to them are little discussed. Still less are the views of the children themselves taken into account, such as of the eight-year-old in the streets of Rio who says: 'I live here because my stepfather doesn't like me. He drinks cachaça (sugar-cane rum) and rows with my mother. He used to hit me, pull my arm, pinch me ... ' (Rocha 1991: 3); or the teenager who says:

> I was thirteen when I went to live on the streets because my mother didn't have the means to support me. She used to go out to sell fish and leave me in the house all day, and I used to spend the whole day very very hungry. Then I made a lot of friends and went out stealing. They'd nick things and I'd be the lookout. Then we went home and divided up the stuff and sold it.
>
> (Dimenstein 1991: 23).

The more progressive among the NGOs are applying to children, including street children, new approaches to social development. Here empowerment is the process which all individuals can undergo to take control over their lives and circumstances. Despite the problems which some see in this approach (Ennew 1994), at least it does recognise children as individuals able to act rationally.

FEAR

At the opposite end of the spectrum of perceptions of street children lies the notion of the 'marginal' or vagabond. In place of the innocent child or the deserving poor we find the dangerous delinquent.

This is a view which has a long history. In the Middle Ages in Europe and the early days of the Russian revolution, gangs of roaming children were a daily sight (Agnelli 1986). From the time of the Black Death in Europe, 'vagabondage' was a route from serfdom to liberty. Later in the seventeenth century, many wandering people with no means of subsistence came to carry the label: 'ex-soldiers, pedlars, jugglers, palm-readers, drovers, tramping artisans, freemasons, begging university undergraduates ... ' (Hill 1996: 49). Many of these 'vagabonds' were youngsters. Still later, during the Industrial Revolution in Britain in the nineteenth century, street children were a commonplace in cities. The perception of danger and thus the feeling of fear was most acute from those who felt such wanderers to be a threat to their livelihoods. As De Swaan notes, 'The problem of the poor is to stay alive; the problem of poverty is a problem for the rich ... ' (1988: 14).

Ironically, it is those close at hand who are more likely to take the view that children are a threat. From a distance, it is easy to recognise that street children are not getting the best deal in life and to understand how petty crime is one of the very few options available for survival on the streets. Pity at a distance can rapidly turn into fear at closer quarters. When a perceived threat of loss of property and/or violence against individuals overcomes sympathy, then the way that street children are viewed becomes quite different. Street children become a social problem to be alleviated not because the children themselves deserve a better chance in life but because they are a threat to the functioning of the rest of society. It is a short step then to considering street children as outside society and therefore not eligible for the same rights and respect that other citizens expect. Once street children are seen to be on the margins of society, the reactions to them can also become less than human. Violence (social and physical) has become a commonplace response to street children in Brazil.

Both reactions – that of pity or that of fear – are directed at an amorphous mass of 'street children' irrespective of their individual circumstances or of their relation to the life of the street.

RESPONSES TO STREET CHILDREN

We have examined the dominant perceptions of street children and the emotions they elicit. Those who pity street children would argue for removing them to the 'safe haven' of the family, and those who fear them will remove them in other ways (through incarceration, and increasingly through other violent means). Neither response necessarily reflects the needs of the child or addresses the circumstance which led the child into the street in the first place. Responses to street children, to put it crudely, appear to be more a function of the responders' needs than anything to do with children's needs.

The diverse nature of the reactions to the growing numbers of children in Brazilian cities provides examples of both kinds of response. The role of the state in relation to street children in Brazil has been the most dramatic and perhaps the most ambiguous. From the end of World War II until the end of the 1980s, the main national guidelines for 'assistance' for children in Brazil included the following two features (Galheigo 1996):

> The Minors' Code. This legislation stayed in effect until 1990. Both its first version of 1927 and its revised version of 1979 were strongly influenced by the adults' penal code and systematised the coercive actions which the State should adopt in the regulation of offences.
> The establishment of institutions for the containment of 'delinquent and abandoned' children and adolescents.

Set up in accordance with the same logic as the adult's penitentiary system, such institutions were the main and most widespread governmental initiatives for assisting children. Formally established at national level in the 1940s, they had as a principle (made clear in the Minors' Code) that 'minors' (delinquent or abandoned) needed to go through a process of resocialisation based on coercion so that their 'distortions' might be straightened out and that they might then be re-integrated into society. In 1964, with the establishment of the so-called 'National Policy for the Well-being of the Minor', the military government in power sought to diminish the coercive action, putting emphasis instead on the adoption of assistance programs. Nevertheless, as the national body for the control and orientation of this 'new policy' was the National Foundation for the Well-being of the Minor (FUNABEM), which was in charge of collecting and containing the 'minors', the same repressive approach was kept.[4]

In the 1980s there was a remarkable, if slow, change in the Brazilian laws concerning children. The Child and Adolescent Statute became law in 1990. Essentially the Statute turned children from 'minors' with no rights to 'citizens' with rights.[5] It was the result of intensive lobbying and a broad coalition of Brazilian NGOs, children's rights activists, lawyers and international organisations. It was also part of a broader process of democratisation in the country after more than 20 years of military rule. In the period since the early 1980s, political reforms and the growth of civil society created an environment which was more conducive to these changes (Paoli 1992).

Of course, legislation is only one step, and the experiences of children still may have little resonance with what is down on paper. The fear of the delinquent seems to lie more in civil society than (at least officially) in the state. The public reaction to street children in Brazilian cities has taken on gruesome proportions. The number of children killed by 'death squads' in Brazil reached an average of approximately one a day in 1989. Estimates are that there are over 400 murders a month of all types in Rio de Janeiro alone, of which a significant proportion are of children. This compares with 160 murders a month in New York and 15 in London (Dimenstein 1991).

The death squads and their documented connection with the police force are not new. During the twenty years of military dictatorship from 1964, death squads operating against 'political' targets were common. What is new is the increasing numbers of young so-called marginals who are becoming the victims of death squads.

In 1990, the human rights organisation, Amnesty International, published a report on the violent activities of the police and 'death squads'. It documents a level of brutality which defies belief; many cases of arbitrary arrest, torture and murder. Some 80 per cent of violent deaths are of males aged between 15 and 18. A growing number of murder victims are children under the age of 15 (Amnesty International 1990).

A common theme of these and other publications is that criminality is on the increase while the judicial system is in crisis and cannot cope with what appears to be a massive social problem. The police and, increasingly, ordinary citizens have taken the law into their own hands to 'clean up the streets'. It is scepticism over conventional means of controlling the threat to social stability which explains the apparent contradiction between the emergence of favourable legislation and an increase in violence against street children. There is a chilling case from Rio de Janeiro where a boy was strangled and a note left on his body which read: 'I killed you because you don't go to school and have no future'. What is behind these private death squads? The long-time campaigner, Bishop Mauro Morelli, Bishop of Duque de Caxias in Rio de Janeiro, explains it this way:

> In my understanding, the killing of children in our streets, behind the killings there is a concept of cleaning, these children make the place disgusting and dangerous. Children running in the streets are seen as a danger to society and to the tourist business. The killings are done by what we call extermination groups, formed by policemen and bandits and paid by businessmen and industrialists who are looking for safety and protection of their own goods and property.
>
> (cited in Hewitt 1992: 55)

In other words, fear seems to be the motive. Fear of theft and of damage to person and property. In recent years the perception that street children are dangerous criminals has spread. As Valdemer de Oliveira Neto, a human rights campaigner, explains:

> The death squads say they are cleaning up the city, helping society by killing criminals. But there is another aspect, the population supports the death squads' activities. Criminality is so big that the population has no other way of getting security.
>
> (cited in Hewitt 1992: 55)

Such fear is breeding increasingly violent responses. Just as the government programme of institutionalisation was an attempt to throw perceived dangers behind locked doors, so too the public response to street children has been to strip them of their humanity and then stamp them out.

Fortunately, there is more to the story. As a reaction to the state of affairs, many small but significant initiatives have got off the ground in recent years. They are frequently run by a small number of individuals on a shoe-string budget. A few examples will give a feel of the possibilities. While the emphasis of government policies has been on institutionalisation and containment, non-governmental initiatives, frequently with the backing of international agencies such as UNICEF, have emphasised a community-based approach. This has involved both emergency help such as food and shelter, but also education projects and work schemes. These latter two are of interest because they function around the existing activities of street children rather than trying to remove them to new surroundings.

Recife, which has one of Brazil's largest concentrations of street children, is in the poor northeast region of Brazil. Here the incidence of violence against children is perhaps the worst in the country. A lawyer, Ana Vasconcelos, appalled by the lives of girls as young as ten who were forced by poverty to become prostitutes, opened the Passage House. The house, which can only cope with some sixty of Recife's estimated 30,000 street girls, provides temporary shelter and food but also counselling training and basic information.

The Salesian Centre for Minors, sponsored by the Catholic Church, was aimed at shanty towns in general but in practice 80 per cent of its participants are street children. The Centre acted as an employment agency providing messenger boys, carriers, market packers and so on to different businesses. As a result nearly 1,000 youngsters have full working documents, rights and benefits, plus an official minimum wage. These are all unusual not only for working children but also for many adults working in the informal sectors. As one observer has noted on this project:

> By its insistence on dignified wages and working conditions from the companies, and on high quality work and dependability from the [children], this programme demonstrated that it is not necessary to accept either economic exploitation or condescending charity as the price for creating employment for poor youngsters. The model was sufficiently replicable that the programme expanded its work to several Brazilian cities.
>
> (Myers 1988: 130)

Nieuwenhuys (1993, 1994) notes a shift in the activities of NGOs working with street children from charitable concern to the encouragement of political mobilisation, participation, self-reliance and empowerment. Organisations which have their roots in the popular movements of the 1960s and 1970s (particularly in Latin America) have been a catalyst for new interventions by NGOs based on more than charitable concerns. One important outcome has been the recognition that street children want to take more control of their own lives through working (earning) and through taking their own decisions.

A Brazilian example of this new approach is the National Movement of Street Boys and Girls (MNMMR). It is an autonomous non-governmental popular organisation, founded in 1985, composed of volunteers, and aimed at supporting children and adolescents in their own struggle for securing and defending their rights as citizens (SEJUP 1994).

One of the important issues to be raised by MNMMR is that of self-advocacy, in other words, the idea that children are not just the passive accepters of 'help' from adults. The movement, itself led by youngsters, promoted a national conference of street children from all over Brazil in the capital Brasília in 1986. Four hundred and thirty-two delegates aged between 8 and 16 years of age spent three days sharing their experiences of work, education, violence, family, political organisation and health. Themes were chosen by the children and covereed education, their exploitation as workers and violence within their families and at the hands of the police (Fyfe 1989; Brazil Network 1992).

Despite the activities of many non-governmental organisations, it is the National Movement of Street Boys and Girls which brought international attention to the problems of street children and thereby jolted the authorities into action. In the last few years there has been unprecedented media coverage of street children both in Brazil and internationally. However, as Jan Rocha notes in the introduction to Gilberto Dimenstein's condemning book, *Brazil: War on Children*:

> These are all good signs, but the indignation that should be caused by the killing, rejection or abandonment of any child is still lacking. Action is being taken because the world has thrown up its hands in horror, not because that horror is shared by the authorities, the congressmen, the judiciary, or the general public. Furthermore, the reasons behind the existence of so many street children are still only being timidly discussed.
>
> (Rocha 1991: 15)

CONCLUSION

This chapter has tried to show – with examples from Brazil – the way in which western cultural constructions of childhood and of the place children occupy within the family, give rise to an undifferentiated notion of 'street children' and of abandonment. Furthermore, the pity and fear these same constructions generate influence many of the responses to children. Despite recent interventions of a different nature – for example from NGOs – the persistence of these perceptions prevents the emergence of solutions which take on the root causes of the problems of street children.

NOTES

1 In recent years, with the belief that 'the idea of development stands like a ruin in the intellectual landscape' (Sachs 1992a: 1), the key terms of the debates have also been questioned. Developing and developed countries, the North and the South, are unsatisfactory solutions to the problem of communicating difficult and contradictory notions. Western is an equally deficient term, being, as it is, nebulous and residual. Here it is intended as a shorthand for those industrialised countries which share a broad similarity of cultural meanings.
2 A recent manual on working with street children (Ennew 1994:36) emphasises searching questions you should ask yourself about your perceptions of children and about the children you will be working with before embarking on any project.
3 Directors Bebenco (1981); Nair (1988).
4 Note that legislation was almost exclusively concerned with delinquent children. The state turned a blind eye to child labour (and continues to do so as far as we can tell).
5 The process whereby the Statute came into being is well documented in Fausto and Cervini (1992) and Swift (1991).

VII

RELIGION, CULTURE AND POLITICS

20

RELIGION AND POLITICAL TRANSFORMATION

Jeff Haynes

INTRODUCTION

Religion has had considerable impact upon politics in many regions of the world over the last few decades. The belief that societies would inevitably secularise as they modernised has not been well-founded. Technological development and other aspects of modernisation have left many people with a feeling of loss rather than achievement. By undermining 'traditional' value systems and allocating opportunities in highly unequal ways within and among nations, modernisation can produce a deep sense of alienation and stimulate a search for an identity that will give life some meaning and purpose. In addition, the rise of a global consumerist culture can lead to an awareness of relative deprivation that people believe they can deal with more effectively if they present their claims as a group. One result of these developments has been a wave of popular religiosity, which has had far-reaching implications for social integration, political stability and international security.

This chapter provides a global perspective on the relation between religion, politics, conflict and identity, although much of the analysis is concerned with the Third World. Using a range of cases from various parts of the world, it examines the complex ways in which religious values, beliefs and norms stimulate and affect political developments and vice versa; the social conditions which give rise to religious movements as well as how such movements are promoted and sustained over time; the relations between religious leaders and followers; and the links between social mobilisation and the pursuit of particularist objectives.

The chapter contends that the defining characteristic of the relationship of religion and politics in the 1990s is the increasing disaffection and dissatisfaction with established, hierarchical and institutionalised religious bodies. Contemporary religious movements seek instead to find God through personal searching rather than through the mediation of institutions. They also focus on the role of communities in generating positive change in members' lives through the application of group effort. In this regard, the chapter argues that religion's interaction with political issues carries an important message of societal resurgence and regeneration, which may challenge the authority of political leaders and economic élites.

The first part of the chapter provides an overview of the relationship between religion and modernisation. It surveys the contradictory effects of modernisation on social values in different cultural and religious settings. Given the uneven impact of modernisation in Third World countries, the relationship between religion and politics has always been a close one. Political power is underpinned by religious beliefs and practices, while political concerns permeate to the heart of the religious sphere. As a result, attempts in many countries to separate politics from religion have largely been unsuccessful, especially as economic crises and global restructuring undermine previous arrangements for promoting social and political cohesion.

The second part of the chapter examines the political significance of identity in the context of religion. Precisely how religious conflicts relate to development is not clear, although, as Hettne (1995: 6) suggests, it is highly likely that 'the differential outcome of both growth and stagnation (has) an impact'. What is clear is that social, political and economic change in the Third World comes about as a result of a complex interaction of both domestic and external developments over time. The contemporary political importance of religion is an integral facet of wider 'identity crises', i.e. serious threats to national integration and to the process of economic development, which focus on the existence of 'cleavages' within societies. The concept of a cleavage refers to the alignment of the population around social dimensions which are conducive to conflict.

The final part of the chapter presents a typology of religious movements in order to demonstrate the political significance of religion as a global phenomenon. Four types of movements are highlighted according to whether religion is used as a vehicle of opposition or as an ideology of community development. Groups which link religion to the pursuit of community development are categorised as community-orientated, while oppositional movements are classified as culturalist, fundamentalist and syncretistic. Threats from powerful outsider groups or from unwelcome symptoms of modernisation largely sustain the oppositional movements; community movements, on the other hand, derive their *raison d'être* from state failures in social welfare development.

RELIGION AND MODERNISATION

One of the most resilient ideas about societal development after World War II was that nations would inevitably secularise as they modernised. The idea of modernisation was strongly linked to urbanisation, industrialisation and to an accompanying rationalisation of previously 'irrational' views, such as religious beliefs and ethnic separatism. Loss of religious faith and secularisation dovetailed with the idea that technological development and the application of science to overcome perennial social problems of poverty, environmental degradation, hunger, and disease would result in long-term human progress. With the decline in the belief in the efficacy of technological development to cure all human ills came a wave of popular religiosity with political ramifications. Examples include: the Iranian Islamic revolution of 1978–80; Christian fundamentalists' involvement in political and social issues in the United States; the recent growth of Protestant evangelical sects in Central and South America which helped to elect two 'born again' presidents in Guatemala; internecine conflict between Hindus and Muslims in India, between Buddhists and Hindus in Sri Lanka, and between Muslims and Christians in former Yugoslavia; the emergence in India of Sikh separatists in Punjab and of Muslim militants in Jammu–Kashmir; religious syncretistic groups in sub-Saharan Africa and elsewhere whose aim is community protection; and the impact of Jewish fundamentalist groups on Israel's political configurations, especially in relation to the Arab Palestinians.

To analyse and explain this wave of apparently unconnected developments we need to confront at the outset an issue consistently ignored in political analysis: how do religious values, norms, and beliefs stimulate and affect political developments and vice versa? For example, historical analysis would point to the close relationship over time between the top hierarchy of the Roman Catholic church and successive less-than-democratic governments in Latin America; yet over the last 20 years (i.e. during periods of dictatorial rule), some Church officials emerged as champions of democracy, vocal in opposing military dictatorships. Senior members of the Roman Catholic hierarchy, on the other hand, retained their roles within the ruling triumvirate along with senior military figures and big landowners and capitalists. How do we explain the contemporary divergence of views between senior Catholic figures and many priests on the ground in Latin America? A similar process occurred among followers of

Islam throughout the Muslim world (i.e. some fifty countries stretching from Morocco to Indonesia). Senior Islamic figures remained close to secular rulers, while political challenges to the status quo were led and co-ordinated by lower- and middle-ranking Muslims. A similar type of schism was observable in Thailand and Myanmar (Burma), where senior Buddhists were often supportive of military(-supported) regimes, while junior figures attacked them for their corruption and political incompetence. A common denominator in these events was senior religious figures' close relationships with secular political and economic élites. Those closest to the people, on the other hand, those involved in religious issues at community level, found themselves responding to popular pressures for change which cut across horizontal class stratifications, vertical ethnic or regional differences, and the urban-rural divide. What emerged was a serious rift between rulers and ruled, where religion was often a focal point for demands for change.

In this chapter I am using the term 'religion' in two distinct, yet related, ways. First, in a material sense, it refers to religious establishments (i.e. institutions and officials) as well as to social groups and movements whose *raisons d'être* are to be found within religious concerns. Examples include the conservative Roman Catholic organisation, Opus Dei, the reformist Islamic Salvation Front of Algeria (FIS), and the Hindu-chauvinist Bharatiya Jana Party of India. Second, in a spiritual sense, religion pertains to models of social and individual behaviour that help believers to organise their everyday lives. In this sense, religion is to do with the idea of *transcendence*, i.e. it relates to supernatural realities; with *sacredness*, i.e. as a system of language and practice that organises the world in terms of what is deemed holy; and with *ultimacy*, i.e. it relates people to the ultimate conditions of existence.

Because of the importance placed here on the explanatory value of the role of modernisation, it may be appropriate at the outset to say a little about it. Throughout the Third World, with the important exception of post-revolutionary states such as China and Iran, the general direction in which social change has taken place is usually referred to as either 'modernisation' or 'Westernisation'. That is, social change is understood to lead to significant shifts in the behaviour and prevailing choices of social actors, with such particularistic traits as ethnicity or caste losing importance in relation to more generalistic attributes such as nationalism. Growth of formal organisations (e.g. political parties) and procedures (e.g. 'the rule of law'), it is claimed, reduces the central role of clientelism and patronage. In short, some believe that the advent of social change corresponding to a presumed process of modernisation would lead to a general jettisoning of older, traditional values and the adoption of other, initially alien, practices. In many respects, however, the adoption of Western traits in many Third World states is rather skin-deep: Western suits for men rather than traditional dress; the trappings of statehood – flag, constitution, legislature, etc.; a Western lingua franca, and so on. The important point is that social change will not be even throughout a society; social and political conflicts are highly likely owing to the patchy adoption of modern practices. Social change destabilises, creating a dichotomy between those who seek to benefit from wholesale change and those who prefer the status quo. New social strata arise whose position in the new order is decidedly ambiguous. Examples include recent rural–urban migrants in Middle Eastern, African, Latin American and other Third World societies, who find themselves between two worlds, often without an effective or appropriate set of anchoring values. Such people are particularly open to political appeals based on religious precepts (Haynes 1993).

Generally, religion is an important source of basic value orientations. It may have a powerful impact upon politics within a state or region, especially in the context of ethnicity, culture or fundamentalism. Ethnicity relates to the shared characteristics of a racial or cultural group. Religious belief may reinforce ethnic consciousness and inter-ethnic conflict, especially in the Third World (but not only there: think of Northern Ireland or former Yugoslavia). Religious fundamentalism, on the other hand, connotes a 'set of strategies, by which beleaguered believers attempt to preserve their distinctive identity as a people or group' in response to a real or imagined attack from those who apparently threaten to draw them into a 'syncretistic, areligious, or irreligious cultural milieu' (Marty and Scott Appleby 1993: 3). Sometimes such defensiveness may develop into a political offensive which seeks to alter the prevailing social, political and, on occasions, economic realities of state–society relations.

THE POLITICAL SIGNIFICANCE OF IDENTITY

Precisely how ethnic and religious conflicts relate to development is not clear, although 'the differential outcome of both growth and stagnation must obviously have an impact' (Hettne 1995: 6). Social, political and economic change in the Third World comes about as a result of a complex interaction of both domestic and external developments over time. The contemporary political importance of ethnicity and religion is what Habermas (1976) regards as an integral facet of wider Third World 'identity crises', i.e. serious threats to national integration and to the process of economic development, which focus on the existence of 'cleavages' within societies.

IDENTITY AND POLITICS

Development is often thought to entail a cultural transformation in the direction of national identity and tradition. Growing economic advancement and political sophistication are assumed to foster a sense of life satisfaction and to lead to generally positive attitudes about the prevailing social and political environment. The orientations and sentiments people have towards politics in general and towards existing political arrangements in particular are formulated within the context of the views they have of themselves and their concept of their own identity. As already noted, two of the most important in the Third World are ethnicity and religion. As Kamrava (1993: 164) notes, 'it is their sense of identity which largely determines how people behave politically and in turn view their own political environment'. The absence of widely accepted, enduring arrays of norms and social values may make it difficult for many people in the Third World, subject to dramatic change over the last two decades, to form cogent opinions about what exactly their identity is, whether at the personal level or nationally. The identity crisis which follows often focuses on unsatisfactory governments, many of which in the Third World are not only fragmented and incoherent but also sectarian and highly changeable.

The social and political characteristics of many countries in the Third World have not only failed to result in a strong sense of national identity and life satisfaction but have also fostered feelings of disappointment and identity crisis. Political repression, rapid industrialisation, the growth of urban-based populations, economic dislocation, and social change prompt people in the Third World to question not only their predominant social and political values but also their identity as part of an often putative nation. It would not be an exaggeration to say that identity crisis has become a ubiquitous facet of politics in much of the Third World.

The extent and intensity of cleavages within Third World countries may have a strong impact upon a country's political stability and the formulation of national identity. Sometimes the pursuit of national identity takes the extreme form of 'ethnic cleansing', whereby the desire for ethnic purity is the 'justification' for large-scale atrocities towards so-called internal enemies, that is, groups with a different ethnic or religious identity. As we have recently seen in former Yugoslavia, Rwanda and Burundi, such atrocities exacerbate wider conflicts aiming at the domination of one ethnic or religious community by another.

A TYPOLOGY OF POLITICAL RELIGION

> Attempts to salvage the secularisation model have interpreted evidence of burgeoning religiosity in many contemporary political events to mean that we are witnessing merely a fundamentalist, antimodernist backlash against science, industrialisation, and liberal Western values ... Religious fervour is often

> dismissed as ethnic hostility ... typically explained away as an isolated exception to unremitting trends of secularisation and seldom recognised as part of a larger global phenomenon.
>
> (Sahliyeh 1990: 19).

The quotation suggests two areas where religion is of particular importance in understanding political and social developments: ethnicity issues and 'religious fundamentalism'. Yet this is only part of the story: we also need to be aware of the political importance of religious syncretism and of community-orientated religious groups (whose position may be bolstered by a national religious hierarchy's institutional voice of opposition during dictatorship), in order to understand fully what has been happening in recent times in the sphere of religious–political interaction.

Each of the four categories of religious movements discussed below has two factors in common. First, leaders of each utilise religious precepts to present a message of hope and a programme of action to putative followers, which may have a political impact. Second, such religious movements tend to be inherently oppositional in character; their leaders capitalise upon pre-existing dissatisfaction with the status quo in order to focus and direct organised societal opposition. It is important to note, however, that not all of the four groups target the governing regime in an overtly politicised manner. Fundamentalist and culturalist groups have as their *raison d'être* an inherent antipathy to government; community-orientated and syncretistic groups, on the other hand, tend to be more diffuse in character, often rurally-based and more concerned with self-help issues rather than emphasising straightforward opposition to government policies.

CULTURALIST GROUPS

Culturalist groups emerge when a community sharing both religious and ethnic affinities perceives itself as a powerless and repressed minority within a state dominated by outsiders. The mobilisation of the opposition group's culture (of which religion is an important part) is directed towards achieving self-control, autonomy or self-government. Examples include Sikhs in India, southern Sudanese non-Muslim peoples (such as the Dinka and the Nuer fighting both Islamisation and arabisation), Tibetan Vajrayana Buddhists in China, Muslim Palestinians living in both the Gaza Strip and in Israel's West Bank, Bosnian Muslims in former Yugoslavia, radical Muslims in Britain, and followers of the American radical, Louis Farrakhan, and his organisation, the Nation of Islam. In each case, the religion followed by the ethnic minority provides part of the ideological basis for action against representatives of a dominant culture whom the minority perceives as aiming to undermine or to eliminate their individuality.

Political culture is an important variable in analyses of culturalist groups, as it suggests underlying beliefs, values and opinions which a people holds dear. It is often easy to discern close links between religion and ethnicity. Sometimes, indeed, it is practically impossible to separate out defining characteristics of a group's cultural composition when religious belief is an integral part of ethnicity. Both are highly important components of a people's self-identity. For example, it would be very difficult indeed to isolate the different cultural components – religious and non-religious – of what it means to be a Sikh, a Jew, a Tibetan, a Somali, an East Timorese, or a loyalist Protestant or a nationalist Roman Catholic Ulsterman or woman.

SYNCRETISTIC GROUPS

A second type of religious entity, found predominantly among certain rural dwellers in parts of the Third World, especially sub-Saharan Africa, is religious syncretistic *groups*, i.e. those involving a fusion or blending of religions (Haynes 1996). They typically feature a number of elements found in more traditional forms of religious association, such as ancestor worship, healing and shamanistic practices. Sometimes ethnic differentiation forms an aspect of syncretism. A syncretistic community uses both religious and social beliefs to build group solidarity in the face of a threat from outside forces – often, but not invariably, the state. Examples include the cult of Olivorismo in the Dominican Republic and, according to some, Sendero Luminoso in Peru, whose ideology, a variant of Maoism, also utilises aspects of indigenous (i.e. pre-Christian) cultural-religious belief to attract peasants in Ayacucho; the *napramas* of north-eastern Mozambique, who combine traditional and Roman Catholic beliefs and were temporarily successful in defeating the then South African-supported guerrilla movement, the Mozambique National Resistance (Renamo) in the early 1990s; and the two 'Alices' – Lakwena and Lenshina – who led syncretistic movements, in Uganda and Zambia respectively, involving a fusion of mainstream Christian faith and traditional beliefs, against their governments in pursuit of regional autonomy.

Syncretistic religious movements are widely found in sub-Saharan Africa. During the colonial era many flourished in the rural areas in the context of widespread dissatisfaction with aspects of colonial rule. On occasion, erstwhile foes – such as the Shona and the Ndebele in colonial Rhodesia (Zimbabwe) – combined to resist British colonialism. Religious identification was an important facet of such organisation. Spirit mediums used 'medicines' to enhance warriors' martial efforts. They created a national network of shrines to provide an agency for the transmission and co-ordination of information and activities, a structure which was re-established during the independence war of the 1970s. The use of medicine also helped galvanise the anti-colonial Maji-Maji rebellion of 1905–7 in German-controlled Tanganyika. The diviner and prophet, Kinjikitili, gave his followers medicine which was supposed to render them invulnerable to bullets. He anointed local leaders with the *maji* ('water') which helped to create solidarity among about twenty different ethnic groups and encouraged them to fight together in a common anti-European cause. In northern Uganda, the cult of Yakan amongst the Lugbara, which also centred on the use of magic medicine, galvanised the Lugbara in their short war against Europeans in 1919 (Allen 1991: 379–80). The list of such religio-political movements could be extended; the point, however, is already hopefully clear: many cults arose, led by prophets, stimulated by colonialism and the social changes to which it led. They employed local religious beliefs as a basis for anti-European protest and opposition. After colonialism, such groups endured as a result of the effects of an unsatisfactory and, in many cases, partial modernisation.

RELIGIOUS FUNDAMENTALISM

Religious fundamentalists, feeling their way of life under threat, aim to reform society in accordance with religious tenets; to change the laws, morality, social norms and sometimes the political configurations of their country. They seek to create a traditionally orientated, less modern(ised) society. Fundamentalists tend to live in population centres – or are at least closely linked with one another by electronic media. Fundamentalists fight against governments because the latter's jurisdiction encompasses areas which the former hold as integral to the building of an appropriate society, including education, employment policy (of men rather than women) and the nature of society's moral climate. Fundamentalists struggle against both 'nominal' co-religionists whom they perceive as lax in their religious duties and against members of opposing religions whom they perceive as evil, even satanic. Examples of fundamentalist groups are to be found among followers of Christianity, Islam and Judaism (the Abrahamic 'religions of the book') and, some would argue, among Hindus and Buddhists too.

The character and impact of fundamentalist doctrines is located within a nexus of moral and social issues revolving around state–society interactions. The main progenitor of recent fundamentalist movements has been a perception on the part of both leaders and followers that their rulers are performing inadequately and, often, corruptly. Religious fundamentalism is often (but not always: Buddhist and Hindu 'fundamentalism' are exceptions) strongly related to a critical reading of religious texts, and the relating of 'God's words' to believers' perception of reality. The significance of this from a political perspective is that it supplies already restive peoples with a ready 'manifesto' of social change leading to a more desirable goal, which their leaders use both to berate their secular rulers and to propose a programme for radical reform of the status quo.

COMMUNITY-ORIENTATED GROUPS

Community-orientated groups utilise aspects of their religious faith to inspire themselves primarily towards self-help improvements in their lives; this may or may not involve overt conflict with government. Especially prominent in this category are local community groups, mostly Roman Catholic in inspiration, which have mushroomed over the last twenty-five years in Latin America, the Philippines, Haiti and parts of sub-Saharan Africa. Many – but not all – derive their ideas from the tenets of radical 'liberation theology', a set of ideas with a core of belief that true religious struggle involves striving to change the political here and now. In addition, because of the oppression associated with the dictatorships which were common in Latin America until the 1980s, national religious hierarchies – such as the Catholic church in Chile – may emerge as a highly significant source of opposition which is capable of offering a degree of sustained resistance, seeking to protect local communities from the depredations of oppressive government.

In recent times, the impact of the spread of Protestantism in the region has been to facilitate the growth of evangelical community groups which function as conduits of solidarity and mobilisation. The origins of the Catholic Basic Christian Communities (BCCs) can be traced back further, to the moves towards popular community development which developed from the early 1960s, encouraged by radicalised clergy at the grassroots. Such priests organised their followers for self-help and spiritual purposes, guided by a vision of the Christian promise of redemption which directly linked the temporal sphere with the spiritual. Social change in the present was seen as integral to people's long-range spiritual redemption. Concretely, this meant the full participation of ordinary people in the shaping of their own lives. Profound dependence and passivity had to be replaced by full participation and self-determination in the economic and political spheres. To achieve these goals, radical priests and nuns became spokespeople for a broad political programme with two main aims: participatory democracy and practical development to deliver desirable social goods, including electricity, schools, health posts, clean water, roads and latrines. BCCs occasionally produced leaders for mass movements, such as trade unions and the Brazilian Labour Party, which were important in the process of popular mobilisation that ultimately helped to undermine the credibility and viability of the country's military dictatorship, forcing it to hand over power to elected civilians in 1985 (Medhurst 1989: 25).

In conclusion, it is necessary to emphasise that the four broad categories of religio-political entity which we have identified are not mutually exclusive. For example, some fundamentalist groups may also be community-orientated, while a number of culturalist groups may also be syncretistic. The purpose of differentiating between them in what is inevitably a somewhat ideal fashion is to seek to identify the nature of their relationship with other religious or ethnic groups and with government. By separating the four types of religious groups, it is possible to arrive at some conclusions relating to the way in which each of the four copes with the stresses and strains of modernisation well as their potential for conflict with others.

CONCLUSION

Over the last twenty years or so, religion has had considerable impact upon politics in many regions of the world. Confidence that the growth and spread of urbanisation, education, economic development, scientific rationality and social mobility would combine to diminish significantly the socio-political position of religion was not well founded. Two broad trends have been observable: religion used as a vehicle of opposition or as an ideology of community self-interest. In the first category are the culturalist, fundamentalist and, in part, the syncretistic, religious entities. Threats emanating either from powerful outsider groups or from unwelcome symptoms of modernisation (breakdown of moral behaviour, over-liberalisation in education and social habits) galvanised religious reactions. Second, the failure of governments to push through their programmes of social improvement led to the founding of local community groups which developed a religious ideology of solidarity and development often without much help from religious professionals.

The developments described above suggest that one of the most resilient ideas about societal development after World War II – that nations would inevitably secularise as they modernised – was misplaced. It was understood that modernisation, involving urbanisation, industrialisation and rationalisation of previously 'irrational' views, including religion, would lead to the development of a new kind of society. Loss of religious faith and secularisation dovetailed with the idea that technological development and the application of science to overcome perennial social problems of poverty, environmental degradation, hunger and disease would result in long term human progress. What became clear was that technological development and other aspects of modernisation left many people with a feeling of loss rather than achievement. One result was a wave of popular religiosity which often had political ramifications.

To analyse and explain what became a virtual global development, it was necessary to look at different manifestations of burgeoning religiosity. Two areas where religion was of particular importance in understanding political and social developments in the contemporary era were in relation to issues of ethnicity and to the growth of religious fundamentalisms. These essentially oppositional manifestations were complemented by the emergence of both community-orientated religious groups and of religious syncretism. While it was suggested that syncretism is a common factor in virtually all organised religions, the importance of religious syncretism in the Third World in the post-colonial era was in the context of failures of central government to oversee local communities' protection, economic development or social cohesion.

When such a loss of faith in central government was writ large, i.e. it galvanised large portions of discrete culturalist groups, then religion often became a central tenet of anti-centre opposition. Hopes of ethnic co-operation largely gave way to fears of endemic ethnic conflict, as one of the features of the modern era was the apparent fracturing of the state system, which appeared solid until the demise of the Cold War led to a plethora of inter-nation conflicts within states.

Religious fundamentalism may be divided into two categories: 'religions of the book' and nationalist-oriented derivatives of Hinduism and Buddhism. Scriptural revelations relating to political, moral and social issues form the corpus of fundamentalist demands. Sometimes these are deeply conservative (American Protestant evangelicals), sometimes they are reformist or revolutionary (many Islamist groups), sometimes they offer an essentially moralistic blueprint for social change (Protestant evangelicals in Latin America), and sometimes they are xenophobic, racist and reactionary (Jewish groups, now banned, such as Kach and Kahane Chai). Hindu and Buddhist 'fundamentalisms', in the absence of a definitive set of scriptural norms and hence goals, assume nationalist dimensions when religious revivalism pertains to the rebirth of national identity and vigour denied in the past, zealots consider, by unwelcome cultural dilution.

A notable feature of the development of religious praxis since the early 1960s was crystallised by the emergence of a popularly driven community religiosity. This might be either conservative or reformist in thrust. The factor which such groups had in common was that religious professionals were respected but were not assumed to have the final word on religious praxis. The development of sets of community-orientated religious beliefs helped to develop mobilising ideologies of opposition and self-expression. All of the groups examined in this paper have in common a disaffection and dissatisfaction with established, hierarchical, institutionalised religious bodies; a desire to find God through personal searching rather than through the mediation of institutions; and a focus on communities' ability to make beneficial changes to members' lives through the application of group effort. This desire to 'go it alone', not to be beholden to 'superior' bodies, marks above all the relationship of religion and politics in the 1990s.

The demise of communism as a mobilising ideology leaves the ideological cupboard rather bare. Religion in all its flexibility offers a rational alternative to those for whom modernisation has either failed or is in some way unattractive. Its interaction with political issues over the medium term is likely to be of especial importance, carrying a serious and seminal message of societal resurgence and regeneration in relation to both political leaders and economic élites.

21

RELIGION AND DEVELOPMENT[1]

Parvati Raghuram

INTRODUCTION

There is no consensus on the meanings or aims of either of the key terms around which this chapter revolves, i.e. religion and development. However, both religion and development have played a significant part in the consciousness and experiences of those living in the world. While the literature on development has grown rapidly, discussions on religion have remained limited. In this chapter, I have attempted to explore the links between these two concepts.

Development, by definition, has involved an idea of change, but the objectives of such change have altered through the decades, varying from growth to welfare, basic needs, equity and empowerment (Moser 1993). The institutionalisation of development, especially during the post-war period, has led to a rhetoric and an industry based around ideas of development (Escobar 1995). However, the idea of progressive change was not new to the 'development machinery' but was embedded in both secular thought, for example, during the French Revolution, and religious ideologies, especially Christianity (Tenbruck 1990). Further, the rise of development discourse is inextricably linked with the rise of capitalism, of modernity and the predominance of the West (Mehmet 1995).

Development is also part of a global process, and has incorporated more than just the countries of the Third World (Slater 1992). Different development paradigms have incorporated the countries of the First and Third Worlds in different ways, as the linkages and the power relations between them have been viewed differently. This has varied from replication of patterns of growth seen in the First World in the countries of the Third, as per modernisation theory, to the more dynamic and exploitative linkages suggested by Wallerstein (1980). These development paradigms have also varied in the ways in which they have incorporated cultural elements, among them, religion.

Religion has been studied and understood in many ways (Durkheim 1957; Gellner 1992). For many, religion relates primarily to belief systems with a commitment to some normative values and some social order. It is also set within systems of authority. This authority may be transcendental but is usually mediated through tangible entities such as priests. Religion thus offers social meanings, providing explanations of the inexplicable.

Such functionalist explanations of religion contrast with accounts of religion as practice. Early analyses of religion focused on performative elements such as rituals and was embedded within detailed anthropological writings (Evans-Pritchard 1937). For instance, writings on African religions focused on magic, and literature on Chinese religions concentrated on practices such as ancestor worship, reducing both religions to performance. These religions were, therefore, considered to be primitive, as they focused on practice rather than thought and were seen to

Plate 21.1 Growing up in the desert offers few distractions for Algerian/Saharan refugee children of Smara Camp in the Tindouff region. This Koran teacher keeps a group of youngsters busy during the school holidays.

Source: Photographer, A. Hollman, 1998. Reproduced with kind permission of UNHCR.

be involved solely in mediating with the inexplicable.[2]

The literature on the relationship between religion and development is disparate. The greatest interest in religion and development has arisen from the explicitness of religious belief in the functioning of political systems in the Third World (Haynes 1993). The significance of religion in Third World politics gained publicity after the Iranian Revolution in 1979. The installation of an Islamic State challenged analysts who had forecast the imminent decline of religion and increasing secularisation of nations through processes of development. The Revolution led to a rethinking of the relationship between religion, economics and politics, but most analyses, however, remained focused on two axes of this triad, i.e. religion and politics.

Underlying this neglect of the relationship between religion and economic development has been the rationale for much of development itself, i.e. the implicit assumption that the economic is separated from the cultural, especially, the religious. Economic development is often viewed as being essentially secular.

In this chapter I examine these assumptions by tracing the implicit and explicit ways in which religion has been implicated in development processes. While this chapter focuses on the concept of economic development in the post-war era, I recognise that both religious ideologies and their linkages with progressive change have been ongoing processes. For instance, early Christian thought has been spread worldwide by missionaries for the past two thousand years. However, until recently, the rate of acculturation through processes of cultural exchange, such as the intermingling of different religions, was comparatively slow. Hence, there was more time for receiving cultures to appropriate the tenets of new religions *within* the framework of other meaningful narratives within their lifeworld. Like other exchanges of ideas, linkages between religions led to some form of syncretism, but this was characterised by selectivity and adaptation. For example, Christianity in India became imbricated with caste structures and with practices of dowry, both of which contain elements of inequality which remain antithetical to broad Christian principles of 'equality'. At the same time such associations have not been equitable. Through colonialism, the cultural exchange became imbued with the power differentials between the ruler and the ruled and the exchange became exceedingly uneven, and became a part of the process of cultural imperialism (Sreberny-Mohammadi 1996). The process of such exchanges has accelerated through cultural exchanges framed in development processes and through processes of globalisation.

This chapter traces some of the discourses where religion has been linked with development.[3] The 'development machine' has had differing definitions and aspirations during this half century. These have been embedded within, and represented through, the 'development discourse', the ways in which development is written about, spoken about and narrated (Crush 1995b). This discourse makes certain assumptions regarding the construction of the world and the relationship between different parts of it. As development discourses have been revised, theorisations of the linkages between development and religion have also altered. I begin by examining the ideas of religion as the passive ground within which ideas of economic development can take root. This was usually related to particular conceptions of development as economic growth, but there were early variations of this, as among the Quakers. The next section examines how development has influenced religious thought, institutions and processes. A more dynamic exploration of the linkages has been undertaken in the following section where both religion and ideas of development are seen to influence each other and I have used here the example of liberation theology. Finally, I explore recent ideas of development and the resurgence of religion in this rhetoric.

RELIGION AS SEED BED

The most explicit discussions of religion and the economy have utilised functionalist definitions of religion by outlining the philosophies which facilitate or hinder development. They have largely been influenced by the writings of Weber and his analysis of the efficacy of religion in promoting capitalist development (Weber 1958).

Weber identified key characteristics of Calvinist Protestant ethics which he deemed to be the ideal setting for capitalist development. He wrote prolifically on the relationship between the economic and the religious and his writing is often considered to lay the groundwork for most other analyses of this relationship. Perhaps, one of his greatest contributions has been his identification of aspects of religion which he felt influenced ideas of development, two of which are discussed below.

One of the most cited Protestant contributions to economic development has been its emphasis on individualism, which made it possible to focus on profit. The conception of the individualistic contract between a person and God, demonstrated by the withdrawal of practices such as confession, also laid the basis for economic, political and legal individualism (Turner 1991). Economic individualism was inimical to feudal organisation and promoted capitalist organisation. However, the role of individualism in capitalism has been contradictory, as state control of the individual became necessary to perpetuate capitalist economic development.

Individualism, the idea that knowledge, thoughts and actions are contained within individual minds, also laid the basis for another key Protestant contribution to capitalist development, i.e. universal rationalism. Protestantism was embedded with normative ideas, ensconced within particular views of morality which influenced daily social as well as economic practices. Protestantism encouraged practitioners to be rationalistic, where rationalism was defined by values which were seen to be universal. Such rationality is positioned within the idea of a 'single, correct, God's eye-view of reason which transcends (goes beyond) the way human beings think' (Thrift 1996: 15). Formal rationality was, for Weber, a product of industrialisation and was a Western thought. It differed from practical rationality, which has a problem-solving orientation, and where the problems are defined by one's individual experiences (Ritzer 1995). Formal rationalistic thought was a foundation for the modernisation paradigm for development.[4]

This paradigm presupposed a definition of development as economic growth following a linear pattern, where countries of the Third World are expected to reproduce the economic trajectories of countries of the First World.[5] Modernisation was embedded within a Protestant concept of universal rationality leading to universally defined goals. A key feature in the operation of modern society was economic rationality, with economic growth as its goal. Calvinistic Protestant ethics, which promoted economic growth, was contrasted with other religions which impeded such development. Other religions often operated as a barrier to economic growth, a psychological impediment to development. As such, religion was placed with other factors such as familialism and collectivism as a hindrance to development (So 1990). These social factors implied a lack of human resource in the 'underdeveloped' society.

For example, Watt in the first chapter of his book *Islamic Fundamentalism and Modernity*, refers to the static nature of Muslim beliefs as the first characteristic of an Islamic world-view:

> For Muslims unchangingness is both an ideal for human individuals and societies, and also a perception of the actual nature of humanity and its environment. Unchangingness is an all-pervading assumption which colours most aspects of the standard world-view, and this justifies giving it a prominent place in the presentation. Moreover, it is something which a Westerner finds it difficult to appreciate without a deliberate effort of thought. The idea of development is part of our general intellectual outlook ... It is thus very difficult for the Westerner to appreciate the outlook of those in whose thinking there is no place for development, progress or social advance and improvement.
>
> (Watt 1988: 3)

Similar claims were also made of other religions. For example, the 'other-worldliness' of Confucianism was considered by Weber to have resulted in the lack of development of many parts of Eastern Asia.[6] The recent rapid economic growth of some of these countries has, however, led to a reinterpretation of Confucianism. The emergence of the 'Four Tiger economies' of East Asia has revitalised interest in the contributions of East Asian religions to these countries' current developmental success. For instance, Kim (1994) has explored the significance of linkages between status-orientation in Confucianism and profit-orientation in capitalism in Korea. He suggests that the need to achieve social status is a key Confucian ideology, which was further strengthened through discriminatory colonial policies which prevented the colonised from acquiring status-conferring posts. Hence, the Koreans' impetus to industrialisation came not only from the desire to accumulate wealth, but also to achieve status, a product of embedded religious ideologies as well as historical trajectories.

Analyses of religion as inimical to, or as an initiator of, change continue to be important in today's development discourse. For example, the analysis of Japanese developmental success and its globality has focused on two fundamental characteristics of Japanese society. The first has been its syncretism, its cross-referencing and revalidation of structures and ideas. Syncretism has permitted Japan to retain those aspects of its religion and social organisation which it deemed most useful to achieving its goals. The systems of cross-referencing also ensured that social and economic goals remained in tandem and not contradictory. Secondly, Robertson (1992), ascribes some of Japan's ability to globalise to Shintoism's purification through ritual by the removal of that which is not wanted. Japan thus underwent a process of conscious economic and religious change.

The power relations between religious groups have also influenced economic development. For instance, in the late nineteenth and early twentieth century, people from 'minority' religions became key economic actors in Britain. One example of this was the way in which 150,000 Jews who immigrated from Eastern Europe between 1891 and 1911, changed the face of the textile industry in Britain during this period. By 1911, one-third of all those employed in the clothing industry were Jews. Thus, religious differences interacted with other differences and sometimes provided the impetus to produce certain trajectories of economic growth.

While the contribution of religious thought to progressive economic change was largely viewed in the context of economic growth, even early writings recognised that, for many, development was not merely growth. Religious groups, such as the Quakers, defined rationality not merely as profit maximisation through hard work but also included values such as 'social conscience'. The ethic of responsibility, which some Quaker businesses adopted, set alternative standards of personal and social development. They provided a good working environment and took social responsibility for their employees, which suggests that development as economic growth was tempered with ideas of welfare even in early capitalism.

To summarise, in the above discussion, religion was seen as the seedbed for development. However, this analysis assumes that development operates through voluntarism and represents an element of human agency, whereas religion is part of the structure of the society which is to develop. In the next section, I look at some examples where the relationship between development and religion appears more complex.

DEVELOPMENT AS INSTIGATOR OF RELIGIOUS CHANGE

While the discussion so far has examined the influence of religion on development, development processes have also resulted in the reconceptualisation of key structures of thought, thus influencing religion. For instance, in many African religions, time and space are only defined through people's experience of it. Thus, time in the future has little meaning and can only be grounded within cyclical patterns such as the seasons or through the life-course of human beings (Mbiti 1969). The future is only seen as a short extension of the 'now'. Hence, the ideas of time as linear and as bifurcated from the experiencing of it are both concepts which have been introduced through the development discourse. However, its influence on religious thought still remains relatively unexplored.

Development discourses have resulted in two distinct patterns of religious change. In the twentieth century, the power relationship between the 'economy' and 'religion' has largely been one where society has been taught to value economic rationality. Both capitalist and socialist development models have resulted in the growing secularisation of cultures. In capitalism, this has partly been achieved by placing value on individual choice. By emphasising the private nature of religious beliefs, the separation of religion from economics and politics was made possible, leading to a process of secularisation of the state.[7] Thus, Protestantism, as the seedbed of capitalism, laid the foundations for its own demise.

Marxist models of development have also given primacy to the economic and treated religion, like other aspects of culture, as part of the 'superstructure' and determined by the economic. Marxism, thus, sees secularism as one of its goals. By placing faith in a 'universal rationalism', which is outside of religious influences, growing numbers have dissociated themselves from professed religion in many parts of the world (Wilson 1982).

However, under capitalist development, the power of the market has meant that particular forms of political and social organisation particularly conducive to the growth of capitalism, have been promoted. Thus, economic development has often been accompanied by religious conversion. Such conversions have been part of world history, but have altered in their conceptualisation, when they have been linked with development discourse.[8] Conversions have often been carried out in the name of eradicating 'irrational behaviour' grounded in 'traditional religions'. Missionaries trying to bring about such conversions focus on the performative elements of the religions they want to displace, on aspects such as rituals, to highlight their irrationality. They draw upon normative ideas embedded within modernist thought and hence define difference as 'irrational'. Recent evangelists have also legitimised their work on the idea of allowing people 'free choice'. Some religious organisations have also become involved in charity, often cloaking their missionary aims under a guise of development activity.

DYNAMIC RELATIONSHIPS BETWEEN RELIGION AND DEVELOPMENT

The most dynamic linkage between definitions of development and religion is offered by the Liberation Theologists. They work with a definition of development as not merely growth but also welfare. In this, there are similarities with the Quaker philosophy at the turn of the century in Britain. However, unlike the Quakers, Liberation Theologists focus not merely on welfare, but on issues of equality. By emphasising equality, they move away from a key element of the Protestant ethic, individualism and the profit motive. They operate within a different set of values, which nevertheless remained normative. At the same time, the redefinition of the economic goals has, in some cases, led to a critical appraisal of their theology itself. Thus, the relationship between development theory and religion has been more dynamic.

Liberation Theology emerged to meet social needs in Latin America; hence the religion was embedded within praxis. It aimed to overcome social inequalities and challenged the individuation of Protestant thought. It also understood development as a function of unequal exchange embedded within unequal power relations, and stressed the need to alter the balance of power in order for development to occur. Thus, it emphasised structural inequalities and gave them precedence over the idea of agency, which was central to modernisation theories of development. They supported ideas of development as equity, rather than growth.[9]

The beliefs of Liberation Theologists were similar to normative theoretical expositions of Marxists, though they were more liberal than the Marxists, as they extended their focus to include all 'have-nots', whether oppressed by race, gender or class. They also differed from Marxists in that, at least for some Liberation Theologists, the emphasis was on learning and responding to the problems of development. Thus, their praxis was not just revolution, but was progressive change. The Liberation Theologists aimed for both religious learning and a process of conscientisation to occur simultaneously. They considered this a form of empowerment, and were among the pioneers in the new social movements. They often had close links with '*dependistas*'. However, they were essentially an episcopal organisation, and their developmental practice was embedded within Christian ideas. Just as developmental ideas were embedded within Christianity, so, also, they used their ideas of development to bring about institutional changes to the organisation of the Church. This was, however, less easily achieved, as most came into conflict with episcopal authorities. For example, Liberation Theologists did not all take the same stance or achieve the same success in different countries. In Nicaragua, for instance, they aligned themselves with conservative elements and achieved less success than in countries like Argentina. Finally, both the increasingly neoliberal trajectory of economies worldwide and the increasingly conservative papacy of Pope John Paul II have dealt a blow to Liberation Theologists.

Liberation Theology had much in common with other social movements in other parts of the world such as environmentalism and Black Consciousness (Beyer 1994; Manzo 1995). Such movements have been seen as an expression of localism, in the context of globalising processes.

Globalisation has been seen as a phase of late capitalist development and a result of modernisation. Globalisation is embedded within notions of globality, of sameness across the globe, of a universalising culture which is predominantly Western but incorporates non-Western elements. Globalisation as process is exemplified, for instance, by the incorporation of certain religious principles, rituals or aesthetics within Western societies. In some cases, practices such as yoga and meditation have been abstracted from their religious moorings and have been redefined within the customs of a 'global lifestyle'. Conversions of a small minority of Christians to Buddhism could also be considered evidence of the same process. Simultaneously, certain world religions have utilised the globalisation processes, economic and technological, to expand their base, to make a bid to become a universal religion. The spread of Protestant Christianity within much of Central and Latin America has been attributed to these processes.

At the same time, contradictory processes such as the growth of religious ethnic and territorial particularism are also evident. This has been fuelled by a number of factors, particularly the contradictions inherent within globalisation. Economically, globalisation processes contain within their structures catalysts for inequality (Beyer 1994). Thus, at the same time that there is a rhetoric of universalism, there has also been a growth in inequalities, with growing numbers being excluded from the benefits of these global processes. Religion has been one of many sites where protest against this exclusion has been expressed. Revival of religion has been portrayed as an expression of socio-cultural particularism against the universalising tendencies of globalisation.[10] Religion may be intertwined with other forms of identification such as ethnicity and nationhood. Religious movements have also varied in their trajectories. Some, such as the Islamic movement, have mobilised themselves through their rejection of the separation of religion from other forms of social organisation, and, as such, a rejection of secularised development. Towards this, they have vested political and legal power with religious heads. The most commonly cited example of this has been the resurgence of Islam and of the concept of '*ummah*' expressed through the Islamicisation of nations such as Iran and Egypt.

Such movements accept ideas of rationality, but argue that rationality is particularistic, rather than universal. Development critiques have also taken on board more particularistic notions of rationality, notions which have been influenced by feminist and post-colonial critiques of one transcendent rationality (Haraway 1989; Escobar 1995). Feminists, for instance, assert that, as knowledge is individually constructed through people's own experiences and from individual subject positions, so rationality must be particular, rather than universal. Rationality is defined by individual experiences of time and space and so operates with contingent contextualised norms. While development per se is recognised as desirable, the goals of such development are recognised as being different according to people's own value systems and their expectations. Development, which, hitherto, had faith in normative processes and ideals, is now redefined. New social movements, the language of localism and empowerment, describe this alternative vision. However, these are also embedded within some universal liberal values such as human rights.

Alternative analyses of the revival of Islam see religion as a universalising doctrine, oppositional to the polycentricity, relativisation and secularism of the post-modern (Turner 1991). Thus, Islam is seen to offer an alternative vision of modernity, an alternative world order, and a means to achieve that vision. In these readings, Islam is not anti-modern, but is replete with ideas of development, albeit a different definition of development (Pasha and Samatar 1996). Such an analysis may be extended to revivalism among other religions as well.

This contradiction between particularistic and universal elements of religion lays the basis for alternative views of the role of religion in future development. Early capitalism had a project and was therefore embedded in an ethic, a value system. Late capitalism is less clear about its direction, it differentiates between experiences, but is less value-laden (Beyer 1994). Resurgence of religion has been interpreted as a rejection of particularism, especially a secularised version of it, which allows plurality of moral outlooks and a return to 'moral' values. Religion, like development, provides the vision of change, of improvement.

CONCLUSION

The relationship between religion and development has been complex. Development, as a process, has brought together people of different cultures and religions. This has resulted in an exchange of ideas, an exchange which has been active and has been embedded within its own politics and power differences. At the same time as religion has defined development goals, these goals have also influenced religious beliefs and practices.

Neither development nor religion is a fixed set of beliefs or signs. Both are thus always contested and, therefore, are being redefined, both in their universal goals and in their particular spatial and temporal manifestations. Any significant analysis, therefore, needs to view both religion and development as a set of interlinking processes which engage with each other to produce changes. These processes may be contradictory or may be mutually supportive. Ultimately, both religion and development remain implicit and explicit narratives with the power to alter the lives of people around the globe.

NOTES

1 I would like to thank Eleonore Kofman and Irene Hardill for their comments on this paper. I would also like to thank my late father for having kindled an interest in this field.
2 However, more recent analyses have returned to ideas of religion as a system of signification, a form of communication in a globalising world (Beyer 1994).
3 Most of the current discussion is focused around the Abrahamic religions.
4 This view of rationality was also linked with the dichotomy between reason and nature, a dualism which was reproduced through other binary categories such as rational/emotional, public/private, market /home, man/woman and urban/rural (Scott 1995). Development operated through and prioritised the first of the binary categories, leading to a neglect of both women and the rural in development theory and praxis.
5 More sophisticated analyses of modernisation recognised that development only offered a field of possibilities (Nettl and Robertson 1968).
6 Certain parallels may also be found with current discussions on the growth of industrial districts in modern-day Europe. For instance, some writers suggest that the growth of such districts is tied with the civic/secular society in Northern Italy (Putnam 1993), while others conversely suggest that structures of authority and the social relationships which operate through Catholicism have led to the growth of Le Choletais as an industrial district in France (Hardill *et al.* 1995). Hence, the search for social and cultural embeddedness of economic growth has not been restricted to studies of countries of the Third World (Granovetter 1985).
7 The extent to which secularisation results from this division between the public (civil) and private (religious, family) is arguable. It must also be recognised that the boundaries between private and public are themselves contested and unstable. One example of this is the way in which behaviour within the family is increasingly being regulated through public codes.
8 For instance, the New Tribes Mission, a Christian Evangelical Group operating from Derbyshire, actively convert 'tribal' people, as they believe in the imminent return of Jesus when the last tribal has been converted. Although, along with Christian thought, the NTM have also exported Western ways of life, this has not been the purpose of religious conversion. Their specific aim has been to baptise, rather than bring about social change. This contrasts with the aims of social reformers working with social development, who have called on ideas of universal relevance and validity of certain social principles to justify their work.
9 Many Liberation Theologists rejected the terminology of development, arguing that development did not reflect the aspirations of the poor (Guttierez 1988). However, Liberation Theology retained an idea of progressive change, albeit redefining the goals of such change.
10 It must be recognised that the distinction between universalism and particularism is unstable.

22

PAYING THE PRICE OF FEMININITY

Women and the New Hinduism

Dina Abbott

INTRODUCTION

One of the most tragic hallmarks of this century is the immeasurable human suffering and misery that has been created by artificial ethnic and religious partitions of people who in fact share a common nationality. India is a classic example of this. Here is a country where, throughout history, there have been chapters of discontent between various religious and ethnic groups, none so apparent as during the struggle for Independence. Today, the hatred created then continues to be reflected in separatist politics as various ethnic and religious groups from the North to the South demand independent rule. Everyone, it seems, is busy planning ethnic states whilst forgetting the cost of national disintegration.

Already two Prime Ministers, Indira Gandhi and Rajiv Gandhi have lost their lives at the hands of separatists. Another, V.P. Singh, was toppled through mass protests and agitation by upper-caste Hindus who did not care for his positive action and equal opportunity policies for the lower socio-economic and caste groups (officially known as scheduled castes and tribes). Newspapers continuously report incidents, sometimes violent and brutal, between Hindus, Sikhs, Muslims and so on. In fact, at times, the violence between groups has reached such critical points that this conflict no longer remains a national issue but becomes an international concern.[1]

Clearly then, from the time of Independence in 1947 to the present, for some Indians, there has been a process of disassociation from a national identity and an increasing association with an ethnic or religious identity. For me, this raises several questions about how this has happened and what processes are involved in bringing about such changes. I would also like to know who is caught up in these processes and what type of individuals are on the receiving and the giving end of escalating conflict. But space restrictions only allow me to focus on three questions which may, however, act as starting points in enabling me to understand what is happening. First then, who and what is involved in the processes of redefining national identities for those biased by ethnicity and religion? Second, how is this change to be achieved? And finally, what are the consequences of redefined identities for individuals caught up in the processes of change?

These are, of course, very large and ambitious questions, particularly because 'processes' and 'identities' are very large issues. To make these more manageable, I have thus decided to focus only on one explanatory factor, that is the role of political agents in the process of identity formulation. I have done this because political agents such as the *Sangh Pariwar* (Pariwar or family) have played a highly visible and pro-active role in attempting to create a new Indian identity based on religion, i.e. (their definition of) Hinduism. Focusing on Pariwar politics will help me to deal with my first question which was about *who and what is involved in the process of redefining identities*.

THE PARIWAR AND THE REDEFINITION OF NATIONAL IDENTITIES

Let me begin by explaining what the Pariwar is, and by defining its political make-up. Within the Indian economic context of oppressive poverty, growing population, lessening resources and resultant political instability, the Pariwar is a group of Indian fascistic parties which offers a 'way out' for millions of disaffected Hindus. It does this by advocating Hindu religious supremacy over all others, particularly Muslims. At the centre of this are

Plate 22.1 An RSS poster of Bharat Mata (Mother India) spreading luminously across and beyond the South Asian continent, erasing national boundaries from Afghanistan to China and Russia.

Source: Reproduced with kind permission from *Manushi: A Journal about Women and Society*, (1993) 76: 19 (New Delhi, India).

parties such as the BJP, the RSS, the Vishwa Hindu Parishad (VHP), Bajran Dal and other similar extremist groups.[2] Its message is clear, that, historically, India has belonged to the Hindus. It is therefore a land of the Hindus (*Hindustan*), and non-Hindus, especially Muslims, must acknowledge this and bow to Hindu supremacy or get out of the country.

These messages are reminiscent of Hitler's ideas on race supremacy and the persecution of the Jews.[3] Thus, it is of little surprise that Muslims feel very threatened in India today as the Pariwar continues to campaign ferociously for Aryan revivalism. Even though the Government has made attempts to contain the situation (for example by officially banning the BJP), both the BJP and other Pariwar parties continue to receive financial and moral support from those who hold influential positions in Government, business and other powerful institutions. And it would appear that the Pariwar, particularly the BJP, continues to gather strength, as illustrated by the volatile Indian elections of spring 1996, when the BJP came to power, even if for just 24 hours (Reuter 1996). This so frightened opposition parties that they hastily formed a coalition to retrieve the image of a secular India (Goldenberg 1996a, 1996b).

In appealing to its large number of followers, the Pariwar uses a number of diverse methods. For instance, it calls on unemployed youths (of whom there are many in India) to spearhead Hindu nationalism. This is done by a network of *shakas* (party branches, organised mainly by the RSS) all over India where:

> Every evening 40–50 men (known as *Kersevaks* i.e. defenders of the faith), dressed in khaki-shorts, meet at the training ground and perform a series of physical exercises and martial training with long sticks (lathis). After the training they gather in the assembly hall in front of a map of Akhanda Bharat (undivided India, i.e. pre-partition India), images of God Ram, saffron flags (saffron being the Hindu colour) and a statue of the founder of the *Organiser*. Here they collectively perform the patriotic Sanskrit prayer of the RSS, promising lifelong selfless service to the regeneration of a pure Hindu Nation, standing in rows with their hands stretched

> in front of their chests in a sort of a military salute ...
>
> (Blom Hansen 1994c)

The Pariwar further ritualises Hindu revivalism by organising long marches (*yatras*), often through Muslim areas. These marches are also usually led by the *kersevaks* and, despite the religiosity surrounding them, they are meant as a show of strength and present a real threat and danger to non-Hindus. Even if State Governments try to ban these marches, this only provides the parties with further ammunition to point out the state's bias against the Hindus and favouring of the minorities. Other methods of popularising appeal include mass meetings, disruption of Muslim events (particularly prayer meetings), and, in a country that has a high illiteracy rate, propaganda through clever use of imagery, poster campaigns and tape-recordings. It is quite common that, in appealing to the Hindus, the messages given to the masses are overtly those of hatred towards the non-Hindus, in particular Muslims.

In spite of the above masculinist tactics, it is admitted that women are the Pariwar's most useful and primary resource, because it is women who can directly influence children, families, and therefore society. Yet it has not been easy for the Pariwar to mobilise women. This is partly because, even if there is a long way to go yet, Indian women have become more and more visible in the public realm. In fact, women in India are increasingly challenging dominant construction of their femininity, sexuality and social space allocations that have been imposed upon them for decades. This protest has found a voice though a rapidly growing Indian women's movement in recent years. In fact, since the 1970s, India has witnessed the growth of a large women's movement. The difference between this recent movement and others that went before is that the new movement has emerged 'from below'. Unlike the élite middle-class pre-independence movement that faded away, the women's movement today is led by poor women, usually from lower-caste/class backgrounds, and is thus able to attract large numbers. Classic examples are the Self-Employed Women's Association (SEWA) and the Working Women's Forum (WWF) (see Abbott 1997; Chakravaty 1972; Jayawardena 1986 for further details).

Yet the Pariwar, in its attempts at Aryan revivalism, seeks to control women's emancipation. The Pariwar's messages are distinctly about gender hierarchies and patriarchy. If anything, the Pariwar would ideally like Hindu women to revert back to old Aryan dictates laid down in the *Manu Smitri* in the first century BC, that is to carry out *sewa* (service) to all, particularly the male members of the family and society. In fact, it is a woman's *dhram* (sacred duty) to obey her father, brother, husband and sons above all else. This belief was aptly demonstrated in the recent Miss World contest when the Pariwar led huge demonstration against the 'sullying' of womanhood by foreigners, and in fact outweighed any feminist attempts at this. Not to be outdone, some Pariwar women also offered to burn themselves if the competition continued (Goldenberg 1996c).

There is, then, a dilemma for the Pariwar. On the one hand, women are strategically placed to spread its messages within society, and, on the other, modernisation and urbanisation have allowed women to challenge patriarchal values such as those manifest in arranged marriages, dowry systems, inheritance laws and so on.

The Pariwar is also worried about the votes it may lose from millions of poor, lower-caste Hindus who make up the masses in India, particularly because it preaches a revival of traditional caste hierarchies. Here too the Pariwar faces a dilemma because several thousand of India's lower-caste Hindus have rejected a religion that subordinates them, in favour of the (supposedly) more egalitarian principles of Islam, Buddhism and Christianity. Like the women, lower-caste groups have also formed strong organisations and are increasingly demanding equal opportunities and freedom from caste-chains.

The way the Pariwar has attempted to deal with this type of dilemma brings me to my second question, i.e. '*how is this change to be achieved?*' How is it that currently, despite what I have said above, both women and lower-caste Hindus are increasingly giving their support to the Pariwar?

ACHIEVING THE CHANGE IN NATIONAL IDENTITIES

At a general level, the Pariwar has achieved this change by playing on and redefining the meaning of Indian nationalism itself. Nationalism, during the struggle for independence, was about a united India, and Gandhi roamed the country (and eventually lost his life) in an attempt to break down the caste and religious barriers which divided his people. The meaning of nationalism for the Pariwar, however, is radically different. To quote Jafferlot (1993: 517), for the Pariwar: 'Hindu nationalism relies on the assumption that the Hindu community embodies the Indian nation and consequently the minorities should give up their specific cultural features in order to get integrated in an 'Hindian' nation ... '. The Pariwar's version of nationalism is, then, defined by two aims, i.e. first, to unite all Hindus, and second, to wipe out the threat posed by non-Hindus.

The first aim is again potentially problematic because there are several million Hindus in India, all coming from highly diverse ethnic, caste, language and regional backgrounds. In fact, if anything, Hindus all over the country have been demanding autonomy on the very basis of these differences and regional loyalties (e.g. the Nagas in Nagaland). For the Pariwar, it is these very differences which are the real 'weakness' of India as a nation. If these differences can be minimised, then India can become a strong nation once again. Hindus will then march as a united people against the rest of the world, much in the same way as the Pariwar imagines Germany and Japan to have done.

To create a united front, all Hindus must prioritise their loyalty to the nation and the Motherland (*Bharat Mata*). And whilst Hindus are capable of doing this, Muslims clearly cannot (see also Panday 1991). To quote from one RSS propaganda tape-recording:

> Muslims who value their religion more than their nation, they can never be nationalists. And Hindus who treat religion as a personal matter can never be communalists ... because those who give pride to their religion over their nation cannot be communalists ...
>
> (Kishwar 1993a: 5)

The mouthpiece of the RSS, *The Organiser*, thus constantly has 'grievance lists' of Muslim 'crimes'. These include a series of accusations such as:

> Muslims are traitors and harbour pro-Pakistan sentiments ... they betrayed Mahatma Gandhi and severed the two arms of Mother India (East and West Pakistan). ... Muslim polygamous practices means that they have a higher birthrate and Hindus are in danger of being swamped by Muslims ...
>
> (Kishwar 1993a: 7)

What this twisted version of nationalism has done, therefore, is allowed both women and lower-caste Hindus to feel that they have an important role to play in building the nation's future. And this role requires them to put aside their 'quibbles' (as the Pariwar sees them) about emancipation and entitlement in order to focus on the ultimate goal, that of national Hindu unity.

Thus, the Pariwar has been able to draw scores of women from all classes into their ranks. In fact, middle- and upper-caste/class women are specially targeted, for example, through the RSS women's wings of the *sakhas,* known as the Rashtriya Sevika Samiti (Patriotic Association of Voluntary Women). Here, like the men, women receive physical training, including martial arts. They also receive extensive moral lessons (*sanskars*) on their duties as women, mothers and carers of the family. To get around the dilemma or conflict of women's public-sphere activity and the revivalism of the domestic-based Aryan womanhood, the Pariwar cleverly argues that:

> Being a mother and looking after the family is at the core of national life, of the reproduction of cultural values, of Hindu culture. Women's performance in the public sphere should not be an impediment to motherhood. Education and work must be encouraged, but should, simultaneously be adapted and subordinated to the supreme goal of motherhood, whose rationale is derived from the nationalist discourse. To raise children as patriotic citizens in the nation state defined by Hindu culture. ...
>
> (Blom Hansen 1994a: 5)

Recruitment of new mothers is, therefore, important, but causes further problems, because of the reluctance of young women, who would rather be doing things other than attending the *sakhas*. A major problem, however, is societal disapproval of young women's social mobility outside the home. To get around this, both the *sakhas* and the associated political parties are presented as extended families where women can deepen their support networks. The RSS, for instance, promises to take care of these 'heroines of the nation (*kersevika*)' and protect them from sexual and other deviance (Basu *et al.* 1993; Blom Hansen 1994a; Sarkar 1991). In fact, the women are constantly reminded of, and likened to, brave and pure women in history and Hindu mythology such as the legendary warrior Queen Jansi ki Rani and Sita, the wife of Rama.

However, motherhood also causes problems. First, it restricts the mother's public involvement and second, her loyalty, which is usually primarily focused on her child rather than national duty. In order to get round this problem, women who represent the Pariwar in the public arena are those who can be presented as pure, de-sexualised holy beings (*sanyashins*) who have renounced family and societal ties in order to serve only the Motherland. Women can thus display a duty to their cause above everything else. One such woman is the (in)famous Rithambra. Rithambra makes speeches and tape-recordings about nationalism which are then sold very cheaply or given away to masses of Hindus all over India. Rithambra's messages of Hindu unity and Muslim hatred are highly stylised and cannot but make an impression. As Sarkar puts it:

> One remembers the voice only too well: high-pitched, shrill, breathless, delivering a non-stop harangue with no modulation. The voice seems always almost about to crack under the sheer weight of passion. The overwhelming and constant impression is one of immediacy, urgency, passion, spontaneity … an inspired voice speaking from the gut.
>
> (Sarkar 1991: 2058)

Thus, in yet another apparent contradiction, under the guise of religiosity, nationalism and purity, Rithambra transmits dangerous, 'satanic' messages of hatred and evil to millions of Indians across India.

In a relatively new development, lower-caste women have also been drawn into the movement recently. Here too heroines are being created and special 'cadres' have been active in addressing mass gatherings, explaining to lower-castes why they should support new Hinduism. As yet, very little is known (at least academically) about how this has been achieved. However, my guess is that an ideology that promises Hindu supremacy offers a place even to those Hindus who are at the very bottom rung of the societal ladder, and, what is more, promises them a superiority over large numbers of non-Hindus.

The Pariwar's nationalist ideology, then, has been able to achieve change by carefully developing strategies to cope with dilemmas created by modernisation in the case of both women and lower-caste Hindus. What is fast becoming shockingly clear is that both these groups are increasingly giving support to Pariwar politics. This brings me to my third and final question: *what are the consequences of redefined identities for individuals caught up in the processes of change?*

CONSEQUENCES OF REDEFINED IDENTITIES

Both men and women are heavily caught up in the processes of creating a new Hinduism, overtly or otherwise. Some take an active part in this process. Others are at the receiving end of it and suffer equally as their homes are burnt down or they are attacked each time the Pariwar marches through their neighbourhoods. Often, however, women bear the severe brunt of the aftermath of violence. For instance, it is they who suffer the social ostracism of rape or are left to look after fatherless children and pick up the pieces of their devastated lives.

In another example, an analysis of the violence that erupted in the Punjab following Indira Gandhi's murder on 31 October 1984 showed that this was not an eruption of spontaneous anger, but an organised systematic attack on Sikhs. In a four-day orgy of brutality, mutilation and massacre, uncountable numbers of Sikh men and boys were killed in public places, on public transport, and in their homes on the pretext of revenging Indira Gandhi's death. Hundreds of Sikh women saw their men and children being hacked to death or being burnt alive in front of them. The rioting crowds, however, appeared to distinguish between the killing of men and women. Whilst the men were murdered openly, women of all ages were raped and molested. Their homes were burnt and thousands of women, traumatised by the loss of their homes and their loved ones, sought refuge in Delhi's relief camps. The women who were lucky enough to have reached these camps have since been forgotten by everyone, and whilst the Indian Government has made gestures about compensation, this is unlikely to happen for a very long time (see Kishwar 1984).

What this violence did was make Sikhs, who have served India well for centuries, become identified as a separate part of it. Sikh women, who have always been represented as brave respectable women in Indian history and legend, now became objects of humiliation for the Hindus. As one person stated:

> As long as people thought in concrete terms of Sikhs they knew personally, they felt that these individuals had to be protected ... but the moment they began to think of themselves as members of the Hindu community or as nationalist patriotic Indians, their ability to see Sikhs as fellow Indians seemed to vanish ...
>
> (Kishwar 1984: 23)

Thus several Sikh women who were caught up in the violence now find themselves destitute and homeless.

For the Muslim men and women the story takes on additional dimensions. There is little doubt that the two Bombay riots that followed the Ayodya incident (December 1992 and January 1993) were organised by Pariwar members (Kishwar 1993b). The scale of brutality and the extent of violence against the Muslims saw a new type of development.

First, the attacks on Muslim families did not just focus on poor settlements and slum areas alone, but also included richer middle-class Muslim homes. Second, the attackers were not only young unemployed youths, but included men of all ages and from all walks of life. Even young girls and women took active parts in demolishing Muslim homes and lives. In one example, hundreds of Hindu women lay down in front of fire engines to stop them reaching an area where several Muslim homes had been set on fire. In another, during the Bombay riots, several young girls and women joined in looting Muslim shops and they also joined men in stoning Muslim men, women and children (Kishwar 1993b: 38). Third, the nature of direct violence on Muslim women was more intense. Here, women were also murdered alongside their men. Seema, a victim who was saved from death, recalls how her neighbourhood women whom she had known for years turned against her:

> They began to hit us. We pleaded with our neighbours to give us shelter, but no one let us find refuge in their house. They said you are Muslim and ours is a Hindu neighbourhood. Muslims deserve to be treated like this. The houses of other Muslim families had also been broken. Even the roof had been blasted. Hindu women from the neighbourhood said to the gangsters, 'beat them up, pierce their eyes ... '
>
> (Kishwar 1993b: 30)

From these accounts it is clear that the consequences of the Pariwar's politics are horrendous for Sikh and Muslim women and men. But I will argue that Hindu women are also losing out. For instance, reverting to old laws can only mean the withdrawal of many women into their homes, a revival of the dowry system, denial of inheritance rights and the loss of any gains made so far. In fact, there is such a move to control womanhood in Hindu India today that once again, in 1987, Roop Kanwar, an eighteen-year-old widow, was burnt alive on her husband's funeral pyre in front of several thousand 'worshippers', including the police (Kishwar and Vanita 1987; Vaid and Sangari 1991). The 'satification' of Roop Kanwar is now being held up as a model to young Hindu women, even if these women may reject this development.[4]

What is equally worrying is that Pariwar politics has led to an undermining of the radical women's movement which has given hope and led to empowerment (little as it may be) to millions of women in poverty. As Sarkar (1991) argues, the Pariwar is essentially about

Plate 22.2 Celebrating the space of the Sati Sthal, where Roop Kanwar was burned to death on her husband's funeral pyre on 4 September 1987, in Deorala, Sikar district, Rajasthan. She was eighteen years old.

Source: Reproduced with kind permission from *Manushi: A Journal about Women and Society*, (1987) 42–3: 17 (New Delhi, India).

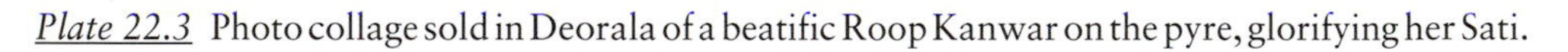

Plate 22.3 Photo collage sold in Deorala of a beatific Roop Kanwar on the pyre, glorifying her Sati.

Source: Reproduced with kind permission from *Manushi: A Journal about Women and Society*, (1987) 42–3: 27 (New Delhi, India).

hierarchical politics and women's place therein is definitely subordinate. For instance, women's mobilisation in radical organisations is horizontal, with leadership emerging at grassroots level. Mobilisation in the *kersevika* movement of the RSS is, however, vertical and in reality led by men. What the women's wing of the RSS does, is, therefore, commanded by the men who belong to it, even if women feel that by dressing up in semi-militaristic gear they have somehow elevated their status within the organisations. The result is that Hindu women, in an increasing number of cases, are breaking down ties with women from other religions. Pariwar politics has created an atmosphere in which women no longer trust or support each other as they have done in organisations such as SEWA and the WWF. Whilst these groups have made an all-out effort to break down caste and religious barriers between women, the Pariwar has created violence between women as is exemplified in Seema's story above.

To answer my third question then, the consequences for individuals wittingly or unwittingly caught up in the processes of creating a new-found Hindu identity are serious. There are the apparent and immediate costs of violence and conflict to individuals and to national unity. But there are the hidden, long-term consequences of dismissing the real problems of poverty and economic decline. These include the growth of segregational politics of the type the Pariwar preaches and the growth of patriarchal and fascistic ideology based on hierarchies and subordination, which can only result in further violence. For women, as this chapter has shown, no matter how the Pariwar manages to disguise the contradictions in its messages of Aryan femininity in a modern world, this can only spell a step backwards. But in a final note, it is clear from the opposition that the BJP received in the Spring elections that, despite its growing strength, there is a strong challenge from millions of men and women who still wish to preserve a secular India.

NOTES

1 For example the Ayodhya incident (December 1993) when thousands of young Hindu men marched into the city to destroy an ancient mosque (the Babri Masjid) with demands to replace it with a temple. The youths were openly incited to carry out acts of violence against Muslims by leaders of the main Hindu fascistic parties the Bhartiya Janata Party (BJP) and the Rashtrya Swayamsevak Sangh (RSS). In the rioting that followed the destruction of the mosque, thousands of Muslims lost their homes and livelihoods. Others were raped, tortured and murdered. For a fuller coverage see Abbott (1996) and Kishwar (1993b).

2 Of these, the BJP and the RSS have the largest following, particularly in Maharastra and Bombay. It is difficult to know the exact membership figures because these are closely guarded secrets. However, Blom Hansen (1993) suggests that the RSS alone had approximately 2 million highly organised activists all over India at the time of his research. But from the crowds that the Parwiar attracts to its events, it is clear that the overall following is massive and growing every day both in rural and urban India (see Blom Hansen 1994b).

3 It is sometimes suggested (for example Chapman 1996; Thapar 1986) that in ancient times the Aryans from the North (probably the Russians) conquered the original inhabitants of India, the Dravidians. The Aryans were a fair race whist the Dravidians were dark. Thus the Aryans named the Dravidians *suddras* (dark people) and deemed that the *suddras* were born to serve their masters. It is believed that this may have been the origin of the caste system where birth dictates one's rightful position in society. Many lower-caste Hindus in India still occupy a similar ranking at the bottom of the societal hierarchy as historically the *suddras* have done. The Aryans also brought with them patriarchal and other hierarchical systems which were embodied in the Aryan religion, Hinduism, which is symbolised by a *swashka*. Hitler may have borrowed Aryan ideas of race supremacy and hierarchy as he did the *swastika*, although he reversed it (Quinn 1994: 130–9).

4 For instance, in her interviews with young RSS women, Sarkar (1991) found that these women completely denounced the *sati* development, arguing that this should never occur in a modern India.

VIII

CULTURE AS PRODUCT: CULTURE AS PLEASURE

23

WHOSE GAME IS IT ANYWAY?

West Indies cricket and post-colonial cultural globalism

Hilary McD. Beckles

INTRODUCTION

Colonialism, with all its subaltern protest and revolt and nationalist agendas, was sustained in large measure by its unifying cultural actions and ideas (see Panday 1982; Fanon 1967; Spivak 1986; Mazrui 1990; Said 1993). In the Caribbean instance this dynamic, paradoxical, and sometimes contradictory process provided it with an advanced multiculturalism as well as a working agenda of ethnic democratisation. There is consensus, furthermore, that cricket culture was the site, a complex field of action, where West Indians articulated in a sophisticated way their anti-colonial philosophies within the political parameters of an aggressive egalitarian, multi-ethnic, nationalist project (Patterson 1969; Cashman 1989; Stoddart 1989; Beckles 1995; Beckles and Stoddart 1995). Yet, specific contests in the cultural relations of Empire, as well as post-colonialism, such as those found in the history of English versus West Indian cricket, have not received detailed systematic attention.

CRICKET AND THE CARIBBEAN

By the beginning of the twentieth century West Indians of all classes and ethnicities had made a substantial social investment in cricket. The ideological centre of the game, nonetheless, like all other major cultural institutions, continued to be occupied by a native white élite whose hegemonic domination was contested by marginalised and excluded groups. Promoted as a symbol of social élitism, high culture and colonial respectability, cricket ritualised the politics of an endemic struggle for social inclusion and by extension the contestations of imperial white supremacy ideology (Cummings 1990; Mandle 1973; Manning 1981; Searle 1990).

C.L.R. James' dialectical analysis of this process suggests that West Indian cricket, as a metaphor of the social relation of Empire and post-colonialism, must be understood as a contradictory action of subversion and accommodation. For him, cricket culture in the West Indies possessed:

(a) an internal social contest for class and ethnic equality;
(b) an intense nationalist desire to challenge at the level of ritual and defeat the imperial centre; and
(c) a deep-seated recognition of cultural affinity and identification with England that promotes sentiments of shared experience and solidarity.

Since James, many researchers have arrived at the same broad conclusion, and have demonstrated that, in general, English versus West Indian cricket contests have long carried this ideological baggage that continues to provide social meaning for specific match encounters (James 1963; for a critique see Tiffin 1981).

Internal social contests undoubtedly facilitated the rapid indigenisation of the game. Its subsequent emergence as a theatre for anti-colonial struggle further secured its status as unchallenged popular culture during the inter-war years. It remains the single cultural action that gives social shape to the idea of nationalist identity within the fragmented nation-building experience of the region. In recent times these relations of culture and political process have been challenged by a post--colonial capitalism that seeks to redefine the cultural meaning of cricket and untangle its links with earlier nationalist paradigms in a discourse of global cultural commodification. The economic collapse of new nation-states in the region and their dependence for sustainability upon north Atlantic financial institutions have facilitated the fracture and global irrelevance of independence projects. The ideological reassertion of market and indirect political authority by Euro-American multinational corporations in these states has resulted in a widespread conception of globalisation as a process of recolonisation.

When, in 1994, the English cricket team arrived in the West Indies at Barbados to play a test match, the encounter released a frenzy of ideological debates that placed in clearer perspective an understanding of developments related to globalisation. The match highlighted the multi-layered cultural meaning of post-colonial cricket by confronting nationalist sentiments as they relate to the legitimacy, relevance and integrity of an indigenous cosmology with respect to cultural ownership and territoriality. In addition, the game publicly raised issues pertaining to the critical features of post-colonial identity, citizenship and cultural rights. It was not a minor social and intellectual eruption, but one that confronted citizens in large numbers at several levels of everyday life.

WEST INDIES VERSUS ENGLAND – POST-COLONIAL IDENTITIES, CITIZENSHIP AND CULTURAL RIGHTS

The background is important and should be recalled. In a document released by the Central Bank of Barbados entitled 'Review of 1994 and the Prospects for 1995', Barbadians read about the 'good' news that followed the 'bad' news they and other West Indians had earlier experienced. The document contained a financial statement released to the press, part of the Bank's normal communication with citizens, concerning indices of national economic performance. It stated:

> Economic recovery in Barbados gained momentum during 1994. Real GDP expanded by an estimated 4 per cent, compared with about 0.6 per cent in 1993. The export sectors together grew twice as quickly as the rest of the economy ... Tourism was again the main engine of growth, expanding by 11 per cent ... The upturn in the UK market has been attributed to an influx of cricket fans, cheaper fares, additional charter flights and intensified marketing.
>
> (Central Bank of Barbados 1994: 1–2)

In the final section of the release, subtitled 'Prospects for 1995', it is stated that 1995 is expected to be a less buoyant year. Overall economic activity is projected to grow by only about 2 per cent as the major foreign exchange earning sectors are likely to be less dynamic than in 1994. Although tourism will again be the leading sector, 'activity will not be as strong in the absence of the special event which took place in 1994' (Central Bank of Barbados 1994: 5–6).

The good news was the enormous financial success of the special event, the influx of cricket fans from England occasioned by a tourism marketing blitz of the West Indies vs England test match scheduled for Kensington Oval, Bridgetown, April 8–13. Over 6,000 English spectators, delivered to the island by a host of charter companies, descended upon the Bridgetown Oval, in support of their embattled national team. The bad news was that the English team, 3–0 down in the five match series, won the game, defeating the West Indies in a convincing display before a massive 'home' crowd. The English victory was historic in proportion. The last occasion that England had defeated the West Indies at Kensington Oval was in 1935.

This kind of spectator migration and

participation had never happened on this scale before. It was a first for English fans, West Indian supporters, cricketers, administrators, tourism officials and the overall culture of West Indies vs English contest. On the surface all was well. Occupying the high ground of civic exchange, 'natives' and 'foreigners' were embraced by the theatrical grip and magic of a cultural practice that had assumed new dimensions and ideological expression. Beneath the surface, however, a new world was being created; ideological representation, aspects of postmodern philosophical discourses, and social reactions to post-colonial cultural globalism were being examined, fashioned and legitimised. It was cricket at its ideological best.

On Friday, 15 April, two days after the 'special event', a letter over the name of P. Francis, appeared in the *Barbados Advocate* with the headline 'When Kensington looked like Lords'. A photograph showing 'a sea of English supporters' over a caption which asked 'Is this really Kensington Oval?' accompanied the letter. In the opening paragraph the author sets a surreal stage for a discussion on cultural possession and celebration, identity and the ethnic politics of heritage:

> Last week-end I went to Kensington Oval to watch the Fourth Cable and Wireless Test Match. Upon arrival, I fell into a trance! I dreamt that all the stands had been painted in various shades of white ... Then someone pinched me and I came to the realisation that I had not been dreaming – this was real! It is estimated that on the first day, approximately 80 per cent of the spectators were English supporters; on the second day over 70 per cent, and on the third over 65 per cent. By the third day many of them, with the help of the sun, had turned from lily white to brown or lobster-red.
>
> (Francis 1994)

On being 'pinched' into consciousness the author asked the question, 'what does all this mean?'

The search for meaning, it is argued here, should begin with a recognition that fundamental changes are taking place in post-colonial Caribbean society, and that these changes have origins within North Atlantic capitalism as well as in the cultural contradictions endemic to the anti-colonial project. The constituent elements within these processes are identified as (i) the globalisation of post-colonial mass culture by international capital; (ii) the willingness of a former imperialist society to consume the cultural products of former colonies as quality offerings; (iii) the post-colonial social acceptance that cultural identity and heritage ownership transcend national political boundaries; (iv) the inability of West Indian nation states to achieve sustainable development; and (v) the extreme vulnerability of these new states with respect to neo-imperial cultural and economic penetration and domination.

The arrival in Barbados of 6,000 English cricket supporters in early April 1994, to witness and celebrate an encounter on the cricket field, was in essence a technology-aided, post-colonial, multicultural communication. Barbados was (re)claimed by these tourists as an idyllic and privileged part of their own heritage; and cricket, the Barbadians believe, is England's lasting gift to them – a nexus more binding than the discarded chains of earlier slave plantations. These cricket fans, as tourists, were going 'back' home to the 'old country' – their own 'South', as the Americans would say – to a place where many cultural things remained in a primal and immensely recognisable form.

In Barbados, English fans were 'away at home', engaging 'their' clever natives in an orgy of heritagemania that continues to be misunderstood by both sides as an ideological contest. The journey to Barbados by the English had little to do with the tourism of the foreigner. As far as this 6,000 was concerned, they could have been at Old Trafford in Manchester or at Lords in London. As a colonial community, Barbados, or as the natives call it 'Bimshire', never bit the English hand, and remained a shire for sure. What for the English was a quarter-day trip across the Atlantic was no more than an internal journey on the cultural terrain of a holistic heritage. It was essentially a reassertion of cultural right, and a subversion of retreating nationalist self-definitions. But it was also an accommodation of sort, and a recognition that things had changed in order to remain the same.

Cultural globalisation, however, is not without its contradictions. Hall (1981) has argued that it is always contested since it necessitates a powerful struggle for space, and because many people have multiple cultural identities which are continuously being reconstructed under the impact of diverse political discourses. During the five-day cricket encounter, Barbadians were divided along much the same cerebral lines as they have been for the past 30 years. But it was not just an additional opening. It was one that moved closer to the heart of the matter. It exposed the spirit of a people whose location at a centre of the periphery has shaped all around and within them. The nerve-ends of cultural ownership and dispossession were touched and the reaction was predictable.

The Kensington Oval is designed to accommodate 11,000 spectators comfortably. During the course of a well-hyped and exciting test match capacity crowds of locals have given the venue its well-known reputation as a place of ease and peace. Six thousand English fans reconfigured the clientele and loud cries of dispossession were heard long before the match commenced. Locals claimed that their cricket, a product shaped and refined by their sweat and tears over a hundred years of a degrading apprenticeship, was being appropriated by international capital – with local agents – and removed from their reach. Talk of cultural disenfranchisement filled the airwaves as angry, excluded spectators recounted pitiful tales of trying to purchase tickets to enter the Oval. Anger led to despair among traditional consumers who in an act of desperation called for a national boycott of the game.

Calypsonians, politicians, intellectuals and other shapers of public opinion, were called upon to assist in resisting the process of perceived cultural alienation. Tourism, they said, was at it again. When, in the 1970s, for example, elements within the Government section of the Barbados Tourism industry indicated a desire to close certain beaches to the public in order to improve the comfort and consumption of tourists, a local calypsonian (The Mighty Gabby) produced a song whose chorus rang out: 'Dat beach is mine, I can bade dey anytime, despite what you say, I gonna bade dey anyway'. The song became a national call to ideological arms, the most popular of its generation, and any policy of hotels' privatisation of public beaches the Government may have had was buried in the sand of popular protest. Barbadians now spoke about winning the battle at sea and losing the war on land. Tempers flared and the politics of cultural-heritage tourism raged.

The letter by P. Francis had located the predicament of local cricket supporters within a developmentalist paradigm that juxtaposed social exclusion and financial globalisation:

> He [the local spectator] has been denied the opportunity to witness a test match being played in his own backyard. Why? Because the almighty dollar (or pound, in this case) is more important than the desire of locals to participate in a West Indian affair – an occasion to render not just moral, but physical support to our fine cricketers ... There is something wrong in our society if we believe that our welfare is always inextricably entwined in something foreign – currency or otherwise. Various sectors of the economy will certainly benefit substantially from the influx of visitors to our shores, but at whose expense?
>
> (Francis 1994)

The dispossessed local spectator, Francis tells us, had also gained the sympathy of some dispossessors, many of them expressing 'surprise at the extent' of the displacement of local spectators.

The size of English market demand for the event was reported in the local press as evidence of 'the intensified sports-tourism drive by the Barbados Tourism Authority (BTA) paying off' (see Blenman 1994). During the 18 months leading up to the tour, the BTA's marketing efforts were spread across the United Kingdom. This included co-operative efforts with travel agents, schools and cricket clubs. Some tour operators, such as Airtours, which has established a strong presence in Barbados, did some individual marketing. The Fred Rumsey Travel Group also played an important role in market preparation. The BTA claimed that it 'spent considerable sums of money' in promoting 'the island as a cricketing destination' and the returns on that investment were finally being realised (*Barbados Advocate* 1994).

By the end of March, reports were appearing in the local press that only the lowest category of tickets were available for sale to residents. The implication was that the prime seats were presold in England by agents who had bulk purchased from the Barbados Cricket Association (BCA), acting in collaboration with the West Indies Cricket Board of Control (WICBC). When the BCA announced on 8 February that tickets would go on sale at its premises, those who entered the market were told that the comfortable seats had been presold to English agents. Some Barbadians reported phoning to England, desperately trying to purchase tickets. A black market developed and ticket prices were inflated by over 400 per cent.

EXILED FROM THE OVAL: 'RACE' AND CLASS AT PLAY

The global demand for access, highlighted by English consumption, drove many locals out of the market and led to a redesigning of the traditional ethnic and class character of the clientele. The BCA, shot onto the export market, confronted local consumers with a new policy that required ticket purchase for all five days. Pre-match panic set in, and many in the financially secure income groups purchased the remaining match tickets while spectators with less cash waited eagerly to see how events would unfold. Strapped for cash for most of its turbulent history, the BCA argued that its position was based upon rational market responses that would work to the overall benefit of Barbados and West Indies cricket. Working-class patrons, the traditional core of cricket spectators, were placed in an unfavourable relationship to the market, and scrambled for access to the recently covered, but still very basic, poor-man's stand – known as the 'Bleachers'. A scenario of race and class emerged with respect to consumption, an indication of the pervasiveness of the globalisation process.

Barbadians were convinced that for most of the match they were outnumbered by English tourists. Pires' column in the *Daily Nation*, 15 April, entitled 'Spot the Bajan: New Game at Kensington' began: 'You could have played spot the Bajan at the Kensington Oval during the fourth test – and lost'. 'The ground', he said, 'was swamped with pot-bellied, lobster-pink, tattooed Englishmen who hung Union Jacks upside-down and sang football songs and made England feel like they were playing at Wembley' (Pires 1994). All told, there were eighty-nine Union Jacks waving on the first day compared with twelve Barbadian flags and three West Indies flags. 'When the English batsmen came on there was a deafening roar', Pires states, 'and Bajans were hushed in Barbados' (Pires 1994).

Sports columnist for *The Barbados Advocate*, Barry A. Wilkinson, presented attendance figures that support those of Francis. He reported being 'extremely surprised at the large numbers of Englishmen'; it was 'the first time' he had seen 'so many white patrons watching a match at Kensington'. On Friday, 8 April, the first day of the test, he informed readers that 'the crowd appeared to be approximately 70 per cent English'. During the weekend more locals came on stream and the ratio fell to 'an estimated 60 per cent' on the Saturday, rising again to 'about 65 per cent' on Sunday. Monday was the rest-day, and on Tuesday, when England's grip on the game intensified, it went up again to about 75 per cent, where it remained the following day.

Barbadians, Wilkinson concluded, 'had been deprived of an opportunity to see their West Indies team in action' (Wilkinson 1994). The *Sunday Sun*, April 10, agreed that 'The British greatly outnumbered their Barbadian counterparts'. Malcolm Marshall, legendary West Indies fast bowler, commented in his *Daily Nation* column that 'West Indians were outnumbered by English supporters', and Oliver Jackman, distinguished Barbadian diplomat, lamented the 'prodigious display of rapaciousness on the part of the WICBC, in cahoots with the BCA', that 'kept thousands of Barbadian cricket lovers out of Kensington' (Marshall 1994; Jackman 1994).

The game was played, however, in an atmosphere which, though charged by some unease due to the unfamiliarity of developments, was cordial and, according to police reports, largely incident-free. Eric Lewis, satirical writer for the *Barbados Advocate*, stated that he was not surprised by the 'historic moment' which witnessed more 'foreigners' at the Oval than locals. The reason he gave was that on the first day of the match he saw 'a woman driving an old car', a phenomenal event in Barbados, where 'only we men drive vehicles which are tetanus traps' (Lewis 1994).

The local and international press expressed fascination with the event and photographers especially revelled in capturing the colourful social event. Roving microphones recorded the opinions of spectators and the print media made splendid copy of narratives. When asked about the atmosphere in the packed stands, Tony Grandison, a Barbadian, said: 'It's all right and I think that we all have a healthy respect for each other'. He added that he appreciated the concerns Barbadian cricket fans had as a result of their inability to purchase tickets on their home ground. William Blades, another Barbadian, was also not too bothered by the British takeover: 'There are more British people sitting where I am than local people but I still know that we (West Indies) will win the game'. Asked how he felt about the extra loud cheers for England's achievement on the field, Blades said: 'They've come to support their team and we have come to support ours and that's the way it is' (*Daily Nation* 1994).

Briton David Rake said the 'invasion' of the famous cricket ground by his countrymen was 'no big thing', since Britons of Caribbean origin always came out in their numbers to support the West Indies team: 'It reminds me of the 1970s. We English always used to be outnumbered whether we were playing at Lords or Old Trafford,' Rake said. An ardent West Indies fan, a black Englishman from Leeds, said he was not happy about the English predominance in the stands:

> They shouldn't have allowed them to buy all of the tickets in advance like that. Later this year or whenever Pakistan or India come here they will be encouraging local fans to come out and support cricket, but for now it is as though they are telling blacks we are not important.
>
> (*Daily Nation*, 11 April 1994)

The massive English presence, noted Tony Cozier, the distinguished Barbadian cricket commentator and analyst, 'emphasised how beneficial sport can be to our tourism – and, consequently, our shaky economies'. The 'presence of thousands of free-spending England supporters', he added 'sent Central Bank's indices soaring' (Cozier 1994). *The Weekend Nation's* Marilyn Sealy described the event as 'the million-dollar test match' (Sealy 1994). WICBC's executive secretary, Steve Camacho, is reported to have said that 'it is undoubtedly the best performance ever at Kensington'. This was confirmed by the executive secretary of the BCA, Basil Matthews (Sealy 1994).

'Local businesses and small entrepreneurs', the press reported, 'cashed in on the crowds by setting up several stalls outside the Oval and by throwing their bars open to the celebrating Brits' (Sealy 1994). Hired-car operations, restaurants, duty-free stores, hotels, liquor shops, food vendors, souvenir stalls and banks, all reported record sales for the month of April, while the 'outcry of hundreds of disappointed locals who were unable to purchase tickets' was drowned in the flow of gallons of rum and beverages (Sealy 1994). Cozier explained the predicament of local cricket supporters. While he sympathised with Barbadians who 'consider a seat at the cricket as part of their birthright', and were angered that English tourists 'occupied so much space', he also recognised that they 'cannot have their cake and eat it too. You either want tourism or you don't' (Cozier 1994).

Popular perceptions of alienation, furthermore, were deepened by the maintenance of WICBC policy provision that matches should not be televised for viewers in countries where the match is being played. Locals, then, were not only 'blocked out' but 'blacked out' with respect to the consumption of their cultural event. Transnational television networks, hooked up by satellite facilities, beamed the match into households across the world. English supporters in England could also watch every moment of the fascinating encounter. Only Barbadian homes, however, that were equipped with satellite dishes could participate in this global viewership. Those without a 'dish', or a 'ticket', resorted to archaic transistor technology and agonised in the embrace of radio.

The Englishman, then, had an all-round advantage. He could tan on the benches at the

Oval in Barbados or sit in his armchair with a can of beer before his set in England. West Indian cricket culture had become a global commodity, but not readily available to the villager who claimed to have produced it. The international division of labour had taken another twist. Barbadians could, however, watch five minutes of televised highlights of the previous day's play during the evening news, compliments of multinational media carriers such as Sky Television and Transworld Communications.

The editorial leader of the *Daily Nation* on Monday, 18 April, was not satisfied with these developments. The WICBC was painted by the paper as 'the last remaining bastion of power without responsibility'. It is a 'very strange decision', the editor stated, that the locals should be refused televised viewing facilities when the match was presold in England. The editorial continued:

> But the whole thing does not just end at counting the millions the Board makes by the sensible policy of pre-selling much of the Oval. There are people living here in Barbados who, as a direct consequence of the Board's decision, will be unable to attend the game. Their exclusion is foreseeable ... We have certainly reached the stage now where, as a community, we will not meekly be shut out of what's going on ... We expect that the BCA representative on the WICBC will appreciate our argument and use their influence to ensure that never again will the WICBC sell out the seats and still black out its potential patrons.
>
> (*Daily Nation*, 18 April 1994)

Barry A. Wilkinson also ended his column in *The Advocate*, Friday 15 April, with a warning to cricket officials. 'Do not repeat this folly in the future', he said. 'Our players need their supporters just as much as England or any other team would need theirs' (Wilkinson 1994).

In all of this, West Indians had little time to reflect on the feelings of English crowds who, since the 1950s, were called upon to absorb their unashamed support for the West Indies at English test match venues. The behaviour of the West Indies cohort at England's Kennington Oval is now legendary; located in Brixton – the heart of London's black community – it has witnessed West Indian calypsonians and revellers basking in English defeats. The English stiffened their upper lips, complained about the noise, but generally tolerated the West Indian bacchanal. The tables were now turned, and the 'cricket chicken' had come home to roost. The English, after years of exposure to the calypso culture, were now behaving like West Indians at the rendezvous of their victory. The 'Empire' had struck back.

THE BATTLE BEGINS: CULTURAL REPRESENTATIONS

Discouraging headlines in the Barbadian press greeted the English tourists – and their national team which had been 'demolished' in the previous match in Trinidad (all bowled out for 46 runs). 'West Indies set to keep England sliding'; 'Into the Lion's Den'; and 'England not good enough' screamed the *Barbados Advocate* in headlines on 8 April – the first day of the test. On 14 April, however, these headlines gave way to 'West Indies Surrender', 'England Break the Jinx', 'English Resilience and Character Paid Off' and 'England – Simply Triumphant!' Gayle Alleyne began a column in the *Daily Nation* on 14 April: 'A White Flag went up over Kensington Oval at 2.54 p.m. yesterday as the West Indies' cricket fortress surrendered to the persistent enemy' (see *Daily Nation* 14 April 1994; *Barbados Advocate* 18 April 1994).

The English media, not surprisingly, did not take the charge of the West Indian press lying down. They had their own style of communications warfare, fashioned by centuries of ideological terrorism in colonial parts, to direct against representatives of West Indian cultural claims. The battle of the subtext had begun with weapons more precise and destructive than those used on the field of play. The 'smart missiles' used by English journalists in the assault at Kensington struck at the centres of West Indian cultural sensibilities. They targeted its post-colonial sense of historical identity, moral authority and the intellectual worth of its cultural heritage. In short, the 'thinking bombs' were designed to obliterate the West Indian voice which spoke possessively and eloquently of its culture now projected so confidently as global fare.

The attack was led by Ian Wooldridge, an English journalist on the London *Daily Mail*. His account of the tour up to the Barbados test was designed to prepare English supporters for the trip to Barbados. It was also published in the Barbados press soon after their arrival. Wooldridge was less than subtle in his rejection of West Indian notions of cultural sophistication. In a Naipaulian sort of way he suggested that the West Indies – outside its persisting and proliferating criminal sub-cultures – is a place of nothingness. He insists upon recognition of a 'boundaries dispute' between West Indies and English cricket culture, and indicated the side and direction from which the barbarians were coming (see *Daily Nation* 13 April 1994).

The tourist brochure spiel, Wooldridge argues, that promoted the Barbados test was designed to inveigle English fans to the island, since it did not tell them that the 'Caribbean is a very tough place', 'born of imperial and free-lance piracy, funded by slavery and now riddled with drugs'. Neither did it tell them that the 'West Indies are beating all hell out of a thoroughly honourable England team' because the islands' team is made up of 'back street' boys for whom cricket is the only 'passport to a better life'. 'There isn't much literature here,' he states, and the West Indian male can pull himself 'out of the back streets' only by being 'pretty smart' at cricket (*Daily Nation* 13 April 1994).

During the week of the test, the 'Wooldridge Affair', as it became known, dominated the media. Barbadians of all shades took umbrage at Wooldridge's pejorative tone and insinuations. Writer and literary critic, John Wickham, in his *Daily Nation* column 'People and Things', dealt with the issue at length. For him, the Wooldridge intervention was 'loaded with transparent innuendo designed to induce a sense of inferiority in the West Indian psyche' (*Daily Nation*, 13 April 1994).

Joel Garner, the former West Indies test cricket star, suggests that the background to the matter has to do with a persistent ideological posture adopted by some in England who believe that 'they are better than the rest'. 'Having savoured the pleasantries of our island, often this is how we are repaid', Garner stated. 'We should encourage our visitors', he added, 'to interact with the locals for there is a wider interest at stake'. In summary, Garner informed readers that 'there is a cultural gap to be bridged' and that 'much of the fear born out of ignorance' can be abated and removed by mutual celebration of cricket's achievement (Garner 1994). According to Louis Brathwaite, sports analyst for *The Barbados Advocate*:

> Wooldridge was trying to scare off his countrymen from coming to this island paradise, but despite his pathetic efforts thousands of them are watching the current test ... They have been mixing without any fear with people from throughout the Caribbean and farther away, and are not only thrilled by the cricket action, but have been eating and drinking as heartily as any local person.
>
> (Brathwaite 1994)

Sports analyst for the Caribbean Broadcasting Corporation (CBC), and writer for the *Daily Nation*, Andy Thornhill, advised Wooldridge that when he sees West Indian cricketers he should not think in terms of emissaries of a sub-cultural criminality, but he should see 'a reflection of the rest of us. They are the mirror images that conceived Nobel laureates like Derek Walcott and Sir Arthur Lewis, musicians like Sparrow and Marley and other luminaries like Lamming and Jones'. Thornhill had no doubt that the West Indies would win the match and considered Wooldridge's missile part of a strategy 'to distract West Indians from their current assignment, side-track them in their quest to impose yet another 'blackwash' on the English (Thornhill 1994).

England's surprising devastation of the West Indies side cut deeper than the 208 runs chasm of victory. When, in 1950, West Indians defeated England at Lords, their first test victory against England on English turf, West Indians held a mass fête on the grounds. The bacchanal was led by Lord Kitchener and Lord Relator, whose calypso 'Cricket Lovely Cricket' is now legend. It was a moment of liberation for the West Indian – a conquest in the nascent anti-colonial struggle. England in Barbados repaid in full all the items on the cultural invoice. Now, unencumbered, the sounds of celebration filled the island under the canopy of dozens of Union Jacks carried by 6,000 plus skimpily-clad, sun-baked bodies.

Kensington (1994) and Lords (1950) were therefore linked on a terrain of cultural heritage made possible by the accumulative logic of post-colonial global capitalism. There was an understanding that the ritual on both sides had become, in fact, one cultural expression. When the 'white flag went up over Kensington' it meant that the 'West Indies' cricket fortress surrendered to the present enemy' (Alleyne 1994). The battle was over, and the victors celebrated in customary style:

> Spraying champagne and beer, a buoyant sea of Union Jack-waving Englishmen converged on the pavilion steps to pay homage to their newest heroes. 'Swing Low, Sweet Chariot', bellowed the jubilant fans who then crafted a new version of 'Jingle Bells': 'Jingle bells, jingle bells,/ England all the way,/ Oh what fun it is to beat the Windies today'. Hugging, shouting, jumping and, a few, even kissing total strangers, they savoured the rare and historic occasion.
>
> (Alleyne 1994)

Simon Daw, an English supporter, managed somehow to say to a journalist: 'I had a marvellous five days at Kensington – and I will be back' (Alleyne 1994). John Toppin, a rejected Barbadian summed it up: 'The fat lady is singing the Beatles now' (Alleyne 1994).

Former Minister of Sports in the Barbados Government, the legendary West Indies fast bowler, Wes Hall, was philosophical and looked at a bigger canvas: 'We have lost a test match, but the sports-tourism fusion has won. Thousands of sports enthusiasts will remember Barbados and tell their family and friends about the wonderful time they had here' (Alleyne 1994). Peter Roebuck, former English cricketer, now journalist, writing in the Barbados *Sunday Sun*, examined the entire process and event – the selling of West Indies cricket as cultural tourism – and summarised, with a touch of irony, 'W.I. cricket must watch cultural yorker' (*Roebuck*, 10 April 1996).

CONCLUSION

Big business it was too! The one-million-dollar match became a metaphor for the ideological representation of post-colonial cultural globalisation. 'What does it all mean?' Francis had asked (Francis 1994). The answer, it seems, is blowing somewhere between the winds of satellite television, the Empire decentred, global tendencies towards cultural homogenisation and, finally, the integration and dispossession of the villager in the global information village. Culture, it seems, no longer has a constitutional owner, as possessors and owners do not occupy the same physical space. Cricket heritage, as an economic commodity, belongs to a 'cultural nation' and the consumer is demanding this ideological value for money.

It's no sporting matter, this advancing post-colonial cultural globalisation. In the West Indies its values are most keenly displayed and contested in the cricket arena. At all levels of the game, radical changes are taking place. The overall effect is that the established official paradigm of cricket and national society is fractured and retreating, gradually being replaced by something less understood and arguably more undesired. An emerging new paradigm is rarely understood as such, and the distance between its form and its informing ideas often generates considerable public anxiety. One major energy source for this movement is the recent trend of cricket commodification by transnationals; another is the cultural assertion of post-colonial societies in response to the crisis of modernity. Together, they demonstrate the ease with which the local can become the global, and why thinking globally but acting locally has become the new institutional creed. But, once again, cricket functioned as a mirror within which to view the paradoxes, ambiguities and contradictions of change within the West Indian circumstance.

24

BOLLYWOOD VERSUS HOLLYWOOD[1]

Battle of the dream factories

Heather Tyrrell

INTRODUCTION

Theorisation around cinema and globalisation has largely been structured in terms of a basic opposition between Western commercial and culturally imperialist cinema, and Third World non-commercial, indigenous, politicised cinema. Much criticism of Hollywood and much support for alternative cinemas have been based on this understood opposition. 'Bollywood', North Indian popular commercial cinema, is an anomalous case which forces us to re-think the global map of cultural consumption and challenge the assumptions generally made concerning world cinema.

While India is not the only non-Western country with a commercial, popular, indigenous cinema – the cinemas of Hong Kong, China, Mexico and Brazil could be similarly described – its film industry is at this time experiencing rapid changes which make it a particularly pertinent subject for examination. ' "Bollywood" has become widespread [sic] nomenclature for the Indian movie industry in recent times' (Reuben 1993: 3) and amalgamates two names: 'Hollywood', and 'Bombay' (India's commercial hub, now renamed Mumbai). But is Bollywood named in imitation of Hollywood, or as a challenge to it? For many years commentators have assumed the former, but if Bollywood were simply a substitute for Western film while economic barriers prevented the import of the original, once those barriers collapsed it would be expected that Bollywood would collapse too.

However, Indian film culture has not been undermined or devalued by the recent influx of Western product as some expected, and multinational companies have not succeeded in dominating the prized Indian market. The role and the importance of popular Indian cinema culture has been misinterpreted or underestimated by external commentators, perhaps precisely because it does not fit easily into the theoretical model developed around the dichotomy of First World and Third World cinema.

This chapter will begin by situating Bollywood within and against the theories formulated around Third World film known as Third Cinema theory.[2] I will then go on to relate theory to practice by looking at three aspects of the Bollywood film industry: first, the current volatile period of change in India's film and media culture, as the international film industry attempts to enter the Indian market; second, Bollywood itself as an international film industry, in terms of production, distribution and exhibition; and third, oppositions to Bollywood as a dominant cultural force in India. By focusing on these areas I hope to demonstrate why Bollywood is fertile theoretical ground for Development Studies and Cultural Studies alike, and may force us to rethink how Third World popular culture is read.

BOLLYWOOD AND THIRD CINEMA

'Third Cinema' is a term coined originally by Argentine film-makers Fernando Solanas and Octavio Gettino (Allen 1996: 53), and generally applied to the theory of cinemas opposed to imperialism and colonialism. Bollywood, as a commercial popular cinema, has a problematic relationship to theories of Third Cinema, which assume a non-commercial, minority cinema as their subject.

In discussions of world cinema, the mainstream is generally taken to be North American and European cinema, with others as oppositional, marginal, and most significantly, non-commercial. Bollywood, the most prolific film industry in the world, and one with an international commercial market, challenges this assumption. Bollywood films are not solely politically motivated, nor are they entirely devoid of nationalist/anti-colonialist content. They are at once 'escapist' and ideologically loaded.

In *Questions of Third Cinema*, Jim Pines and Paul Willemen (1989) talk about Third World films as 'physical acts of collective self-defence and resistance'. Bollywood can be read both as defending itself and Indian values against the West, and as a dangerous courier of Western values to the Indian audience, and is read in both these ways by the Indian popular film press. A constant process of negotiation between East and West takes place in Bollywood films, operating both in terms of style (narrative continuity, *mise-en-scène*, acting styles), and in terms of content (the values and ideas expressed in the films). Indian cinematic style negotiates the cinematic traditions of Classical Hollywood, while its content addresses the ideological heritage of colonisation; just as, in the 'picturisation' of a single film song, hero and heroine oscillate between Eastern and Western dress in a rapid series of costume swaps as they dance and mime to music which is itself a hybrid of Eastern and Western styles.

But does this negotiation, and its often overt anti-Western agenda, qualify Bollywood as Third Cinema? A cinema does not automatically qualify for the title because it is produced in and for the Third World. Argentine film-makers Fernando Solanas and Octavio Gettino, defined any 'big spectacle cinema' financed by big monopoly capital as First Cinema, 'likely to respond to the aspirations of big capital'. Third Cinema was 'democratic, national, popular cinema' (quoted by Allen 1996: 53). But both these statements can equally be applied to Bollywood, which, despite its prolific commercial profile, has always been refused industry status by the Indian government, and which, historically, received subsidies from Nehru's government to pursue an explicitly anti-colonial agenda.

Fidel Castro fiercely criticised Hollywood in his closing speech at the 1985 Havana Film Festival:

> They are poisoning the human mind in incredible doses through commercial cinematography, grossly commercial. [Third world cinema must be supported, because] if we do not survive culturally we will not survive economically or politically.
>
> (Quoted by Allen 1996: 53)

Compare this speech with an article by Shah Rukh Khan, India's top film star, in 1996, defending Bollywood's commercial film industry in an introduction to a feature on 100 years of Indian cinema in *Movie International* magazine.

> I'd like to stress we are part of world cinema and we are making films – films we like, not for film festivals ... Mark my words one day Indian cinema will rule the world. Once we get the technology we are going to kill them.
>
> (Khan 1996: 44)

Khan's military metaphors are directed explicitly against the West, and not only against Hollywood and commercial cinema, but also against the independent, alternative cinema of 'film festivals' – cinema that could, in many cases, be described as 'Third Cinema'.

Bollywood seems both diametrically opposed to, and fiercely aligned to, Third Cinema. This confusion arises because commercialism has been exclusively identified with the West in cultural criticism, without taking into account a non-Western, even anti-Western commercialism. A cinema which is

Plate 24.1 Front cover of *Movie International* magazine, February 1996 issue, showing (clockwise from top right) Shah Rukh Khan, Madhuri Dixit, Shammi Kapoor, Sharmila Tagore, Rekha, Raj Kapoor, Amitabh Bachchan, Dilip Kumar and (centre) Madhubala.

Source: By kind permission of *Movie International*.

both commercial and concerned with 'decoding ... the deemed superiority of the West' (Gerima 1989, quoted in Allen 1996: 55) problematises established theoretical oppositions of East and West. Some of the strategies of Third Cinema can be applied to it, but so can some of the criticisms levelled at Hollywood. Vijay Mishra in his essay *The Texts of 'Mother India'* (1989) argues that Bollywood cannot be seen as Third Cinema, despite its 'defiantly subversive' stance (Mishra 1989: 136), because it is ultimately conforming: 'popular Indian Cinema is so conservative and culture specific as to make a radical post colonial Indian Cinema impossible' (Mishra 1989: 134–5).

Third Cinema is commonly perceived as 'serious' cinema, challenging in an aesthetic as well as a political sense. Bollywood films generally include light-hearted song-and-dance numbers, causing Tim Allen to dismiss them from the Third Cinema equation in his dossier on Third Cinema: 'In India serious films are not generally very popular at all. Most cinemas show jolly musicals ... ' (Allen 1996: 55).

However, as Mira Reym Binford says in her essay *Innovation and Imitation in Indian Cinema*:

> the obligatory song-and-dance sequences of the Indian mainstream film are a striking example of indigenously based aesthetic principles [with remote antecedents in the traditional Sanskrit drama] shaping the use of imported technology.
>
> (Binford 1988: 79)

These very song-and-dance sequences are a form of opposition to Western cultural imperialism. Also, Bollywood films are not musicals alone; they are an 'Omnibus' (Vasudevan 1995) or a 'Masala' (Binford 1988) form, combining melodrama, action, comedy, social commentary and romance, violently juxtaposing intensely tragic scenes with jolly song and dance numbers, jolting the viewer from one extreme of feeling to another (an aesthetic similarly inherited from Sanskrit theatre).

However, if Bollywood has not developed, stylistically, as 'serious' Third Cinema, neither has its style much connection with Hollywood. Indian cinema has developed a film language which has little or nothing in common with the codes of classical Hollywood cinema and, ironically, this has caused some critics to dismiss Bollywood as escapist. Modes of presentation termed escapist according to the classical Hollywood mode, like the song-and-dance number, are, however, used to play on 'deep tensions – between wealth and poverty, old and new, hope and fear' in Indian films (Barnouw and Krishnaswamy, quoted in Tyrrell 1995: 2). For example, in the 1996 film *Army*, a song-and-dance routine breaks out in a prison compound, and prisoners sing, while cartwheeling about the exercise yard, that poverty is so extreme in Indian society outside the prison walls that they are better off in jail, under a death sentence, because a death sentence hangs over them even outside prison.

Ironically, while, from the outside, Bollywood is popularly viewed as a more escapist cinema than even Western commercial cinema, it has absorbed within it as successful commercial product a number of challenging and 'serious' films that in the West achieved only a small, independent distribution. Shekhar Kapoor's *Bandit Queen* (1995) was among the top ten grossing films of 1996 in India, over a year after its small-scale, independent release in the UK, and made $1 million in its first week of Indian release. The harrowing film is based on the life of outlaw Phoolan Devi, and confronts head-on the abuse of women in Indian society: Bollywood's aesthetic evidently cannot be dismissed as 'frivolous' if a film this 'serious' can achieve such enormous commercial success.

'HOLLYWOOD RAISES HELL IN BOLLYWOOD'[3]

Hollywood/Bollywood relations are at a moment of crux, as the lifting of the ban on dubbing foreign films into Hindi in 1992 has left Hollywood free to enter the Indian market. However, audiences have shown little interest in Western imported film product; the barriers against the West are revealed as cultural, not simply economic, and 'Hindi films' have, effectively, 'triumphed over Hollywood in India' (Rajadhyaksha 1996). Media coverage taken from the British and American film press, of the attempts of Hollywood to dominate the last remaining world market, chart some of the assumptions made, and broken down, before and during the current surprising impasse for Hollywood in India. In comparison, the discourses around East and West, film and culture, that are used in the Indian press, are just as dismissive, even hostile, towards the West, and believe just as confidently in the greater merit of their own cinema.

The *Guardian*'s film critic, Derek Malcolm, warned that 'a giant culture clash [was] looming' in India, as 'Spielberg's *Jurassic Park*, dubbed into Hindi, [had] given a fright to the massive Indian film industry' (1995: 12–13). An article in *The Sunday Times* in June 1995, 'Hollywood Raises Hell in Bollywood' (Lees 1995) predicted doom for the Indian film industry following the release of action movies such as Sylvester Stallone's *Cliffhanger* (1993), with its higher production values, and, as Lees quotes Indian sources as saying, 'machine guns instead of rifles'. However, the reception of Stallone's *First Blood* (1982), renamed *Blood* and released in July 1995 in India, was lukewarm. Trade reviews commented that 'the film holds appeal mainly for action film lovers', and judged its publicity and opening 'so so' ('Blood (dubbed)' *Film Information,* 1995: 4).

BOLLYWOOD VS HOLLYWOOD

The reasons for Bollywood's resistance to colonisation by Hollywood are aesthetic and cultural as well as political. The formula for Bollywood films has been jokingly summarised as 'A star, six songs, three dances' (Barnouw and Krishnaswamy 1963: 148), and these Omnibus or Masala films must have the right mix of a diverse range of ingredients to satisfy their audiences. Without them a film 'lacks in entertainment value' (*Film Information,* 1995: 6).

However rigid this formula, adherence to it does not guarantee a film's success. Only one in ten films makes a profit, and whether a film is a hit or a flop depends on the unquantifiable judgement of the Bombay audience, who either fill or desert cinema houses in a film's first week of release. Films which imitate the formula of previous hits sink without a trace, while others appear from nowhere to become blockbusters. As Subhash K. Jha remarks in *g magazine*: 'The vagaries of the box-office have flummoxed film-makers and trade watchers forever' (Jha 1995: 23). If Indian filmmakers are unable to guarantee audiences, Western film product is unlikely to do so.

The market for undubbed Western films in India before 1992 was very small, consisting only of an English speaking middle-class élite, and Western films had far shorter runs than Hindi films. Hollywood first attempted to attract Indian audiences after 1992 by dubbing major American hits into Hindi, but so far only a fraction of the films released have been commercially successful with the Hindi-speaking mass market. *Jurassic Park* (1993), *Speed* (1994) and *Dunston Checks In* (1995) – colloquially translated as 'A Monkey in a

Hotel' – have been box office successes, but others, such as *Schindler's List* (1993), *The Flintstones* (1994) and *Casper* (1995), have 'bombed' (Lall 1996: 14). Even those films which did not 'bomb' achieved only a fraction of the success of domestic Indian films: in the same year that *Jurassic Park* grossed $6 million, *Hum Aapke Hain Koun ... !* (1994) grossed $60 million. Hollywood has not yet discerned a pattern as to which films succeed and which 'flop' in India.

One significant factor in films' successes, which may be too culturally specific for Hollywood to duplicate, is their music. 'Popular music in India is synonymous with film music' (Ranade, quoted by Kumar 1988), and the popular film and music industries in India are interdependent. Not only does Indian popular film depend on music, Indian popular music also depends on film. Peter Manuel (1993), in his book *Cassette Culture*, explains the history of this symbiosis in economic terms; before the cassette revolution in the 1970s, the cinema was the most accessible way to hear popular music for many Indians. Film music is also culturally important; as Sanjeev Prakash (1984) notes in his article *La musique, la danse et le film populaire [Music, Dance and Popular Film]*, film music so pervades Indian culture that it is played even at marriages and religious festivals.

Plate 24.2 Drawing of Amitabh Bachchan by Deepa Mashru, age nine, from Leicester.

The star system too is a formidable force in India, and another factor excluding Western cinema. The earliest Indian films were known as 'mythologicals', portraying the adventures of Hindu gods such as Krishna, and the equation of actors with gods has remained. Many Indian film stars go on to become politicians and national icons, representing quintessential 'Indianness'. Nargis, 'the woman in white', was the personification of 'Mother India' in the 1950s; Amitabh Bachchan has been India's greatest cinema icon for thirty years, and his face has come to be used as a symbol for India itself (as we shall see later). Western stars cannot compete with such quasi-religious iconography. A recent Hindi film, *Rock Dancer* (1995), starring Samantha Fox, a British glamour model turned pop singer, singing all her own songs in Hindi, received very little press attention and no commercial success. Though the urban middle classes knew her name well enough to merit an aside in one film news column, to the mass Hindi film audience, she was an unknown.

Having largely failed to export Western product to India, Hollywood is now investing in Indian studios – putting money into Bollywood, not attempting to replace it with its own product. The Indian view of this seems to be of a cultural victory; as Shah Rukh Khan (1996) expresses in his piece, 'Soon Hollywood will come to us'; but economically this is no great victory for India over the West, since profits from what appears a quintessentially Indian product will now go back to the West.

The Indian cinema box office was not essentially diminished by the rise of video in the 1970s, but now Bollywood must accommodate satellite and cable expansion.[4] Rupert Murdoch's Star network attempted to sell Western programmes in India, but could only attract élite minority audiences; but when an Indian company set up a Hindi satellite channel, Zee TV, they attracted a far larger market, and were the impetus for a whole

Plate 24.3 Publicity poster for *Cliffhanger* in Hindi.

Source: By kind permission of Columbia Tristar/*Film Information*.

industry of Indian satellite and cable channels, which Star have now bought into. As interviews with Zee TV and Star TV spokesmen (they were all men) showed, both Indian and Western companies interpreted this as a victory. The Indian company believed they had beaten Star at their own game and reaped the rewards. Star felt they had finally found a way to infiltrate the Indian market, by using an Indian figurehead company. The successful move of multinational media companies into the Indian market was ultimately demonstrated, however, when the 42nd Annual Filmfare Awards, otherwise known as 'the Indian Oscars', were screened exclusively on Sony Entertainment Television's Hindi Channel in March 1997.[5]

POPULAR DISCOURSES OF HOLLYWOOD/BOLLYWOOD OPPOSITION

Both Hollywood and Bollywood have made their direct opposition explicit in India, and their rivalry has passed into popular cultural vocabulary. The promotion poster for Stallone's *Cliffhanger* (1993) reads 'Hollywood challenges Bollywood'; Hollywood's decision to choose *Cliffhanger* as the vehicle for its challenge was perhaps based on a superficial reading of contemporary Indian film as high in action content, without taking into consideration its juxtaposition with other elements of the 'Masala' mix, such as song and emotional melodrama. *Cliffhanger*'s challenge failed. In contrast, as one Indian trade paper commented, a series of Indian music cassettes entitled 'Bollywood vs Hollywood' have been highly commercially successful.

Within Indian popular culture, the commercial success of Indian cinema has become emblematic of India's resistance to the West, and Bollywood stars have become figureheads in what is viewed as a battle against Western-

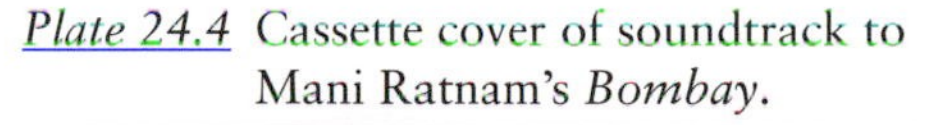

Plate 24.4 Cassette cover of soundtrack to Mani Ratnam's *Bombay*.

Source: By kind permission of Spark International.

isation. Actress Madhuri Dixit, known as Bollywood's 'queen bee', 'drew herself up and lectured the guy on patriotism' when a fan 'offered her a Canadian dollar for an autograph' (Giggles 1996). I have already mentioned the nationalist sentiments expressed by actor Shah Rukh Khan in a *Movie* magazine feature. Another instance is an advert for BPL (an Indian electrical hardware company) which appeared in *g magazine*, a leading Indian English-language film magazine, every month from October 1996 to January 1997. The advert combines a photograph of film star Amitabh Bachchan with discourses around national pride. December's advert concludes:

> Who would have guessed a few centuries ago that India would become a poor, Third-World country? And who knows what India will become in the next century? Who knows what may happen if we believe in ourselves?
> (BPL 1996: 63)

Hollywood's failure to supersede Bollywood reveals that an existing Third World culture can be a crucial factor in halting Western cultural imperialism, even when political and economic barriers are lifted. Barnouw and Krishnaswamy (1963) describe in *The Indian Film* how Hollywood monopolised the world cinema market during the First World War, while other film producers were handicapped by the loss of resources and labour-power to the war effort, and successfully defined the cinematic experience for the rest of the world according to their product, so that, in effect, politics shaped economics shaped culture. However, Hollywood has not defined what makes a film work in India, where, conversely, cultural disparity, rather than any political or economic factor, has slowed Western commercial expansion.

BOLLYWOOD AS INTERNATIONAL FILM INDUSTRY: PRODUCTION, DISTRIBUTION, EXHIBITION

The very mechanics of Bollywood's film industry form a cultural barrier between it and Hollywood; mechanics which have been exported to the diasporic Indian film industries such as the one in the UK. This section will look at how the processes of production, distribution and exhibition of Bollywood films carry their own cultural agenda. I will use the UK as my case study, although it can by no means be taken as a universal model, because my own position within the cinema exhibition industry in Britain has given me a particular insight.

Bollywood is no less an international industry than Hollywood, with a worldwide distribution system which ensures that films are available abroad almost immediately after their release in India. The international market for Indian films includes not only other South Asian nations, and South Asian communities in Britain and North America, but also non-South Asian audiences in the Middle East, Africa, the Caribbean, China, and even Eastern Europe and the former USSR, as Rosie Thomas points out in her article *Indian Cinema: Pleasures and Popularity, an Introduction* (1985: 116). She also says that:

> ... they are frequently consumed more avidly than both Hollywood and indigenous 'alternative' or political cinemas. Such preference suggests that these films are seen to be offering something positively different from Hollywood, and in fact, largely because it has always had its own vast distribution markets, Indian cinema has, throughout its long history, evolved as a form which has resisted the cultural imperialism of Hollywood.
> (Thomas 1985: 116)

Indian film distribution in Britain and North America presents a paradox, however, for while Bollywood enjoys popularity among white Eastern Europeans and Russians, in Britain and North America the industry is rigidly delineated along ethnic lines, so that there exist two parallel universes of Hollywood and Bollywood, with no commercial crossover between them, operating simultaneously in the same nation. The dominant culture is virtually blind to the parallel culture's existence, so that, while Indian films are often shown at a local cinema every week, they are advertised exclusively within the Indian community, and never featured in film listings in the national press. Bollywood culture remains a closed culture, so that one Indian film, *Bombay* (Ratnam 1995), was given its 'national première' (in a 'Western cut' form, minus songs) at the Edinburgh Film Festival in 1995 when it had already been playing to Asian audiences in Britain for six months. I will consider why this mutual blindness has come about below.

CINEMA DISTRIBUTION AND EXHIBITION

Bollywood's use as cultural resistance is reflected in the political economy of the Indian film industry in Britain, which runs according to the Indian, and not the British, European and North American model. India's rejection of the international model for distribution, established predominantly by the US film industry after the First World War, was perhaps a part of the post-independence rejection of the coloniser and the technological trappings of colonisation.

In India, and throughout the international Indian film industry, distributors allocate their territories to agents, and exhibitors pay a flat fee, not a percentage. The size of these fees necessitates large audiences for films to break even, and in the UK inflated ticket prices (£5–6 instead of £3–4), which differentiate Bollywood films from the Western film economy. There have been moves to change the system to a percentage one in India, perhaps prompted by new partnerships with Western distributors and the building of Western-style multiplexes, but this will doubtless take time to trickle down into diasporic industries which are consciously choosing an Indian model in opposition to the Western one, an economic choice which forms part of their culturally embattled stance against the host country's dominant culture.

There exists an international Indian film press, some of which, such as *Movie* and *Cineblitz*, is published in the UK, but inhabits an ambivalent space between India and the Indian Diaspora, referring to India as 'our home'. By doing so, the magazines invite readers to collude in a cultural displacement in which they imagine themselves back in India, not only when watching films, but also when reading about them.

VIDEO

The video boom of the 1970s affected Bolly-

Plate 24.5 Still from Mani Ratnam's *Bombay*.

Source: By kind permission of Spark International.

wood output by provoking more commercial films, and by the 1980s quality had been sacrificed to action, sex and violence, as studios competed with television and video, producing larger quantities of more explicitly commercial product. But although it affected film quality, television and video did not affect cinema audiences in India. It did, however, effectively destroy Diasporic Indian cinema audiences; as more affluent Indians in the West invested in video recorders, Indian cinemas across Britain (of which there were several) were forced to close. The video shop took the place of the cinema as the social hub of Indian Diasporic communities in Britain throughout the 1970s and 1980s. The viewing of Indian film became a more private pursuit for Diasporic Indians, contributing perhaps to the disappearance of Indian film from mainstream British consciousness and the 'mutual blindness' I referred to earlier.

The Indian video industry in the UK is itself now under threat, however, not only from the rise of satellite and cable, which enable easier access to Indian film directly from the home, but also from the steady rise of video piracy. Certification of Indian films in Britain was only introduced as recently as 1994 (perhaps because the videos, in Hindi and with no subtitles, presented a problem of access for the British Board of Film Classification), necessitating an additional expense for video distributors, which made 'official' videos considerably more expensive to hire than pirate versions of the same film. Also, film distributors often delayed the video release of a film which was still attracting large cinema audiences in India, increasing demand for pirate video copies (generally filmed on a camcorder in a cinema auditorium), particularly in the UK, where there was little access to Indian cinema exhibition. The 1994 hit *Hum Aapke Hain Koun ... !*, for example, was not released on video until 1998, four years after its cinematic release. Piracy has estranged the Indian film political economy still further from that of the mainstream in the UK, where video piracy has been more effectively controlled.

It is the delay in releasing popular films on video which has prompted a revival of Indian cinema exhibition in the UK, however, in initiatives frequently led by the video store owners themselves. At first commercial cinemas were rented for single screenings once a month in major cities, but hits such as *Hum Aapke Hain Koun ... !* secured large enough audiences for more frequent screenings, until exclusively Indian cinemas began reopening in large Indian population centres, such as Southall in London, and Leicester. Independent cinemas throughout the UK have also begun to screen Bollywood films as part of their cultural remit to cater to minority groups. While screenings were only in commercial cinemas they remained invisible outside local Indian communities (Indian nationals studying in the UK whom I interviewed in 1995 were unaware it was possible to see Indian cinema in the UK, although regular screenings were taking place only a mile from where they lived). However, the opening of Indian cinemas and screenings in independent cinemas have attracted more mainstream interest, exposing the Indian film industry's 'invisible' economy to mainstream scrutiny for the first time. Nevertheless, until recently distribution of Indian film in the UK remained aligned to the Indian rather than the British model.

SATELLITE/CABLE TELEVISION

Zee TV is not only an all-India phenomenon, it is now available internationally – which alters the East/West media equation quite dramatically. So far, critiques of the phenomenon of globalisation have assumed that now India has opened its doors to the West, its own media will be overwhelmed by the higher technology levels and better production values of Hollywood films, and British and American television. But at the same time as India is being flooded by Western media, the same process is operating in reverse. The same globalisation process has given Indians in the West access to Asian cable networks such as Zee TV and Asianet, with the result that they have switched over from nationally oriented television networks to international, ethnically oriented ones. *Movie International*'s 'Opinion Poll 95' revealed that 54 per cent of readers watched Zee TV, compared with 21 per cent who watched

BBC1, 11 per cent who watched ITV, 7 per cent who watched Sky TV and 3 per cent who watched Channel 4 (ironic, in view of Channel 4's remit to programme for minority audiences).[6] This was an international poll, so the British Asian audience is only a percentage of the respondents, but it should be pointed out that *Movie* is an English-language magazine, so linguistic barriers were not preventing readers from watching other channels. Their viewing preferences were entirely culturally led.

The same technology which allows the Western media to move into foreign markets has enabled the Western 'home' market to divide along ethnic lines. As Zee TV will move into other global markets at the same time as Western cable and satellite networks, including markets where Indian films are already more popular than Western ones, Hollywood's global dominance is by no means a foregone conclusion.

Plate 24.6 Sanjeev Baskar, presenter of BBC 2's *Bollywood or Bust!*.

Source: By kind permission of Ken Green/BBC.

'BOLLYWOOD OR BUST!'

Bollywood's 'parallel universe' situation of cultural invisibility in the UK is already beginning to change in Britain, despite the fact that films and videos themselves remain largely unsubtitled in English. *Bollywood or Bust!*, the BBC's Indian film quiz, has acquired a cult following outside the Indian community; television adverts have begun quoting Bollywood cinematic styles, as in a 1995 advert for Diesel jeans; and items on the Indian film industry have appeared on film programmes such as Channel 4's *Moviewatch* (1996).[7] Films like the British-Asian *Bhaji on the Beach* (1993), and the British-French-Indian collaboration *Salaam Bombay!* (1988) have quoted or parodied the Bollywood film style, bringing a basic awareness of its aesthetic to a wider audience. Indian cinema's visibility in the West has increased enormously over the last two years, coinciding with the Western move into the Indian film market.

In the West, Bollywood is viewed as an alternative to the mainstream, not only by Indians, but increasingly by a wider public, while ironically, in India itself, it has always symbolised adherence to it. The process of cultural influence is working two ways; recent media attention suggests that Western audiences are beginning to become aware of the cultural importance of Bollywood, even if it remains aesthetically impenetrable. Indian product, in its turn, is showing the influence of new Western collaborations, so that the 'parallel universes' are beginning to touch.

OPPOSITIONS TO BOLLYWOOD

To see Bollywood purely in terms of its oppositional relationship to Hollywood, however, is to overlook the similarities between the two industries; it must be remembered that the Bombay film industry's nickname is an imitation of, as well as an opposition to, the West. This section will look at the hegemonic motivation behind Bollywood's attempts to be an 'All-India' cinema, and at voices of opposition to its implied North Indian Hindi nationalism, first by considering a number of critiques of Bollywood, and second by looking at the work of some Indian film-makers popularly seen as 'oppositional'.

Hindi or Bollywood movies have also become known as 'All-India films', with a specifically unificatory agenda. In his essay 'Addressing the Spectator of a "Third World" National Cinema', Ravi Vasudevan (1995) notes the intention of Nehru's post-colonial government to use the Hindi film to unite an ethnically disparate nation, with Hindi as its lingua franca. Ashish Rajadhyaksha and Paul Willemen, in the *Encyclopaedia of Indian Cinema*, explain the term 'All-India film' as 'an all-encompassing entertainment formula designed to overcome regional and linguistic boundaries' which also offered a form of 'cultural leadership' at a time when strong leaders had not emerged 'because of the hiatus between the intelligentsia, to which the leaders belong, and the masses' (Rajadhyaksha and Willemen 1994: 41).

The generic nature of Bollywood as Masala film is not merely an aesthetic, but also a cultural and political consideration. The Hindi film is designed to function as a lingua franca in a country of over 500 languages, where many of the audience understand little or no Hindi; the use of slapstick comedy, melodramatic acting styles and song-and-dance numbers ensures films retain entertainment value for those with limited understanding of Hindi. However, the use of Hindi as a lingua franca is fiercely contested within India, and this is one instance where Bollywood falls down against the 'Third Cinema' model. Regional cinemas oppose Bollywood's domination throughout South Asia: South Indian Tamil and Telegu cinemas now have a higher rate of production than Bombay, so that, effectively, the South Indian film industry is bigger than Bollywood. Bengal too has a fiercely defended regional cinema; so much so that cinemas screening Hindi films are regularly bombed in Calcutta.

Religion, too, forms an area of contention surrounding Indian film. It has been said that religion and film are the two great obsessions of India, and the two together form a volatile mix. There is a long history of tension between Hindus and Muslims in Bollywood, as in the whole of India. Ravi Vasudevan covers the history of this tension throughout 1950s film culture. He notes that the push towards unification within the 'All-India Film' was often accompanied by a Hindu nationalism which marginalised Muslims, including those within the film industry. Muslim actors were required to adopt Hindu pseudonyms, and a pro-Hindu subtext within the films of the 1940s and 1950s led to scenes such as one where two Muslim actresses, Nimmi and Nargis, kiss the feet of Hindu actors Premnath and Raj Kapoor in *Barsaat* (1949); a scene read in overtly racial terms by critics of the period. This tension is still alive: a hit film of 1995, *Bombay*, told the story of the 1992 Hindu/Muslim riots through that of a Muslim woman who runs away to marry a Hindu man. Both Hindus and Muslims protested at the story of inter-racial reconciliation, however, and the film provoked rioters to burn down the cinema in Bombay where the film was due to be screened, as well as the attempted assassinations of the lead actress and the director.

Third Cinema is taken to be cinema by and for the underclass, but Bollywood takes an élite position in relation to India's lower classes. Bollywood is seen to serve the poor of India, as this quote from Manmohan Desai, an Indian film-maker, shows:

> I want people to forget their misery. I want to take them into a dream world where there is no poverty, where there are no beggars, where fate is kind and God is busy looking after his flock.
>
> (Manuel, quoted in Tyrell 1995: 1)

However kind this 'service', it is not part of an empowering strategy of the type Third Cinema embraces. Behind these generous sentiments is the assumption of an 'us and them' stance; 'we, the privileged, must look after the faceless masses who depend on us'. But 'looking after' does not mean offering empowerment: it means offering placatory escapism.

While for many Indians abroad Bollywood becomes synonymous with India itself, many Indians in India take a more oppositional stance towards it. One Indian student I interviewed in Glasgow commented: 'The rate of generation of Hindi movies is really very, very high, but really very few are good enough to watch, on an intellectual level' (Tyrrell 1995: 64).

The belief held by many of India's intellectual élite is that Bollywood films, as escapist fantasies portraying casual violence and sexuality, are a corrupting influence on the masses who consume them. This is only the other side of the same attitude of 'benevolent paternalism' outlined above, however; still the underclass 'cannot' think for themselves. The Indian intellectual élite depict Hindi film's mass audience as passive, undiscriminating and receptive to ideological conditioning; as children in fact. This colours Indian notions of censorship. Peter Manuel quotes actor-director Raj Kapoor as saying:

> The best entertaining film is a film that does not raise any controversy. In a democracy of ours ... one has to be very careful as to what kind of fare to present and how much of truth you can present along with that.
> (quoted in Tyrrell 1995: 16)

Issues of imperialism become covert not overt with increased commercialism, and John A. Lent suggests in *The Asian Film Industry* that films often avoid political issues to appease the censors, who are themselves often eager to appease the government in power for their own betterment (Lent 1990).

The issue of censorship has come to the fore in Indian film press coverage of two films which could be considered 'oppositional' in their relationship to Bollywood. One is Shekhar Kapoor's *Bandit Queen* (1995) and the other is Mira Nair's *Kama Sutra* (1996). Kapoor fought to have his film released in India uncut for a year, eventually compromising and agreeing to 'a few minor cuts' (Somaya 1995). The film was the biggest grossing hit in India in February 1996, the month it was finally released, and May 1996 saw 'the election of the real-life bandit queen, Phoolan Devi, to the Indian parliament' (*Screen International Film and TV Year Book* 1996: 25). Mira Nair, director of *Salaam Bombay!* (1988) – a commercial and critical success – and *Mississippi Masala* (1991), and considered by many to be a proponent of Third Cinema, is currently campaigning against proposed cuts to her provocatively titled new film. The dispute began in December 1996 and her film, released in North America in Winter 1996, and in Britain in Summer 1997, was, like *Bandit Queen*, seen in the UK long before it was seen in India.

Strangely, neither Kapoor nor Nair are viewed as oppositional film-makers in India in the same way as they are in the West. Kapoor's anti-establishment stance has been viewed with some cynicism by the Indian film press since *Bandit Queen* made a sizeable profit in India, and he has been accused of 'selling out' to the West by accepting funding from Channel 4 by film-maker J.P. Dutta, currently completing *Border*, a controversial film on the India-Pakistan war. Mira Nair's earlier film, *Salaam Bombay!* (1988), has been read by Western critics as setting up a conscious opposition to Bollywood (although it was in fact commercially successful in India); the street children who acted in it were discouraged from imitating the 'false' acting styles of Bollywood film stars. However, *Kama Sutra* stars Rekha, one of Bollywood's leading film icons, and the popular Indian film press has given the forthcoming film extensive coverage, treating it no differently to a commercial Bollywood release. This despite the fact that the film is in English, not Hindi. Nair, whether she likes it or not, has been absorbed into the cinema she set herself in opposition to.

Just as the Bollywood aesthetic has been shaped without reference to the West, so India's definition of oppositional cinema sits awkwardly in relation to those made in the West. This disjunction is amplified when considering Satyajit Ray. He is perhaps the best-known Indian film director outside India, but his relationship with Indian cinema is in fact an awkward one. Ray is, more or less, the Western view of Indian Third Cinema, but viewed from within India, his position is considerably more problematic. Ray's book *Our Films Their Films* (1976), is not, as its title would suggest, a defence of Indian cinema and a critique of Western cinema. It is instead Bombay cinema to which Ray places himself in opposition, with chapter headings such as 'What is wrong with Indian films?' (1976: 5). East and West come to stand not only for world divisions, but also for divisions between eastern and western India, the West becoming Bombay.

Ray commented of his own film-making: 'We learned from the Indian cinema what not to do'; and instead took his model from European film-makers such as Jean Renoir. If, as Tim Allen and Simon Blanchard suggest (1996a: 65), 'there has been little interest in his work in India', it is because his films are made according to Western codes, and consequently, it could be argued, for Western audiences. How anti-colonial, how 'Third Cinema' is a film-maker of whom it can be said 'Ray was too sophisticated an artist for most Indians, including Bengalis' (quoted by Allen and Blanchard 1996a: 66). Who is defining 'sophistication' here?

Who, then, are oppositional film-makers according to India's own definition? Subhash K. Jha (1996: 25) traces the beginning of India's 'Parallel Cinema' to the end of the 1960s and the work of directors Mrinal Sen and Basu Chatterjee, but restricts his discussion to films which had some commercial success, arguing that: 'If a work of art fails to communicate itself to a sizeable audience, if its appeal remains restricted only to a handful of intellectuals it cannot be regarded as a success' (Jha 1996: 25). The Parallel Cinema directors he singles out as most significant are M.S. Sathyu, Shyam Benegal, Basuda, Robin Dharmraj (whose 1980 film *Chakra*, he notes, predated Mira Nair's *Salaam Bombay!* by almost ten years in its depiction of slum life), Govind Nihalani, Gautam Ghose, Ketan Mehta, Kundan Shah and Kalpana Lajmi; names little known in the West.

In fact there seems little consensus even within India as to who are the most important film-makers of the Indian 'New' or 'Parallel Cinema'; a value judgement apparently dependent on which part of India the critic is writing from. As already mentioned, an embattled regionality informs much internal debate around Indian cinema, so much so that in fact many critiques which ostensibly discuss East–West post-colonial relations, in fact use them to air grievances against another region of India. Bidyut Sarkar, in his biography *The World of Satyajit Ray*, comments on Ray's Oscar win late in his career: 'The honour could have come to the nation much earlier if the Bombay-dominated film industry had taken a larger view of its responsibility and not sent year after year entries typical of its kind ... ' (Sarkar 1992: 101).

In other words, India could have won an Oscar sooner if more Western-style product had been submitted. Confusingly, Bollywood is at once too Western and not Western enough for its detractors. However, these contradictions make more sense when viewed in the context of India's internal regional conflicts, which reveal the difficulty of defining 'Indianness' itself, and also begin to explain why Indian cinema has absorbed so much 'oppositional' cinema within the 'mainstream'. To survive as a nation, made up as it is of a myriad of conflicting ethnicities, India itself must constantly absorb oppositions. Thus, despite his hostility to the Indian commercial film aesthetic, and his inaccessibility to the mass Indian audience, after Ray's death 'an estimated one million people [can] follow Ray's body to the crematorium' (Allen and Blanchard 1996a: 66).

There may be hidden agendas of Hindi nationalism and of élitism behind Bollywood's 'All-India film', but there is equally a hidden agenda of regionalism behind India's oppositional cinema. According to Solanas and Gettino's definitions of First and Third Cinema discussed at the beginning of this chapter, Bollywood finds itself in the anomalous position of being Third cinema in the First World and First cinema in the Third World. At the same time, however, with equal irony, cinema which sets itself in opposition to Bollywood within India aligns itself, by default, with the West, leaving Bollywood more 'anti-Colonial' than its apparently more radical counterparts. This irony was illustrated to the full at India's International Film Festival in January 1997, in Thiruvananthapuram, on the south coast of India. 'Politics dogged [the festival] from the very start' (Cunha 1997a) when many major Indian film-makers (several of whom have achieved success at Western film festivals in the last year) were excluded from the Panorama of New Indian Cinema, 'virtually all of the rejects being in the Hindi language'.

CONCLUSION

Bollywood is a wild-card in the globalisation process of the media. Its position is constantly shifting: influenced by its diasporic audiences, by Western moves into India, by newly emerging cultural dialogues between East and West, and by new technologies and their implications. Its relationship with the West has undergone radical changes in the last four years, which will no doubt change its future, although quite probably on its own terms rather than those of the West. Bollywood does not see itself as a minority cinema, but claims the right to be taken seriously as a commercial popular cinema. It demonstrates, finally, that the use of culture as a global force, and as a hegemonic force, is not confined to the West alone.

The existence of another economically imperialist international cinema outside Hollywood is in itself no cause for celebration simply because that cinema opposes Hollywood. Problematic issues around Bollywood and Hindu nationalism, élitism, censorship and corruption should not be glossed over. It has been my intention instead to suggest a reappraisal of current dichotomies of thought between East and West, between commercial and oppositional cinema, by highlighting how unstable these positions look when viewed from an entirely different perspective, a perspective taken, as far as possible, from within India.

A reappraisal of Indian cinema may challenge our assumptions not only about First World and Third World cultural politics, but also our assumptions about what constitutes commercial, and what oppositional, or 'art' cinema, for, as I have discussed, what has in the West been seen as 'difficult' independent cinema fare, has in India been consumed by mass audiences with greater enthusiasm than what we understand as overtly commercial Hollywood films.

I have left the issue of quality out of my discussion of Bollywood, largely because I do not presume to make value judgements on a cultural product designed for consumption by a culture relatively alien to my own experience. Bollywood films have, historically, been dismissed as formulaic and poor quality, and their audience, by inference, as unsophisticated. However, not only can the Bollywood audience watch a film for longer, generally, than a Western audience (Hindi films are uniformly three hours long), it is tolerant of, in fact hungry for, film which in the West is considered too 'challenging' for mainstream, commercial audiences. Which begs the question: which is, in fact, the more truly 'sophisticated' cinema audience? Hollywood's, or Bollywood's?

ACKNOWLEDGEMENTS

I would like to thank Professor Richard Johnson and Daniel Saul for their comments on earlier drafts of this chapter, and for their invaluable advice and support.

NOTES

1 The title is borrowed from that of a popular music cassette on sale in India in 1995 (see section on 'Popular discourses of Hollywood/Bollywood opposition').

2 As addressed by critics such as Jim Pines and Paul Willemen (1989) in their book *Questions of Third Cinema*.

3 The title of an article in *The Sunday Times* by Caroline Lees, Bombay (1995).

4 As described in a BBC 2 *Late Show* report on 'Satellite TV' (Beavan 1994), and a Channel 4 documentary, *Satellite Wars* (Fienburgh 1995).

5 Sony Entertainment Television was launched in India in October 1995, and has 'reached a prime position among Hindi channels. It had rights to exclusively cover events like the Michael Jackson concert and Femina Miss India '97' (D'Souza 1997: 5).

6 This question was not included in the 1996 and 1997 *Movie* opinion polls.

7 Sadly, the BBC TV series *Bollywood or Bust!* was discontinued after its 1996–7 run.

25

MIMICKING MAMMON?

What future for the post-communist Russian film industry?

Kate Hudson

INTRODUCTION

Whatever one's view of communism and whatever judgement one may arrive at about the political and economic project of the Soviet Union, the cultural impact of that experiment has been very significant on a global scale, particularly in the field of cinema. Whilst the political and economic impact of the Russian revolution of 1917 has clearly been of massive significance throughout the twentieth century, the revolution also changed artistic and cultural life in quite dramatic ways. In fact, for most of this same period, the film practices encouraged and developed by the Russian revolution have been regarded as some of film's greatest experiments.

The framework within which these developments took place was the nationalisation of the film industry and its evolution within a rigid ideological framework. Its development was chiefly to underpin the economic and political programme of the ruling system and win the hearts and minds of the population to it. With the overthrow of the communist system in Eastern Europe and the Soviet Union in 1989 and 1991 this basis was removed and the film industries of the former communist countries have had to redevelop within the framework of a market system which has introduced a wide range of challenges and conflicting pressures for creative workers and cultural producers. These changes have fallen into two major categories.

First there are those resulting from economic changes: the end of state funding and organisation of production, distribution and creative input, which removed the box office imperative; the end of subsidised performance, which brought about extremely high film-going statistics amongst the population as a whole; the attempts at production for the international market, resulting in an increase in sex and violence in cinematic output; and western attempts at penetration and control of distribution and theatrical space.

Second are those resulting from political changes: freedom from censorship has produced a paradox – freedom of expression has resulted in a much reduced output from the great creative spirits of the communist period; in addition, the lack of state funding has reduced many previously in secure 'art' jobs to a position of penury.

This chapter will explore these changes, primarily in the context of the development of the Soviet film industry and the Russian film industry in these early years of the post-communist period. The chapter will also show these changes to mirror some wider and more fundamental issues in contemporary Russian society and politics: western penetration of the Russian film industry and the perceived undermining of national cultural values is a microcosm of the great political and economic debates within contemporary Russia. Should Russia turn to the west, to western market economics, to western liberal values, and risk being subordinated? Or should it attempt to plough a Russian furrow, preserving the integrity of its national economy, and redeveloping its own national values, and indeed, cultural traditions? These are the great questions, which still remain unresolved for Russia, and thus, for the Russian film industry. In short, will Russia continue to provide some of the world's greatest directors, developing and evolving a new Russian cinema built on the great cultural and artistic achievements of the Soviet period, or will the Russian industry become merely a pale shadow of the worst aspects of its western counterparts, and end up simply mimicking Mammon? This chapter argues that the answer, to a great extent, depends on the outcome of the political struggle in Russia. The best combination for the Russian film industry is economic support from the state, regulation of the industry, and protection from external penetration, combined with freedom from censorship and ideological control.

THE SOVIET EXPERIENCE

Much of the debate that has taken place about developments in eastern Europe and the former Soviet Union since 1989 and 1991 has hinged on the idea that with the demise of communism, history would somehow 'reassert' itself, as if the communist period were some kind of aberration which had no continuity with the previous period, and would have none with the subsequent one: as if 'normality' would re-emerge in Russia and Eastern Europe. Whilst it would be wrong to see these massive changes taking place without some continuity within these countries, there is a sense in which the idea of history reasserting itself is correct with regard to the film industry: the nationalisation of the film industry during the 1920s ended foreign domination of Russian film production and distribution. The end of the Soviet system has immediately reopened the floodgates to foreign penetration and domination and the marginalisation of Russian production because the market is no longer protected by – and production no longer funded by – the state.

It may seem strange, at first thought, that the Soviet Union, which experienced for much of its existence a most repressive political regime, should have produced a national cinema that was, for significant periods, a byword for creative and technical innovation and artistic advance. On closer investigation it is clear that the Soviet period cannot be seen in a uniform fashion. Whilst for political reasons the 1930s to 1950s were a time of cultural repression and artistic stagnation, the 1920s, and 1960s to 1980s were times when Soviet cinema produced many superb works, often pathbreaking in cinematic terms. It can be seen that, when the Soviet political environment was less fraught and oppressive – when censorship and political manipulation were less overwhelming – the positive features of the system – massive state support for the arts – were able to come into play and enable great art and artists to emerge. An understanding of the period of Soviet cinema is essential to an understanding of the reality and the potential of Russian cinema today.

THE EARLY YEARS

It would be wrong to imagine, however, that Russian cinema began with the revolution, with Eisenstein and with the Soviet period. Prior to the revolution, cinema was an enormously popular part of Russian life – for example, in 1917 there were around four thousand cinemas operating in Russia. The first film theatre in Russia was opened in 1896 by the Lumière company, and the film business was dominated by French companies such as Pathé, which held a particularly powerful position prior to the outbreak of the First World War, progressing from being the main distributor in Russia, to being one of the main producers. From 1908, independent Russian production developed to reap the profits of the increasing demand for cinematic entertainment: between 1908 and 1912, 351 films were produced by Russian companies –

139 of these were dramatic, and 212 were newsreels; yet the industry remained dominated by foreign companies. After the revolution took place in 1917, many of the private film companies moved away from Moscow and communist-held regions to southern Russia, where they continued to make films into the early 1920s (Leyda 1983: 17–71).

FILM AND REVOLUTION

Despite this national significance, however, international awareness of Soviet film stems primarily from the recognition of the technical and artistic achievements of Sergei Eisenstein, particularly his political development of montage and what he described as his 'dialectical approach to film form' in the mid-1920s. Through his revolutionary approach to film he was able to develop the medium to serve the political interests of the young Soviet state. Films such as *Strike* (1924), *Battleship Potemkin* (1925) and *October* (1928) are notable examples of this. In particular, the Soviet communists, whom Eisenstein strongly supported, were very aware of the political potential of film and its importance as a mass communicator. In fact, Lenin, the leader of the Bolshevik party, said, 'Of all the arts, for us the cinema is the most important'. Another Bolshevik leader, Leon Trotsky, was more explicit about the political role of art, and therefore of film. He said: 'Bourgeois art is a mirror, and proletarian art is a hammer. We must use the hammer to smash the mirror'.

In this view, bourgeois art or art from the ruling class in society merely reflects the status quo in society and encourages conformity. Proletarian art – art from the working class – is a political tool that must be used to help smash the existing order. In Soviet film one can see how the great directors began to try and use film in innovative ways to change people's perceptions of society, and ultimately through this to help change society itself. It was this conviction which inspired the achievements of the early Soviet directors and established their reputation and significance, technically and artistically, in the film industry on a worldwide scale.

SOCIALIST REALISM

Clearly, in spite of the positive developments in the first decade of the Soviet Union, there were subsequently about thirty years during the Soviet period in which cultural development and artistic expression were extremely difficult. Why was this the case? A crucial question in understanding the negative repressive aspects of the Soviet period in cultural terms, is the development of 'socialist realism'. In its original form, as pursued by Eisenstein, socialist realism meant realism combined with socialist commitment; in many ways in continuity with the great Russian realist cultural tradition of writers like Chekhov. However, it gradually became an official doctrine which dominated all areas of Soviet creative and cultural life from the first Soviet writers' congress in 1934 to the emergence of *glasnost* (the policy of 'openness' associated with the reforms of the Soviet Union in the 1980s) under Gorbachev. The congress defined socialist realism as follows:

> Socialist realism ... demands of the artist the truthful, historically concrete representation of reality in its revolutionary development. Moreover, the truthfulness and historical concreteness of the artistic representation of reality must be linked with the task of ideological transformation and education of workers in the spirit of socialism.
>
> (Kenez 1992: 157)

This official doctrine was opposed to formal experimentation in cinema and other arts and called for narratives and styles capable of conveying the political perspective of the ruling party in simple form to a mass audience of workers and peasants. It turned the innovative and dynamic young Soviet cinema into a ponderous machine repeating and reinforcing stagnant cultural norms.

Given the internationally recognised genius of Eisenstein and the great talents of a whole generation of Soviet directors, how could such a transformation have taken place? In the view of Alec Nove (1989), the dogmatisation of the earlier commitment-based approach was based on a double misunderstanding of Lenin's views. First, on what Lenin is supposed to have written about 'Party literature' in

1905, and second, on the basis of a conversation with the German communist Klara Zetkin, where Lenin is alleged to have said: 'Literature should be understandable by the people' (*ponyatna narodu*). This was interpreted, in Nove's view:

> as meaning that it should be at a level at which people would understand it without difficulty. However, it seems that what he did say was that 'literature should be understood by the people' (*ponyata narodom*), which implies that people should raise their level of understanding, rather than that authors should lower themselves to the existing level.
>
> (Nove 1989: 1)

Whatever the reason, socialist realism was propelled further and further from its origins through the needs of the increasingly repressive Soviet state to give a positive official representation of reality, which coincided less and less with the actual, lived reality of the population. One anecdote which illustrates this disparity is recounted by Nove, as told to him by a colleague who swears he witnessed it himself:

> In Moscow in 1952 a film was shown, *Cavalier of the Golden Star*, in which well-dressed peasants were feasting at a well-stocked table. Sitting in front of him were two peasants. When the film ended, one of them asked: 'Where is all this supposed to be?' The other peasant replied: 'Dunno, probably somewhere in America'.
>
> (Nove 1989: 2)

The film was, of course, set in the Soviet Union.

The Stalin period had a terrible impact on Soviet cultural life, not only through the rigidities of socialist realism, but also through the arrests and killings of many notable creative workers, such as Mandelshtam, Babel and Meyerhold – who were accused of anti-Soviet terrorism for failing to conform to Soviet cultural norms – to name but a few. Others such as Bulgakov were almost totally censored.

THE THAW

After Stalin's death there was a relaxation of censorship, and during Khrushchev's leadership, particularly in the early 1960s, there was a considerable thaw. The publication of Alexander Solzhenitsyn's *One Day in the Life of Ivan Denisovich* was one of the clearest indicators of this, as were the posthumous rehabilitations of Babel and Mandelshtam.

During the years of Khrushchev's thaw, Soviet cinema could be described as having experienced an 'artistic renaissance' (Lawton 1992: 2). During the late 1950s and early 1960s, a new generation energised and revitalised the film industry: with the breaking out of the straitjacket of socialist realism, there was a revival of formalist experimentation, the development of the 'poetic' style of the directors from the southern republics, and the emergence of Andrei Tarkovsky.

CREATIVITY AND CONSERVATISM

Unfortunately, the thaw during this period was brought to an end by the downfall of Khrushchev in 1964 and his succession by more conservative forces in the Communist party leadership, with the eventual emergence of Leonid Brezhnev as the outright leader in the late 1960s. Under Brezhnev, whilst there was some turning back of the clock and less liberalisation in the cultural field than there had been under Khrushchev, there was never a return to the rigours and extremes of the Stalin years.

A number of remarkable films were produced during this period, which then faced obstacles in their distribution. For example, *Teni Zabytykh* (*Shadows of Forgotten Ancestors* 1964), was directed by the Soviet Georgian, Sergei Paradzhanov. Hailed as 'one of the most striking achievements in world cinema since World War II' (Sklar 1993: 447), its style was intensely visual, folkloric and colourful, breaking with conventional narrative forms, highly symbolic and experimental in many of its techniques. *Shadows of Forgotten Ancestors* was very negatively criticised within the Soviet Union, and Paradzhanov found obstacles placed in the way of his work. Eventually he was allowed to make *Sayat Nova* (*The Colour of Pomegranates*, 1969), but this suffered from the Brezhnevite conservatism, and it was re-edited by others prior to its eventual release in 1972. Paradzhanov himself was imprisoned in 1974 for his championing of the ethnic and national values of his native republic, Georgia, which were themes clearly reflected in his films. He was eventually released in 1977.

Andrei Tarkovsky also experienced obstacles in the way of his film-making, which was also mystical and highly symbolic. *Andrei Rublev* (1966) was shelved until a screening was allowed at Cannes in 1969, where it won the International Critics' Prize.

Taken as a whole, however, the 1960s were a period of radical and remarkable renewal of Soviet film. As Anna Lawton has observed: 'Revival of film art in those years brought Soviet cinema to the attention of international audiences and critics and, as in the 1920s, it scored high marks' (Lawton 1992: 3).

GLASNOST

During the 1970s further change took place: economic stagnation and the competition from television led to commercial considerations being taken into account more, and there was a widening of the genre repertoire – of the different types of films being made. Whilst these changes catered more for wider public taste, there were, however, also some superb films made in the 1970s, which were only released under *glasnost*. Indeed, much of the cultural impetus for *glasnost* came from the Soviet artistic and creative community who had a great desire for freedom of expression, rather than any concern about the introduction of the market economy. The great films of the *glasnost* period were either shelved films made in previous years, or recent works of great Soviet directors. Such outstanding achievements as have occurred in the cinematic field since 1991 have usually come from well-established directors of the Soviet period – such as the recent Oscar-winning *Burnt by the Sun* (1994), directed by Nikita Mikhalkov, thus ensuring a significant level of continuity between the culture of the Soviet period and that of the new post-communist Russia.

Gorbachev's policy of *glasnost* or 'openness', was much more systematic with regard to the arts than the general liberalisation under Khrushchev and can be seen as 'a renaissance planned and sustained by the party' (Lawton 1992: 3). Two films of particular note that were released during *glasnost*, in the second half of the 1980s, were Alexander Askoldov's *Komissar* (*The Commissar*, made in 1967), which was set during the Civil War period, and was apparently shelved because the censors objected to a prevision of the Holocaust experienced by one of the characters. *The Com-missar* became a huge international success. *Dolgiye Provody* (*Long Farewells*, made in 1971), by woman director Kira Muratova, was suppressed until 1987 because of its negative portrayal of relations between the sexes.

Perhaps most significant of all was *Monanieba* (*Repentance* 1984), made by Georgian director Tenghiz Abuladze. This film was briefly shelved, but released in 1986, and was the first Soviet film, through surreal comedy, to explore the dictatorial aspects of life during the Stalin period, and break down the wall of silence around the critical cinematic depiction of Soviet history and ideology.

Of the new films made during the glasnost period, a real pathbreaker was *Malenkaya Vera* (*Little Vera* 1988), directed by Vasily Pichul at the age of 28, which dealt with the question of sexuality in an open way, previously not experienced by Soviet audiences. Described by Nicholas Galichenko as 'a caustically depressing ... modern drama of alienation ... [in which] ... the optimism of the Gorbachev era is offset by indifference in hellish working class life' (Galichenko 1991: 111), *Little Vera* drew audiences totalling over 50 million. For Pichul, however, this proved to be a one-off success, for his second film, *Dark Nights on the Black Sea* (1989), was a massive box-office disaster.

Another outstanding film of the *glasnost* period was *Come and See* (1985) made by Elem Klimov, not because it broke any previously existing taboos, but because it confirmed the world-class quality of Soviet film-making. The film depicts Nazi atrocities on a horrendous scale, through the eyes of a child, as he witnesses the burning alive of the entire population of a Byelorussian village; a similar fate was shared by more than 600 Byelorussian villages. The film was enormously popular because it dealt with the reality of the Second World War – devastating and ever-present in the memories of the Soviet population who lost an estimated twenty million citizens in its course – in a deeply moving and profoundly dignified fashion.

CRISIS IN THE FILM INDUSTRY

Whilst *glasnost* provided the artistic freedom that Soviet directors had long hoped for, Gorbachev's other policies led to the breakup of the nationalised, state-supported film industry, both in terms of production and distribution, and also from the point of view of employment security for creative workers. Immediate effects of the breakup of the previous system included a massive influx of American films as import restrictions were lifted and a proliferation of videotheques showing pornographic films. As Vasily Pichul, director of *Little Vera*, commented, unless Russian film-makers could meet these new challenges:

> We'll go under and we'll be reduced to making advertisements, pop promos or television programmes. There's every chance that Russian cinema can become an important part of the cultural life of the country. That possibility exists, and if we don't make use of it, we'll lose it.
>
> (in Sklar 1993: 508)

The responses to these changes have been as varied within the cultural sphere as they have within the political sphere. Russian academic Sergei Serebriany, for example, sees both positive and negative factors arising in the post-Soviet situation for Russian culture today:

> ... liberated from the oppressive care of the state, culture by now has found itself 'liberated' also from its habitual material support and has been mercilessly thrown into the elements of a market economy ... The cinema production in Russia, they say, soon may stop altogether, but now and then we see on the TV our cinema people celebrating their achievements – not to mention the fact that now we can see on our TV a lot of great foreign films, that formerly we could only read about; true we are shown a lot of rubbish too.
>
> (in Saikal and Maley 1995: 161–2)

Serebriany sees both pros and cons as a result of the introduction of market forces into the cultural sphere; yet many writers have been more outspoken in their criticism of the negative factors. Writing even before the final demise of the Soviet Union, but after *glasnost* and *perestroika* had taken their massive toll, Victor Bozhovich commented in 1991:

> The Soviet film and video market is glutted with second-rate American product bought wholesale – and cheap. Whether the market can sustain trash for very much longer is beside the point: the fast buck, not artistic standards rules ... With no protective mechanism to stem the flood of foreign films, Soviet culture is in mortal danger.
>
> (Soviet Weekly 1991: 8 August)

Bozhovich was particularly concerned because, at that time, the Soviet film industry was producing 400 feature films a year, but the vast majority of them were not reaching the screen because they were being elbowed out of the way by American films. New Russian film-makers, possibly with great talents among them, would not be able to emerge in front of the public. Bozhovich also goes on to make an extremely important point which demonstrates his concern, about the future of the film industry not only in terms of artistic production, but also in terms of its significance for the future of the national culture:

> Cinema is not just a money-maker, it is an essential part of our culture. Art has never been, and never will be, able to pay its way. Its life-blood is the material and spiritual resources which society has to provide if national identity and social cohesion are to be preserved.
>
> (Soviet Weekly 1991: 8 August)

Bozhovich convincingly argues for state intervention to protect the domestic film industry, but his concern for the preservation and development of a national culture is perhaps the point that has most resonance in terms of the wider political and cultural debate in Russia today, of which the situation in the film industry is just a microcosm.

The scale of the changes since the introduction of *glasnost* and *perestroika* and the subsequent collapse of the system can be seen clearly through cinema attendance figures. In 1983 in the Russian Federation (the largest republic within the Soviet Union), there were 2,700 cinemas and 80,000 film clubs in towns and villages, which attracted in total around 2,600 million visits in that year alone. The audiences would be watching films of predominantly Russian origin. In 1987, when the effects of *perestroika*, or Gorbachev's economic restructuring, were beginning to be felt, Russians went to see an average of 15 films a year. By 1993 there were only 250 million visits to the remaining 1,600 cinemas. In other words, there had been a 90 per cent decrease in cinema attendance, with cinema audiences in major cities at around 8 to 10 per cent of capacity for each showing. By 1995, that figure had fallen even further to around 3 to 4 per cent.

This decline in attendance has been matched by the decline in the number of Russian films shown on the big screen. In 1994, of every 100 films shown in cinemas, 74 were American, 14 European, and only 8 from Russia and the other former Soviet republics. As Daniil Dondurel, editor in chief of the Russian film journal, *Iskusstvo Kino*, observed:

> Desertion of the cinemas has occurred almost exactly at the same time as the invasion by US films. Audience surveys show that the choice of programmes is attracting a younger and less educated public and is starting to put older age groups off the cinema.
>
> (*Audiovisual Eureka* 1995: 5)

THE CHALLENGE OF THE SMALL SCREEN

In his assessment of the current state of the industry, Dondurel is firmly of the view that the shift is not away from film, but towards the small screen, and indeed, the public has a large choice at its disposal. In the first half of 1994, the six major public channels between them showed 4,650 hours of fiction film and 1,850 hours of documentaries and animations. Every day Russians can choose from 33 hours of non-stop films. 44 per cent of this availability is Russian productions – a very different situation from the programming in cinemas. In addition there are broadcasts from over a hundred regional channels, and on a less legal footing, there are also 'an uncontrolled mass of small, private cable channels practising private broadcasting on a grand scale' (*Audiovisual Eureka* 1995: 5). Small-screen film consumption is then doubled by films seen on video at home.

The issue of video is an enormous one for the film industry and has been the subject of much debate and many attempts to secure legislation. Video pirates offer a choice from around 10,000 stolen titles at prices with which legitimate operators cannot compete. In 1995 the legitimate hiring of a video cassette cost around 5,000 roubles ($1.50 or ECU1.25), the price of two cinema tickets. The collapse of cinema distribution is a major blow to the industry, especially as compounded by video piracy. In Dondurel's view, however, demand for Russian films on television seems to have recovered after a temporary decline. A survey in 1994 showed that for the first time in years, 69 per cent of viewers stated a preference for Russian cinema. The main problem appears to be that the films the public prefers are not those that are currently being produced by the industry. Of the titles mentioned in the survey, only one in ten was a contemporary production. The most popular films are comedies and dramas from the Soviet period.

THE RISE AND FALL OF PRODUCTION

An investigation of the figures on the production side is also very interesting. Again from Dondurel's figures, we can see that until the mid-1980s, Soviet studios produced around 150 full-length feature films a year, with an additional 40 or more films made for television. A monopoly on their distribution was held by Sovexportfilm, which also purchased a similar number of titles abroad for showing within the Soviet Union. Figures for production after the end of state control underline the contradictory impact of the introduction of the free market on the film industry. In 1991, production doubled as newly-established companies invested in the sector, only a tenth of the total production of 375 films being subsidised wholly or partly by public authorities. However, these companies did not have a sufficient understanding of the industry or the market conditions and paid no attention to the growth in competition from foreign productions, the nature of public demand or the terrible audience figures in the film theatres.

Then, with new opportunities for privatisation and investment introduced under the *Gaidar* government in 1992, cinema ceased to be of interest to Russian business. The economic crisis meant that inflation began to destroy production budgets and increases in interest rates made it impossible for either new companies or existing studios to gain credit. It appeared that with films taking two or more years to show a return, banks preferred to make shorter-term investments. After the initial expansion in Russian film-making in 1991, by 1993 Russia produced only 136 feature films. The end of the communist system provided the freedom for film-makers to make whatever films they chose – but market forces actually resulted in a reduction in Russian film-making. The initial flowering of Russian culture promoted by the end of state interference in film content was soon cut back by the economic power of the great American film magnates.

CAN THE RUSSIAN FILM INDUSTRY SURVIVE?

With a situation where production costs are on the increase and it is cheaper to import foreign films, there can be little future for the Russian film industry without state support – a situation not so unfamiliar even to western European film industries in the face of the US monolith. Attitudes have shifted significantly amongst key figures in the Russian film industry themselves since the introduction of the free market; as Dondurel pointed out in 1995: 'More and more directors and producers who had enthusiastically refused government support are now convinced that the industry will only survive with state subsidies' (*Audiovisual Eureka*, 1995: 6).

As Dondurel observes, there should be a workable market for Russian films: Russia has a population of 160 million, and it is surrounded by republics of the former Soviet Union where most people understand Russian.

As this chapter shows, given the cultural traditions and cinematic output and experience of the former Soviet Union, it is absolutely possible for Russian cinema to continue and develop the same quality output, particularly if Russian art remains free from censorship and political control. What is totally clear, however, is that this will not take place if foreign – predominantly US – business interests are allowed to dominate the industry, and flood the Russian market with cheap foreign imports. Video-pirating is more or less the last straw for the beleaguered theatrical side of distribution. The Russian industry cannot compete financially in the free market, even were it to force its film content, indeed, to mimic Mammon, because of the global strength of US interests. What can succeed, however, is a newly regulated and restructured Russian industry with effective state subsidies, which, operating in an artistically open environment, will allow a whole generation of new Russian film-makers to emerge, helping to fulfil the cultural needs not only of the Russian people, but of the world as a whole.

REFERENCES

'Aatank Hi Aatank' (1995) *Film Information* 22, 45 (5 August): 6 (review).

Abbott, D. (1994) 'Women's home-based income generation as a strategy towards poverty survival: Dynamics of the Khannawalli (mealmaking) activity of Bombay', unpublished Ph.D thesis, Open University.

—— (1996) 'Creating conflict: How Hindu fundamentalists seduce the poor, illiterate and the disaffected of India', *Contemporary Politics* 2, 1: 89–96.

——(1997) 'Who else will support us: How poor women organise the unorganisable in India', *Community Development Journal* (special issue on social movements).

Abdalla, I.-S. (1980) quoted in 'Qué es el Tercer Mundo', in *Guia del Tercer Mundo 1981*, Mexico: Periodistas del Tercer Mundo.

Abegglen, J. (1958) *The Japanese Factory: Aspects of its Social Organisation*, Asia Publishing House.

Abu-Lughod, J. (1991) 'Going beyond global babble', in A.D. King (ed.) *Culture, Globalization and the World System*, London: Macmillan.

Adams, S. (1984) *Roche versus Adams*, London: Jonathan Cape.

Adamson, P. (1991) Speech to UNICEF National Committee, Geneva.

—— (1993) 'Charity begins with the truth', *Independent*, 18 May.

Agnelli, S. (1986) *Street Children – A Growing Urban Tragedy, Report for the Independent Commission on International Humanitarian Issues*, London: Weidenfield & Nicholson.

Albrow, M. (1996) *The Global Age*, Cambridge: Polity Press.

Allen, R. (1986) 'Behind the News 2: Bob's not your uncle', *Capital & Class* 30 (Winter): 31–7.

Allen, S. and Barker, D. (1975) 'Introduction: The interdependence of work and marriage', in S. Allen and D. Barker (eds) *Dependence and Exploitation in Work and Marriage*, London: Longman.

Allen, T. (1991) 'Understanding Alice: Uganda's holy spirit movement in context', *Africa* 61, 3: 370–99.

—— (1992a) 'Prospects and dilemmas for industrializing nations', in T. Allen and A. Thomas (eds) *Poverty and Development in the 1990s*, Oxford: Oxford University Press.

—— (1992b) 'Upheaval, affliction and health: A Ugandan case study', in H. Bernstein, B. Crow and H. Johnson (eds) *Rural Livelihoods: Crises and Responses*, Oxford: Open University and Oxford University Press.

—— (1994) 'Ethnicity and tribalism on the Sudan-Ugandan border', in K. Fukui and J. Markakis (eds) *Ethnicity and Conflict in the Horn of Africa*, London: James Currey.

—— (1995) 'Taking Culture Seriously', in T. Allen and A. Thomas (eds) *Poverty and Development in the 1990s*, Oxford and Milton Keynes: Oxford University Press and Open University.

—— (1996) 'The Third Cinema', in *U208 Summer School Student Notes*, Milton Keynes: Open University.

Allen, T. and Blanchard, S. (1996a) 'Satyajit Ray: 'Distant Thunder", in *U208 Summer School Student Notes*, Milton Keynes: Open University.
—— (1996b) 'Mira Nair: "Salaam Bombay!" ', in *U208 Summer School Student Notes*, Milton Keynes: Open University.
Allen, T. and Eade, J. (eds) (1999) *Divided Europeans: Understanding Ethnicities in Conflict*, London: Kluwer.
Allen, T. and Seaton, J. (eds) (1999) *The Media of Conflict*, London: Zed Books.
Alleyne, G. (1994) 'West Indies surrender', *Daily Nation*, 14 April.
'Altered images: Media, myths and misunderstanding' (1987) *Links* 28.
Alvares, C. (1994) *Science, Development and Violence: The Revolt against Modernity*, Delhi: Oxford University Press.
Amin, M. (1989) 'A Vision of the Truth', *Refugees*, 22–5 October.
Amnesty International (1990) 'Brazil: Torture and extrajudicial execution in urban Brazil', London: Amnesty International Briefing.
Anderson, B. (1991, [1983]) *Imagined Communities: Reflections on the Origin and Spread of Nationalism*, London: Verso.
Anderson, M. (1996) 'Do No Harm: Supporting Local Capacities for Peace through Aid', *Local Capacities for Peace Project*, Cambridge, MA: Collaborative for Development Incorporated.
Anderson, M.B. (1985) 'Technology transfer: Implications for women', in C. Overholt, M.B. Anderson, K. Cloud and J.E. Austin (eds) *Gender Roles in Development Projects: A Casebook*, Connecticut: Kumarian Press.
Andrei Rublev (1986) Andrei Tarkovsky, USSR.
Anthias, F. (1992) 'Connecting "race" and ethnic phenomena', *Sociology* 26: 421–38.
Ante Projeto do Grupo de Comunicacao Popular da Vila Nossa Senhora Aparecida (1984) São Miguel.
Anwar, F. (1994) 'Bhaji on the Beach', *Sight and Sound*, February: 47–8 (review).
Anwar, M. (1979) *The Myth of Return: Pakistanis in Britain*, London: Heinemann.
Appadurai, A. (1990) 'Disjuncture and difference in the global cultural economy', in M. Featherstone (ed.) *Global Culture*, London: Sage.
—— (1995) 'The production of locality', in R. Fardon (ed.) *Counterworks: Managing the Diversity of Knowledge*, London and New York: Routledge.
Appleton, H. (ed.) (1995) *Do it Herself: Women and Technical Innovation*, London: IT Publications.
Apthorpe, R. (1996) 'Policy anthropology as expert witness: Review article', *Social Anthropology* 4, 2: 163–79.
Archer, M. (1988) *Culture and Agency: The Place of Culture in Social Theory*, Cambridge: Cambridge University Press.
—— (1990) 'Theory, culture and post-industrial society', in *Global Culture*, special issue of *Theory, Culture and Society* 7, 2–3: 97–119.
Armes, R. (1987) *Third World Film Making and the West*, Berkeley, CA: University of California Press.
Army (1996) Raam Shetty, India.
Armytage, W.H.G. (1968) *Yesterday's Tomorrows: An Historical Survey of Future Societies*, London: Routledge & Kegan Paul.
Arnold, S. (1988) 'Constrained crusaders: British charities and development education', *British Charities and Education*, Education Network Project Occasional Paper no. 1, University of Sussex.
Asia Watch and Women's Rights Projects (1993) *A Modern Form of Slavery: Trafficking of Burmese Women and Girls into Brothels in Thailand*, New York: Human Rights Watch.
Atkinson, P. (1996) 'The Liberian Civil War: Images and Reality', *Contemporary Politics* 2, 1: 79–88.

Atsumi, R. (1980) 'Patterns of personal relationships: A key to understanding Japanese thought and behaviour', *Social Analysis* (Special Issue: Japanese Society: Reappraisals and New Directions) 5–6: 63–78.

Audiovisual Eureka (1995) Brussels, 16/9, 13 (March).

Bahaguna, S. (1979) 'The Himalayas: Towards a programme of reconstruction', in K.M. Gupta and Desh Bandhu (eds) *Man and Forest*, Delhi: Today and Tomorrow Publishers.

Bailey, D.A. (1989) 'The black subject at the centre: Repositioning black photography', *Links* 34: 31–8.

Ballard, R. (1992) 'New clothes for the emperor? The conceptual nakedness of the British race relations industry', *New Community* 18: 481–92.

Bandit Queen (1994) Shekhar Kapoor, India and UK.

Bandyopadhyay, J. and Shiva, V. (1987) 'Chipko: Rekindling India's Forest Culture', *The Ecologist* 17, 1: 26–34

—— (1988) 'Political economy of ecology movements', *Economic and Political Weekly*, 11 June: 1223–32.

Banks, M. (1996) *Ethnicity: Anthropological Constructions*, London: Routledge.

Barbados Advocate (1994) 'Barbados test a good catch: Tourism', 4 April.

Barbados Advocate (1994) 8, 14, 15 and 18 April.

Barbier, E. (1989) *Economics, Natural Resource Scarcity and Development*, London: Earthscan.

Barclay, H. (1964) *Buuri al Lamaab. A suburban village in the Sudan*, Ithaca, NY: Cornell University Press.

Bardhan, P. (1991) 'On the concept of power in economics', *Economics and Politics* 3, 3: 265–77.

Barker, M. (1982) *The New Racism*, London: Junction Books.

Barnard, A. and Spencer, J. (eds) (1996) *Encyclopaedia of Social and Cultural Anthropology*, London: Routledge.

—— (1996) 'Culture', in *Encyclopaedia of Social and Cultural Anthropology*, London: Routledge.

Barnett, C. (1986) *The Audit of War*, London: Macmillan.

—— (1987) *The Pride and the Fall. The Dream and Illusion of Britain as a Great Nation*, London: Free Press.

Barnouw, E. and Krishnaswamy, S. (1963) *The Indian Film*, New York: Columbia University Press.

Barry, K. (1985) *The Prostitution of Sexuality: The Global Exploitation of Women*, New York: New York University Press.

Barsaat (1949) Raj Kapoor, India.

Barth, F. (1969) 'Introduction', in F. Barth (ed.) *Ethnic Groups and Boundaries*, London: George Allen & Unwin.

Barthes, R. (1967) *Elements of Semiology*, trans. A. Lavers and C. Smith, London: Jonathan Cape.

—— (1971) 'The rhetoric of the image', *Working Papers in Cultural Studies* 1: 37–51.

—— (1977) *Image–Music–Text*, trans. S. Heath, New York: Hill & Wang.

—— (1984) *Camera Lucida*, trans. J. Howard, London: Flamingo.

—— (1989) *Mythologies*, trans. A. Lavers, London: Paladin.

Basu, T. *et al* (1993) *Khaki Shorts and Saffron Flags*, Hyderabad: Orient Longman.

Battleship Potemkin (1925) Sergei Eisenstein, USSR.

Bauman, Z. (1973) *Culture as Praxis*, London: Routledge & Kegan Paul.

—— (1995) *Life in Fragments*, Oxford: Blackwell.

Baumann, G. (1996) *Contesting Culture: Discourses of Multi-ethnic London*, Cambridge: Cambridge University Press.

Beckles, H. (ed.) (1995) *An Area of Conquest: Popular Democracy and West Indies Cricket Supremacy*, Kingston: Ian Randle Publishers.

Beckles, H. and Stoddart, B. (eds) (1995) *Liberation Cricket: West Indies Cricket Culture*, Manchester: Manchester University Press.

Befu, H. (1993) 'Nationalism and Nihonjinron', in H. Befu (ed.) *Cultural Nationalism in East Asia: Representation and Identity*, Berkeley, CA: University of California Press.

Bell, D. (1962) *The End of Ideology*, New York: Collier.

—— (1973) *The Coming of Post-industrial Society: A Venture in Social Forecasting*, London: Heinemann.

Ben-Ari, E. (1990) 'Ritual strikes, ceremonial slowdowns: Some thoughts on the management of conflict in large Japanese enterprises', in S.N. Eisenstadt and E. Ben-Ari (eds) *Japanese Models of Conflict Resolution*, London: Kegan Paul.

Benedict, R. (1935) *Patterns of Culture*, London: Routledge.

Benhabib, S. (1990) 'Epistemologies of postmodernism: a rejoinder to Jean-François Lyotard', in L.J. Nicholson (ed.) *Feminism/Postmodernism*, London: Routledge.

Benson, S. (1996) 'Asians have culture, West Indians have problems: Discourses of race and ethnicity in and out of anthropology', in T. Ranger, Y. Samad and O. Stuart (eds), *Culture, Identity and Politics*, Aldershot: Avebury.

Benthall, J. (1993) *Disasters, Relief and the Media*, London: I.B. Tauris & Co.

Berger, P. (1988) 'An East Asian development model?', in P.L. Berger and H.H.M. Hsiao (eds) *In Search of an East Asian Development Model*, New Brunswick and Oxford: Transaction Books.

Bernbaum, G. and Mattheson, M. (1991) 'The British disease: A British tradition?', in R. Moore and J. Ozga (eds) *Curriculum Policy*, Oxford: Pergamon Press.

Beyer, P. (1994) *Religion and Globalization*, London: Sage.

Bhachu, P. (1985) *Twice Migrants: East African Settlers in Britain*, London: Tavistock.

Bhagavan, M.R. (1990) *Technological Advance in the Third World*, London: Zed Books.

Bhaji on the Beach (1993) Gurinder Chadha, UK.

Bhalla, A.S. (ed.) (1991) *Small and Medium Enterprises: Technology Policies and Options*, London: IT Publications.

Bhalla, A.S. and James, D.D. (1991) 'Integrating new technologies with traditional economic activities in developing countries: An evaluative look at "technology blending"', *Journal of Developing Areas* 25: 477–96.

Binford, M.R. (1988) 'Innovation and imitation in Indian cinema', in W. Dissanayake (ed.) *Cinema and Cultural Identity: Reflections on Films from Japan, India, and China*, Lanham: University Press of America.

Black, M. (1992) *A Cause for our Times: Oxfam the First 50 Years*, Oxford: Oxfam.

Blackbourn, D. and Eley, G. (1984) *The Peculiarities of German History*, Oxford: Oxford University Press.

Blanchard, S. (1996) 'Projecting "the good life" – Hollywood and world cinema', in *U208 Summer School Student Notes*, Milton Keynes: Open University.

Blenman, R. (1994) 'Sports tourism paying off', *Daily Nation*, 7 April.

Blom Hansen, T. (1993) 'RSS and the popularisation of Hinduvata', *Economic and Political Weekly of India*, 16 October: 2270–2.

—— (1994a) 'Controlled emancipation: Women and Hindu nationalism', unpublished paper, International Development, Roskible University, Denmark.

—— (1994b) 'The marathaization of Hinduvata: BJP and Shiv Sena in rural Maharastra', paper given at the 13th European South Asian Conference, Toulouse.

—— (1994c) 'Recuperating masculinity: Hindu nationalism, violence and the exorcism of the Muslim other', unpublished paper, International Development Roskible University, Denmark.

—— (1994d) 'The Safforan wave: Hindu nationalism in the 1980s: A case study of Maharastra', unpublished Ph.D thesis, Roskible University, Denmark.

'Blood (dubbed)' (1995) *Film Information* 22, 43 (22 July): 3 (review).

Boddy, J. (1982) 'Womb as oasis: the symbolic context of pharaonic circumcision in rural Northern Sudan', *American Ethnologist* 9, 4: 682–98.
—— (1989) *Wombs and alien spirits: women, men and the zar cult in northern Sudan*, Madison, WI: University of Wisconsin Press.
Bollywood or Bust! (1997) BBC Asian Programmes, UK.
Bombay (1995) Mani Ratnam, India.
Bonacich, E. and Modell, J. (1980) *The Economic Basis of Ethnic Solidarity: Small-Business in the Japanese-American Community*, Berkeley, CA: University of California Press.
Boyd-Barrett, O. (1982) 'Cultural dependency and the mass media', in M. Gurevitch *et al.* (eds) *Culture, Society and the Media*, London: Methuen.
Boyden, J. and Pratt, B. (1985) (eds) *Field Directors' Handbook: An Oxfam Manual for Development Workers*, Oxford: Oxford University Press.
Boyle, C.M. (1993) 'Touching the air: the cultural force of women in Chile', in S.A. Radcliffe and S. Westwood (eds) *'Viva' Women and popular protest in Latin America*, London and New York: Routledge.
BPL (1996) 'Advertisement', *g magazine: glamour, glory, grandeur*, December: 63.
Brake, M. (1982) *Human sexual relations: a reader* Harmondsworth: Penguin Books.
Bramble, B. and Porter, G. (1992) 'Non-governmental organizations and the making of US international environmental policy', in A. Hurrell and B. Kingsbury (eds) *The International Politics of the Environment*, Oxford: Clarendon Press.
Brathwaite, L. (1994) 'Author's attempt to belittle our cricketers', *Sunday Advocate*, 10 April.
Brazil Network (1992) 'Children without a future', September.
Brockett, L. and Murray, A. (1993) 'Sydney's Asian sex workers: AIDS and the geography of a new underclass', *Asian Geography* 12, 1 & 2: 83–95.
Brohman, J. (1996) 'New directions in tourism for Third World development', *Annals of Tourism Research* 23, 1: 48–70.
Bromley, Y. (1974) 'The term *ethnos* and its definition', in Y. Bromley (ed.) *Soviet Ethnology and Anthropology Today*, The Hague: Mouton.
Bujra, J. (1992) 'A Third World in the making: Diversity in pre-capitalist societies', in T. Allen and A. Thomas (eds) *Poverty and Development in the 1990s*, Oxford: Oxford University Press.
Burawoy, M. (1976) 'The functions and reproduction of migrant labor: Comparative material from South Africa and the United States', *American Journal of Sociology* 81, 5: 1050–87.
Burgin, V. (ed.) (1982) *Thinking Photography*, Basingstoke: Macmillan Press.
—— (1986) 'Something about photographic theory', in A.L. Rees and F. Borzello (eds) *The New Art History*, London: Camden Press.
Burnell, P. (1992) *Charity, Politics and the Third World*, Hemel Hempstead: Harvester Wheatsheaf.
Bustamante, J. (1993) 'Undocumented migration from Mexico to the United States: Sociological decoding frameworks and data', paper delivered to the Mexico and the NAFTA: Who will benefit Conference, London: Chamber of Commerce and Industry, 12–14 May: 1–46.
Cain, P.J. and Hopkins, A.G. (1993) *British Imperialism 1688–1914*, London: Longman.
Caldeira, T. (1984) *A Politica dos Outros*, São Paulo: Brasiliense.
Calhoun, C. (1997) *Nationalism*, Minneapolis, MN: University of Minnesota Press.
Campbell, A. (1997) *Western Primitivism: African Ethnicity*, London and Washington: Cassell.
Campbell, B. (1987) 'A feminist sexual politics: now you see it, now you don't', Feminist Review (eds) *Sexuality: A Reader*, London: Virago.
Caplan, P. (ed.) (1987) *The Cultural Construction of Sexuality*, London: Routledge.
Cardoso, R. (1992) 'Popular movements in the context of the consolidation of democracy in Brazil', in A. Escobar and S.E. Alvarez (eds) *The Making of Social Movements in Latin America*, Boulder, CO: Westview.

Carter, S.L. (1993) *The Culture of Disbelief: How American Law and Politics Trivialise Religious Devotion*, New York: Harper.

Cashman, R. (1989) 'Cricket and colonialism: Colonial hegemony and indigenous subversion', in J.A. Mangan (ed.) *Pleasure, Profit, and Proselytism: British Culture and Sport at Home and Abroad*, London: Frank Cass.

Casper (1995) Brad Silberling, US.

Castells, M. (1997) *The Power of Identity*, Oxford: Blackwell.

Castles, S. and Miller, M. (1993) *The Age of Migration: International Population Movements in the Modern World*, London: Macmillan.

Catálogo, Educação e Comunicação Popular (undated) Centro de Evangelizacao Popular de Guaianazes.

Central Bank of Barbados (1994) *Review of 1994 and the Prospects for 1995.*

Centre for Contemporary Cultural Studies (1982) *The Empire Strikes Back*, London: Hutchinson.

Chakravarty, S.S. (1996) *National Identity in Indian Popular Cinema 1947–1987*, Delhi: Oxford University Press.

Chakravaty, R. (1972) 'New Perspectives on the women's movement after 25 years of drift', *Link* 15, 1.

Chalmers, N. J. (1989) *Industrial relations in Japan: The peripheral workforce*, Nissan Institute Series, London: Routledge.

Chambers, R. (1986) *Normal Professionalism, New Paradigms and Development*, IDS Discussion Paper, University of Sussex.

Chambers, R., Pacey, A. and Thrupp, L.A. (eds) (1989) *Farmer First: Innovation and Agricultural Research*, London: Intermediate Technology Publications.

Chaney, D. (1988) 'Photographic truths', *Discourse Social* (Social Discourses) 1, 4: 397–422.

Chant, S. and McIlwaine, C. (1995) 'Gender and export manufacturing in the Philippines: Continuity or change in female employment? The case of the Mactan export processing zone', *Gender, Place and Culture* 2, 2: 147–76.

Chapman, G. (1996) 'Geopolitics of South Asia', unpublished book, University of Lancaster.

Charmley, J. (1995) *Churchill's Grand Alliance: The Anglo-American Special Relationship 1940–57*, London: Jonathan Cape.

Chaui, M. (1987) *Conformismo e Resistencia, aspectos da cultura popular no Brasil*, São Paulo Brasiliense.

Chipasula, F. (1990) 'Friend, ah, you have changed!', in A. Maja-Pearce (ed.) *The Heinemann Book of African Poetry in English*, Oxford: Heinemann.

Christian Aid (1988) *Images of Development, Links between Racism, Poverty and Injustice* (Insight Series), London: Christian Aid.

Clark, R. (1979) *The Japanese Company*, New Haven, CT: Yale University Press.

Cleary, D. (1992) 'Commentary', *International Journal of Sociology and Social Policy* 12, 4–7: 152–4.

Cleasby, A. (1995) *What in the World is Going On?: British Television and Global Affairs*, London: Third World and Environment Broadcasting Project.

Cliffhanger (1993) Renny Harlin, US.

Cohen, A. (1969) *Custom and Politics in Urban Africa*, London: Routledge & Kegan Paul.

Cohen, A.P. (ed.) (1982) *Belonging: Identity and Social Organisation in British Rural Cultures*, Manchester: Manchester University Press.

—— (ed.) (1986) *Symbolising Boundaries: Identity and Diversity in British Cultures*, Manchester: Manchester University Press.

Cohen, S. and Young, J. (eds) (1973) *The Manufacture of News*, London: Constable.

Colby, M.E. (1990) *Environmental Management in Development: The Evolution of Paradigms*, World Bank Discussion Paper no. 80, Washington: World Bank.

Colchester, M. (1994) 'Sustaining the forests: the community-based approach in South and South-East Asia', in D. Ghai (ed.) *Development and Environment: Sustaining People and Nature*, Oxford and Cambridge, MA: Blackwell Publishers/UNRISD.

Coleman, J. (1989) 'Social capital in the creation of human capital', *American Journal of Sociology* 94 Supplement: 95–210.

Como fazer um socio-drama (1978) São Miguel: Centre for Popular Education and Communications.

Constantinides, P. (1985) 'Women heal women: spirit possession and sexual segregation in a Muslim society', *Social Science and Medicine* 21: 685–92.

Corcoran-Nantes, Y. (1993) 'Female consciousness or feminist consciousness?: Women's consciousness raising in community-based struggles in Brazil', in S.A. Radcliffe and S. Westwood (eds) *'Viva' Women and Popular Protest in Latin America*, London and New York: Routledge.

Cornia, G. A., Jolly, R. and Stewart, F. (eds) (1987) *Adjustment with a Human Face, Volume I: Protecting the Vulnerable and Promoting Growth*, Oxford: Oxford University Press.

Counts, D., Brown, J. and Campbell, J. (eds) (1992) *Sanctions and Sanctuary: Cultural Perspectives on the Beating of Wives*, Boulder CO: Westview.

Cozier, T. (1994) 'Lara: The icing on series to savour', *Sunday Sun*, 24 April.

Crang, M. (1998) *Cultural Geography*, London: Routledge.

Crawcour, E.S. (1978) 'The Japanese employment system', *Journal of Japanese Studies* 4, 2.

Crush, J. (ed.) (1995a) *The Power of Development*, London: Routledge.

—— (1995b) 'Imaging development', in J. Crush (ed.) *The Power of Development*, London, Routledge.

Cummings, C. (1990) 'The ideology of West Indian cricket', *Arena Review* 14, 1: 25–32.

Cunha, U.D. (1997a) 'Politics threaten Int'l India Film Fest', *Variety*, 13–19 January: 4.

—— (1997b) 'India's Doordaarshan may lose grip', *Variety*, 10–15 February: 1.

Curran, J. (1996) *Mass Media and Society*, London: Hodder.

D'Souza, J. (1995) 'Rushes [Rock Dancer]', *g magazine: glamour, glory, grandeur*, September: 62.

—— (1997) 'Rushes [Sony Entertainment Television]', *g magazine: glamour, glory, grandeur*, March: 58.

Dahlman, C.J. (1989) 'Technological change in industry in developing countries', *Finance and Development*, June: 13–15.

Daily Mail (1995) 'The Industrious and the Drones', 26 September.

Daily Nation (1994) 13–14 April.

Daly, M. (1978) *Gyn/ecology: the metaethics of radical feminism*, London: The Woman's Press.

Dankelman, I. and Davidson, J. (1988) *Women and Environment in the Third World*, London: Earthscan Publications in association with IUCN.

Das, J.C. and Negi, R.S. (1982) 'The Chipko movement', in K.S. Singh (ed.) *Tribal Movements in India*, vol. 2, Delhi: Manohar, 381–92.

David, R. (1992) 'Filipino workers in Japan: Vulnerability and survival', *Kasarinlan* 6, 3: 9–25.

Davies, M. (ed.) (1994) *Women and Violence: Realities and Responses Worldwide*, London: Zed Books.

Davis, N. (ed.) (1993) *Prostitution: An International Handbook on Trends, Problems, and Policies*, Westport: Greenwood Press.

Dawtrey, A. and Weiner, R. (1995) 'Indie pic pioneers feel global warming', *Variety*, 20–6 November: 1.

De Swaan, A. (1988) *In Care of the State: Health Care, Education and Welfare in Europe and the USA in the Modern Era*, Oxford: Polity Press in association with Basil Blackwell.

De Waal, A. (1994) 'Genocide in Rwanda', *Anthropology Today* 10, 2: 1–2.

Deniozos, D. (1994) 'Steps for the introduction of technology management in developing economies: The role of public governments', *Technovation* 14, 3: 197–203.

Devereux, G. (1967) 'The irrational in sexual research', in G. Devereux (ed.) *From Anxiety to Method in the Behavioural Sciences*, The Hague: Moulton.

Dialogue (1988) 'The Image of Africa', 5 September (1–2), Washington DC: US Information Agency.

Dimenstein, G. (1991) *Brazil: War on Children*, London: Latin American Bureau.

Dissanayake, W. (ed.) (1994) *Colonialism and Nationalism in Asian Cinema*, Bloomington, IN: Indiana University Press.

Doi, T. (1973) *The Anatomy of Dependence*, : Kodansha International.

Dore, R. (1979) *Japanese Factory–British Factory*, Berkeley, CA: University of California Press.

—— (1987) *Taking Japan Seriously: A Confucian Perspective on Leading Economic Issues*, London: Athlone.

Dragadze, T. (1995) 'Politics and anthropology in Russia', *Anthropology Today* 10, 4: 1–3.

Dreyfus, H.L. and Rabinow, P. (1986) 'What is Maturity?', in D. Couzens Hoy (ed.) *Foucault: A Critical Reader*, London: Basil Blackwell.

Dubos, R. (1968) *Man, Medicine, and Environment*, London: Pall Mall Press.

Duffield, M. (1996a) 'The Symphony of the Damned: Racial Discourse, Complex Political Emergencies and Humanitarian Aid', *Disasters* 20, 3: 173–93.

—— (1996b) 'The Political Economy of Internal War: Asset Transfer, Complex Emergencies and International Aid', Macrae and Zwi (eds.) *War and Hunger: Rethinking International Approaches to Complex Emergencies*, London and New Jersey: Zed Books.

—— (1997) 'Evaluating Conflict Resolution: Context, Models and Methodology', Bergen, Norway: discussion paper for the Christian Michelsen Institute.

Duffy, J. (1963) 'Masturbation and clitoridectomy: a nineteenth century view', *JAMA* 3: 166–8.

Duke, K. and Marshall, T. (1995) *Vietnamese Refugees Since 1982*, Home Office Research Study 142, London: HMSO.

Dunn, C. (1994) *The Politics of Prostitution in Thailand and the Philippines: Policies and Practice*, Working Paper 86, Clayton, Australia: Monash University Centre of Southeast Asian Studies.

Dunston Checks In (1995) Ken Kwapis, Canada.

During, S. (ed.) (1993) *The Cultural Studies Reader*, London: Routledge.

Durkheim, E. (1957) *The Elementary Forms of Religious Life*, London: Allen & Unwin.

Eade, J. (1996) 'Ethnicity and the politics of difference: An agenda for the 1990s?', in T. Ranger, Y. Samad and O. Stuart (eds), *Culture, Identity and Politics*, Aldershot: Avebury.

—— (1997) *Living the Global City: Globalization as Social Process*, London and New York: Routledge.

Eccleston, B. (1996a) 'NGOs and competing representations of deforestation as an environmental issue in Malaysia', *Journal of Commonwealth and Comparative Politics* 34: 116–42; repr. in D. Potter (ed.) *NGOs and Environmental Policies: Asia and Africa.*

—— (1996b) 'Does North–South collaboration enhance NGO influence on deforestation policies in Malaysia and Indonesia?', *Journal of Commonwealth and Comparative Politics* 34: 66–89.

The Economist (1996) 'Cultural Explanations', 9 November: 25–30.

ECPAT Australia (1996) *Bulletin* 35 (October).

'Editorial' (1994) *Daily Nation*, 18 April.

Edwards, E. (1992) 'Introduction', in E. Edwards (ed.) *Anthropology and Photography 1860–1920*, London: Yale University Press and Royal Anthropological Institute.

El Dareer, A. (1982) *Women, Why do you Weep? Circumcision and its Consequences*, London: Zed Books.

Eliot, T.S. (1948) *Notes Towards the Definition of Culture*, London: Faber and Faber.

Enloe, C. (1990) *Bananas, Bases and Beaches: Making Feminist Sense of International Politics*, London: Panorama.

—— (1993) 'Images and realities: "Policies" for children in development among non-governmental organisations', paper for the joint seminar on 'Children, NGOs and the State', Open University and London School of Economics, 1 July.

Ennew, J. (1994) *Street Children and Working Children: A Guide to Planning*, London: Save the Children.

Ennew, J. and Milne, B. (1989) *The Next Generation: Lives of Third World Children*, London: Zed Books.

Eriksen, T.H. (1993) *Ethnicity and Nationalism: Anthropological Perspectives*, London: Pluto Press.

Escobar, A. (1995) *Encountering Development: The Making and Unmaking of the Third World*, Princeton NJ: Princeton University Press.

Esteva, G. (1985) 'Regenerating people's space', *Alternatives* 12, 1.

Evans-Pritchard, E.E. (1937) *Witchcraft, Oracles and Magic among the Azande*, Oxford: Clarendon Press.

Fanon, F. (1967) *The Wretched of the Earth*, New York: Viking.

Fausto, A. and Cervini, R. (eds) (1992) *O Trabalho e a Rua: Crianças e adolescentes no Brasil urbano dos anos 80*, São Paulo: Cortez Editora, UNICEF/FLACSO.

Fei, J.C.H. and Ranis, G. (1964) *Development of the Labour Surplus Economy: Theory and Policy*, Homewood: Irvin.

Ferguson, J. (1990) *The Anti-Politics Machine: 'Development', Depoliticisation and Bureaucratic State Power in Lesotho*, Cambridge: Cambridge University Press.

—— (1996) 'Development', in *Encyclopaedia of Social and Cultural Anthropology*, London: Routledge.

Fernandez-Kelly, M.P. and García, A.M. (1989) 'Informalization at the core: Hispanic women, homework, and the advanced capitalist state', in A. Portes, M. Castells and L.A. Benton (eds), *The Informal Economy: Studies in Advanced and Less Developed Countries*, London: Johns Hopkins University Press.

Firestone, S. (1974) *The Dialectic of Sex: The Case for Feminist Revolution*, New York: Morrow.

First Blood (1982) Ted Kotcheff, US.

Fitch, R. (1994) 'Explaining New York City's aberrant economy', *New Left Review* 207: 17–48.

The Flintstones (1994) Brian Levant, US.

Foner, N. (1977) 'The Jamaicans: Cultural and social change among migrants in Britain', in J. Watson (ed.) *Between Two Cultures: Migrants and Minorities in Britain*, Oxford: Blackwell.

FORWARD (1989) *Report on the First National Conference on Female Genital Mutilation: Unsettled Issues for Health and Social Workers in the UK*, London: Foundation for Women's Health Research and Development.

Fowler, R. (1991) *Language in the News: Discourse and Ideology in the Press*, London: Routledge.

Francis, P. (1994) 'When Kensington looked like Lords', *The Barbados Advocate*, 15 April.

Frank, A.G. (1969) *Latin America: Underdevelopment or Revolution*, New York: Monthly Review Press.

Freire, P. (1975) *Cultural Action for Freedom*, Harmondsworth: Penguin.

—— (1978) *The Pedagogy of the Oppressed*, Harmondsworth: Penguin.

Friedman, J. (1994) *Cultural Identity and Global Process*, London: Sage.

'Friendly rivals in the stands' (1994) *Daily Nation*, 11 April.

Fukuyama, F. (1992) *The End of History and the Last Man*, New York: Free Press.

—— (1995) *Trust: The Social Virtues and the Creation of Prosperity*, London: Hamish Hamilton.

Fuller, C.J. (1984) *Servants of the Goddess: the Priests of a South Indian Temple*, Cambridge: Cambridge University Press.

Fyfe, A. (1989) *Child Labour*, Cambridge: Polity Press in association with Basil Blackwell.

Gabayet, L. and Lailson, S. (1992) 'The role of female wage earners in male migration in Guadalajara', in S. Díaz-Briquets and S. Weintraub (eds) *The Effects of Receiving Country Policies on Migration Flows*, Boulder, CO: Westview.

Gaete, R. (1993) *Human Rights and the Limits of Critcal Reason*, London: Dartmouth

Galheigo, S.M. (1996) 'Juvenile policy making, social control and the state in Brazil: A study of laws and policies from 1994 to 1990', University of Sussex.

Galichenko, N. (1991) *Glasnost: Soviet Cinema Responds*, Austin, TX: University of Texas Press.

Galtung, J. and Ruge, M. (1973) 'Structuring and selecting news', in S. Cohen and J. Young (eds) *The Manufacture of News*, London: Constable.

Gamble, A. (1988) *The Free Economy and the Strong State*, London: Macmillan.

Garcia Canclini, N. (1995) *Hybrid Cultures*, Minneapolis, MN: University of Minnesota Press.

Garner, J. (1994) 'On the Wooldridge issue', *Sunday Advocate*, 10 April.

Geertz, C. (1963) 'The integrative revolution: primordial sentiments and civil politics in the new states', in C. Geertz (ed.) *Old Societies and New States: The Quest for Modernity in Asia and Africa*, New York: Free Press.

Gellner, E. (1964) *Thought and Change*, London: Weidenfeld & Nicholson.

—— (1992) *Postmodernism, Reason and Religion*, London: Routledge.

Giddens, A. (1990) *The Consequences of Modernity*, Cambridge: Polity.

—— (1991) *Modernity and Self-Identity: Self and Society in the Late Modern Age*, Cambridge: Polity.

—— (1994a) *Beyond Left and Right*, Cambridge: Polity.

—— (1994b) 'Living in a post-traditional society', in U. Beck, A. Giddens and S. Lash (eds) *Reflexive Modernization*, Cambridge: Polity.

—— (1997) *Sociology*, Cambridge: Polity.

Giggles (1996) 'No Dollars for Ms Dixit', *g magazine: glamour, glory, grandeur*, October: 14.

Gill, P. (1988) 'Conclusion: Helping is not enough', in R. Poulton and M. Harris (eds) *Putting People First: Voluntary Organisations and Third World Development*, Basingstoke: Macmillan.

Gillespie, M. (1995) *Television, Ethnicity and Cultural Change*, London: Routledge.

Glasbergen, P. and Cörvers, R. (1995) 'Environmental problems in an international context', in P. Glasbergen and A. Blowers (eds) *Environmental Policy in an International Context 1: Perspectives on Environmental Problems*, London: Arnold.

Glasgow University Media Group (1976) *Bad News*, London: Routledge.

—— (1980) *More Bad News*, London: Routledge.

—— (1982) *Really Bad News*, London: Routledge.

—— (1985) *War and Peace News*, Milton Keynes: Open University Press.

Glazer, N. and Moynihan, D. (1963) *Beyond the Melting Pot: The Negroes, Puerto Ricans, Jews, Italians and Irish of New York City*, Cambridge, MA: The MIT Press.

—— (1975) 'Introduction', in N. Glazer and D. Moyhihan (eds), *Ethnicity: Theory and Experience*, Cambridge, MA: Harvard University Press.

Gledhill, J. (1994) *Power and its Disguises*, London: Pluto Press.

Glendon, M. A. (1992) *Rights Talk: The Impoverishment of Political Discourse*, New York: Macmillan.

Gluckman, M. (1958) *Analysis of a Social Situation in Modern Zululand*, Manchester: Manchester University Press, first published 1940.

Goddard, V., Llobera, J. and Shore, C. (eds) (1994) *The Anthropology of Europe: Identities and Boundaries in Conflict*, Oxford and Providence, RI: Berg.

Godfrey, M. (1976) *Education, Training, Productivity and Income: A Kenyan Case Study*, Brighton, Institute of Development Studies.

Goldenberg, S. (1996a) 'BJP leader takes office', *Guardian*, 17 May: 16.

—— (1996b) 'India calls on left bloc as BJP cedes power', *Guardian*, 29 May: 15.

—— (1996c) 'Street riots greet Miss World', *Observer*, 24 November: 2.
Goodwin, N. and Guest, J. (1996) 'By-pass operation', *New Statesman and Society*, 19 January: 14–15.
Gordon, A. (1985) *The Evolution of Labor Relations in Japan*, Harvard East Asian monographs, Cambridge, MA: Harvard University.
Gordon, D. (1991) 'Female circumcision and genital operations in Egypt and the Sudan: a dilemma for medical anthropology', *Medical Anthropology Quarterly* 5: 3–14.
Gordon Drabek, A. (1987a) 'Development alternatives: The challenge for NGOs – An overview of the issues', *World Development* 15 (Autumn) (Supplement): ix–xv.
—— (1987b) 'Editor's Preface', *World Development* 15 (Autumn) (Supplement): iv–viii.
Graham-Brown, S. (1988) *Images of Women. The Portrayal of Women in Photography of the Middle East 1860–1950*, London: Quartet Books.
Gramsci, A. (1971) *Selections from the Prison Notebooks*, New York: International Publishers.
Granovetter, M. (1985) 'Economic action and social structure: a theory of embeddedness', *American Journal of Sociology* 91: 481–510.
Greer, G. (1984) *Sex and Destiny: The Politics of Human Fertility*, London: Secker & Warburg.
Gruenbaum, E. (1982) 'The movement against clitoridectomy and infibulation in Sudan: Public health policy and the women's movement', *Medical Anthropology Newsletter*, 13, 2: 4–12.
—— (1988) 'Reproductive ritual and social reproduction: female circumcision and the subordination of women in the Sudan', in N. O'Neill and J. O'Brien (eds) *Economy and Class in Sudan*, Aldershot: Avebury.
The Guardian (1994) 'The First Cut', April 25.
Guha, R. (1989) *The Unquiet Woods: Ecological Change and Peasant Resistance in the Himalayas*, New Delhi: Oxford University Press.
Guignon, C.B. (1983) *Heidegger and the Problem of Knowledge*, New York: Hackett Publishing Company.
Guttierez, G. (1988) *A Theology of Liberation: History, Politics, and Salvation*, New York: Orbis.
Habermas, J. (1976) *Legitimation Crisis*, London: Heinemann.
—— (1978) *L'Espace Public: Archéologies de la Publicité comme Dimension Constitutive de la Société Bourgeoise*, Paris: Payot.
Hacking, I. (1983) *Representing and Intervening: Introductory Topics in the Philosophy of Natural Science*, Cambridge: Cambridge University Press.
Haeri, S. (1994) 'A fate worse than Saudi', *Index on Censorship* 23, 4–5: 49–51.
Hall, C.M. (1994) *Tourism in the Pacific Rim: Development, Impacts and Markets*, Melbourne: Longman.
Hall, M. (1992) 'Sex tourism in South-east Asia', in D. Harrison (ed.) *Tourism and the Less Developed Countries*, London: Belhaven Press.
Hall, S. (1972) 'The Determinations of News Photographs', *Working Papers in Cultural Studies*, Birmingham Centre for Cultural Studies 3 (Autumn): 53–87.
—— (1973) 'A world at one with itself', in S. Cohen and J. Young (eds) *The Manufacture of News*, London: Constable.
—— (1980) 'Encoding/Decoding', in S. Hall *et al.*, *Culture, Media, Language*, London: Hutchinson.
—— (1981) 'Cultural studies: Two paradigms', in T. Bennett (ed.) *Culture, Ideology and Social Process: A Reader*, London: Open University Press.
—— (1982) 'The rediscovery of ideology: The return of the repressed in media studies', in Gurevitch *et al.*, *Culture, Society and Media*, London: Methuen.
—— (1992) 'The question of cultural identity', in S. Hall, D. Held and T. McGrew (eds) *Modernity and Its Futures*, Cambridge: Polity.
Hall, S. and Jefferson, T. (eds) (1976) *Resistance Through Rituals: Youth Sub-Cultures in Post-War Britain*, London: Hutchinson.

Hane, M. (1979) *Peasants, Rebels and Outcasts: The Underside of Modern Japan*, New York: Pantheon.

Hannerz, U. (1990) 'Cosmopolitans and locals in world cultures', in M. Featherstone (ed.) *Global Culture: Nationalism, Globalization and Modernity*, London: Sage.

—— (1991) 'Scenarios for peripheral cultures', in A.D. King (ed.) *Culture, Globalization and the World System*, London: Macmillan.

Hantrakul, S. (1988) 'Prostitution in Thailand', in G. Chandler *et al.* (eds) *Development and Displacement: Women in Southeast Asia*, Melbourne: Monash University.

Haraway, D. (1989) *Primate Visions: Gender, Race and Nature in the World of Modern Science*, London: Routledge.

Hardill, I., Fletcher, D. and Montagne-Villette, S. (1995) 'Small firms' "distinctive capabilities" and the socio-economic milieu: findings from case studies in Le Choletais (France) and the East Midlands' (UK) *Entrepreneurship and Regional Development* 7, 2: 167–86.

Harrell-Bond, B. and Carlson, S. (1996) 'Revisiting the dream of a new world information order', *Contemporary Politics* 2, 1: 97–121.

Harris, N. (1986) *The End of the Third World: Newly Industrializing Countries and the Decline of Ideology*, Harmondsworth: Penguin Books.

—— (1995) *The New Untouchables: Immigration and the New World Worker*, London: Taures.

Harrison, P. and Palmer, R. (1986) *News out of Africa: Biafra to Band Aid*, London: Hilary Shipman.

Hart, A. (1987) 'Consuming compassion: The Live Aid phenomenon', *Links* 28: 15–17.

—— (1989) 'Images of the Third World', *Links* 34: 12–18.

Harvey, D. (1989) *The Condition of Postmodernity*, Cambridge, MA: Blackwell.

Hayes, R.O. (1975) 'Female genital mutilation, fertility control, women's roles, and the patrilineage in modern Sudan: a functional analysis', *American Ethnologist* 4: 617–33.

Haynes, J. (1993) *Religion in Third World Politics*, Buckingham: Open University Press.

—— (1996) *Religion and Politics in Africa*, London: Zed Books.

Hebdige, D. (1988) 'Postscript 1: Vital strategies', in D. Hebdige (ed.) *Hiding in the Light*, London: Comedia and Routledge.

Held, D. and McGrew, A. (1993) 'Globalization and the liberal democratic state', *Government and Opposition* 28, 3: 261–85.

Helmsing, A.H.J. (1991) 'Non-agricultural enterprises in the communal lands of Zimbabwe', in N.D. Mutizwa-Mangiza and A.H.J. Helmsing (eds) *Rural Development and Planning in Zimbabwe*, Aldershot: Avebury.

Henderson, J. (1989) *The Globalization of High Technology: Society, Space, and Semiconductors in the Restructuring of the Modern World* London: Routledge.

Hettne, B. (1995) *Development Theory and the Three Worlds*, Harlow: Longman.

Hewitt, T. (1992) 'Children, Abandonment and Public Action in Brazil', in M. Wuyts, M. Mackintosh and T. Hewitt (eds) *Development Policy and Public Action*, Oxford: Oxford University Press.

Hill, C. (1993) 'Planning for prostitution: An analysis of Thailand's sex industry', in M. Turshed and B. Halcomb (eds) *Women's Lives and Public Policy*, Westport, CT: Greenwood Press.

—— (1996) *Liberty Against the Law: Some Seventeenth Century Controversies*, London: Allen Lane and Penguin.

Hirst, P. (1989) *After Thatcher*, London: Collins.

Hite, S. (1976) *The Hite Report on Female Sexuality*, 2nd edn, New York: Pandora Press.

Hobsbawm, E. and Ranger, T. (eds.) (1983) *The Invention of Tradition*, Cambridge: Cambridge University Press.

Hodgson, D. (1995) 'Combating the organized sexual exploitation of Asian children: Recent developments and prospects', *International Journal of Law and the Family* 9: 23–53.

Hoggart, R. (1957) *The Uses of Literacy*, Harmondsworth: Penguin.

Holingsworth, M. (1986) *The Press and Political Dissent: A Question of Censorship*, London: Pluto Press.
Holmström, M. (1994) 'A cure for loneliness: networks, trust and shared services in Bangalore', paper presented at Workshop on Industrialisation, Organisation, Innovation and Institutions in the South, Vienna Institute for Development and Co-operation, November.
Honey, J. (1987) 'The sinews of society: The public schools as a system', in D.K. Muller, F. Ringer and B. Simon (eds) *The Rise of the Modern Educational System: Structural Change and Social Reproduction 1870–1920*, Cambridge: Cambridge University Press.
Horgan, J. (1986) *Images of Africa: Interim Research Report*, Ireland: School of Communication, National Institute for Higher Education.
Hosken, F.P. (1982) *The Hosken Report: Genital and Social Mutilation of Females*, Lexington, MA: Women's International News Network.
Hum Aapke Hain Koun...! (1994) P. Barjatyta, India.
Huntington, S. (1993) 'The Clash of Civilisations?', *Foreign Affairs* 72: 22–49.
—— (1996) *The Clash of Civilizations and the Remaking of World Order*, New York: Simon & Schuster.
Hutton, W. (1995) *The State We're In*, London: Hodder & Stoughton.
Ignacio Lopes, V.J. (1987) *Como Fazer um Socio-drama*, Serie Radio Popular l, CEMI.
Ignatieff, M. (1998) *The Warrior's Honor: Ethnic War and the Modern Conscience*, London: Chatto & Windus.
'The Image of Africa' (1988) *Dialogue*, September–October: 1–2.
Ishida, T. (1971) *Japanese Society*, London: Random House.
Jackman, O. (1994) 'One for the history books', *Sunday Sun*, 17 April.
Jackson, J.A. (ed.) (1969) *Migration*, Cambridge: Cambridge University Press.
Jackson, P. (1989) *Maps of Meaning*, London: Routledge.
Jafferlot, C. (1993) 'Hindu nationalism: Strategic syncretism in ideology building', *Economic and Political Weekly of India*, 20–27 March: 517–24.
—— (1994a) 'Hindu nationalism and Hindu society', paper given at the 13th European South Asian Conference, Toulouse.
—— (1994b) *The Hindu Nationalist Movement 1925–1993*, London: C. Hurst & Co.
James, C.L.R. (1963) *Beyond a Boundary*, London: Hutchinson.
Jayawardena, K. (1986) *Feminism and Nationalism in the Third World*, London: Zed Books.
Jeffrey, P. (1976) *Migrants and Refugees: Muslim and Christian Families in Bristol*, Cambridge: Cambridge University Press.
Jenkins, R. (1997) *Rethinking Ethnicity: Arguments and Explorations*, London: Sage.
Jenks, C. (1993) *Culture*, London: Routledge.
Jennet, C. and Stewart, R.G. (1989) *Politics of the Future. The Role of Social Movements*, Melbourne: Macmillan.
Jenson, J. (1989) 'The talents of women, the skills of men: Flexible specialisation and women', in S. Wood (ed.) *The Transformation of Work*, London: Unwin Hyman.
Jha, S.K. (1995) 'Flop show', *g magazine: glamour, glory, grandeur*, July: 23–5.
—— (1996) 'The different dozen', *g magazine: glamour, glory, grandeur*, January: 25.
—— (1997) 'Haqueeqat was a defeatist film...Border is not!', *g magazine: glamour, glory, grandeur*, April: 10 (interview with J. P. Dutta).
Johnson, H.E. (1992) 'Rural livelihoods: Action from below', in H. Bernstein, B. Crow and H. Johnson (eds) *Rural Livelihoods: Crises and Responses*, Oxford: Open University and Oxford University Press.
—— (1995) 'Reproduction, exchange relations and food insecurity: Maize production and maize markets in Honduras', unpublished Ph.D thesis, Open University.
Jurassic Park (1993) Steven Spielberg, US.
Kabeer, N. (1990) 'Poverty, purdah and women's survival strategies in rural Bangladesh', in H. Bernstein *et al.* (eds), *The Food Question*, London: Earthscan.

Kaida-Hozumi, M. K. (1989) 'The role of NGOs and the media in development education', M.Phil. dissertation (unpublished) IDS, Sussex University.

Kama Sutra (1996) Mira Nair, India.

Kamata, S. (1982) *Japan in the Passing Lane*, New York: Pantheon.

Kamrava, M. (1993) *Politics and Society in the Third World*, London: Routledge.

Kaplan, R. (1994) 'The coming anarchy', *Atlantic Monthly*, February: 44–76.

Kaplun, M. (1987) 'La comunicación popular, alternativa valida?', *Ensayo*, November.

Kayanja, R. (1996) 'Preventative journalism for sub-Saharan Africa', Geneva: paper presented at World Aid 1996 Conference, October.

Kenez, P. (1992) *Cinema and Soviet Society 1917–1953*, Cambridge: Cambridge University Press.

Kennedy, P. (1988) *The Rise and Fall of the Great Powers*, London: Fontana.

Kenyon, S. M. (1991) *Five Women of Sennar: Culture and Change in Central Sudan*, Oxford: Oxford University Press.

Kerr, C. *et al.* (1962) *Industrialism and Industrial Man*, London: Heinemann.

Khan, S.R. (1996) 'The hundred luminaries: Foreword', *Movie International*, February: 44–5.

Kinzley, W.D. (1991) *Industrial Harmony in Modern Japan: The Invention of a Tradition*, London: Routledge.

Kishwar, M. (1984) 'Gangster rule: The massacre of Sikhs', *Manushi* 25 (November–December): 10–33.

—— (1993a) 'Religion at the service of nationalism', *Manushi* 76 (May–June): 2–20.

—— (1993b) 'Safety is indivisible: The warning from Bombay riots', *Manushi*, double issue 74–5 (January–April): 2–48.

Kishwar, M. and Vanita R. (1987) 'The burning of Roop Kanwar', *Manushi* 42–3 (September–December): 20–6.

Kitching, G. (1982) *Development and Underdevelopment in Historical Perspective: Populism, Nationalism and Industrialization*, London: Methuen.

—— (1989) *Development and Underdevelopment in Historical Perspective*, 2nd edn, London: Routledge.

Kouba, L.J. and Muasher, J. (1985) 'Female circumcision in Africa: an overview', *African Studies Review* 28: 95–110.

Kouchner, B. (1991) *Le Malheur des autres*, Paris: Éditions Odile Jacob.

Kowarick, L. *et al.* (1978) *Cidade, Usos e Abusos*, São Paulo: Brasiliense.

Kroeber, A. L. (1960 [1948]) *Anthropology*, New York: Harcourt Brace.

Kumar, K.J. (1988) 'Youth culture and popular music: the Indian experience', London: Centre for the Study of Communication and Culture, unpublished paper of the International Association of Mass Communication Research, Barcelona, 22–4 July.

Kuper, A. (1994) *The Chosen Primate*, London: Harvard University Press.

Kyong-Dong, K. (1994) 'Confucianism and capitalist development in East Asia', in L. Sklair (ed.) *Capitalism and Development*, London: Routledge.

Lall, B. (1996) 'Focus India: Sunrise Boulevard, Bollywood huffle, cities of joy', *Screen International*, 2 August: 11–15.

—— (1997) 'Murdoch finds new DTH passage to India', *Screen International*, 10–16 January: 1.

Lam, T. and Martin, C.J. (1994) *Vietnamese in the UK: Fifteen Years of Settlement*, Occasional Papers in Sociology and Social Policy 2, London: School of Education, Politics and Social Science, South Bank University.

—— (1997) *The Settlement of the Vietnamese in London: Official Policy and Refugee Responses*, Social Science Research Papers 6, London: School of Education, Politics and Social Science, South Bank University.

Late Show (1994) Bearan, BBC2, UK.

Lawton, A. (1992) *Kinoglasnost: Soviet Cinema in our Time*, Cambridge: Cambridge University Press.

Le Vine, S. and Le Vine, R. (1981) 'Child abuse and neglect in sub-Saharan Africa', in J.E. Korbin (ed.) *Child Abuse and Neglect: Cross-Cultural Perspectives*, Berkeley, CA: University of California Press.

Leach, E. (1954) *Political Systems of Highland Burma: A Study of Kachin Social Structure*, London: Bell.

Lebra, T.S. (1976), *Japanese Patterns of Behavior*, Honolulu: University of Hawaii Press.

Lee-Wright, P. (1990) *Child Slaves*, London: Earthscan Publications.

Lees, C. (1995) 'Hollywood raises hell in Bollywood', *The Sunday Times*, 25 June.

Legassik, M. and Wolpe, H. (1976) 'The Bantustans and capital accumulation in South Africa', *Review of African Political Economy* 7: 87–107.

Legendre, P. (1974) *L'Amour du censeur: Essai sur l'ordre dogmatique*, Paris: Éditions du Seuil.

Leheny, D. (1995) 'A political economy of Asian sex tourism', *Annals of Tourism Research* 22, 2: 367–84.

Lent, J.A. (1990) *The Asian Film Industry*, London: Christopher Helm.

Lerner, D. (1958) *The Passing of Traditional Society: Modernizing the Middle East*, New York: Free Press.

Levinas, E. (1993) *Outside the Subject*, London: Athlone.

Lewis, A. (1954) 'Economic development with unlimited supplies of labour', *The Manchester School* 22: 139–91.

Lewis, E. (1994) 'Historic moment', *Barbados Advocate*, 15 April.

Leyda, J. (1983) *Kino: A History of the Russian and Soviet Film*, London: George Allen & Unwin.

Lidchi, H.J. (1993) 'All in the choosing eye: Charity, representation and the developing world', unpublished Ph.D thesis, The Open University.

—— (1996) 'Projecting the "bad life" – representations and world cinema', in *U208 Summer School Student Notes*, Milton Keynes: Open University.

Lindenbaum, S. (1991) 'Anthropology rediscovers sex', *Social Science and Medicine*, 33: 865–6.

Lindquist, S. (1997) *Exterminate All The Brutes*, London: Granta Books.

Lipset, S.M. (1969) *Political Man*, London: Heinemann.

Lissner, J. (1977) *The Politics of Altruism*, Geneva: Lutheran World Federation.

—— (1981) 'Merchants of misery', *New Internationalist* 100: 23–5.

Lovelock, J. (1991) *Gaia: The Practical Science of Planetary Medicine*, London: Gaia Books Ltd.

Lukes, S. (1974) *Power: A Radical View*, London: Macmillan.

Lull, J. (1995) *Media, Communication, Culture: A Global Approach*, Cambridge: Polity.

Lyons, H. (1981) 'Anthropologists, moralities and relativities: the problem of genital mutilations', *Canadian Review of Sociology and Anthropology* 18: 499–518.

Lyotard, J.-F. (1986) *The Post-Modern Condition: A Report of Knowledge*, Manchester: Manchester University Press.

McClelland, D. (1961) *The Achieving Society*, Princeton, NJ: Princeton University Press.

—— (1963) 'The achievement motive in economic growth', in B.F. Hoselitz and W.E. Moore (eds) *Industrialization and Society*, The Hague: UNESCO and Mouton.

McCormick, D. (1994) 'Industrial district or garment ghetto? The case of Nairobi's mini-manufacturers', paper presented at Workshop on Industrialisation, Organisation, Innovation and Institutions in the South, Vienna Institute for Development and Co-operation, November.

MacGrane, B. (1989) *Beyond Anthropology: Society and the Other*, New York: Columbia University Press.

McGrew, T. (1992) 'A global society?', in S. Hall, D. Held and T. McGrew (eds) *Modernity and its Futures*, Cambridge: Polity.

Machado, Magri and Masago (eds) (1987) *Radio Livres; a reforma agraria no ar*, Brasil: Brasiliense.

Macías, J. (1991) 'Informal education, sociocultural expression, and symbolic meaning in popular immigration music text', *Explorations in Ethnic Studies* 14, 2: 15–31.
MacIntyre (1985) *After Virtue: A Study of Moral Theory*, London: Duckworth.
McQuail, D. (1994) *Mass Communication Theory*, London: Sage.
Macrae, J. (1996) 'The origins of unease: Setting the context of current ethical debates', London: Overseas Development Institute, mimeo.
Macridis, R.C. (ed.) (1992) *Foreign Policy in World Politics*, London: Prentice-Hall.
Malcolm, D. (1995) 'Dangerous Liaisons in Bollywood', *Guardian*, 'Screen', 9 February: 12–13.
Malinowski, B. (1946) *The Dynamics of Culture Change: An Inquiry into Race Relations in Africa*, New Haven, CT: Yale University Press.
Mamdani, M. (1996) *Citizen and Subject: Contemporary Africa and the legacy of late colonialism* London: James Currey.
Mandle, W.F. (1973) 'Cricket and Australian nationalism in the nineteenth century', *Journal of the Royal Australian Historical Society* 59, 4: 225–46.
Manning, F. (1981) 'Celebrating cricket: The symbolic construction of Caribbean politics', *American Ethnologist* 8, 3: 616–32.
Manuel, P. (1993) *Cassette Culture: Popular Music and Technology in North India*, Chicago, IL: University of Chicago Press.
Manzo, K. (1995) 'Black consciousness and the quest for a counter-modernist development', in J. Crush (ed.) *The Power of Development* London: Routledge.
Maquet, J.J.P. (1961) *The Premise of Inequality in Rwanda: A Study of Political Relationships in a Central African Kingdom*, Oxford: Oxford University Press.
March, R.M. (1991) *The Japanese Negotiator: Subtlety and Strategy Beyond Western Logic*, Tokyo: Kodansha International.
Marquand, D. (1988) *The Unprincipled Society*, London: Jonathan Cape.
Marshall, M. (1994) 'Tourism still holds the upper hand', and 'Tetley too bitter for Windies', *Daily Nation*, 14 April.
Martin-Barbero, J. (1993) *Communication, Culture and Hegemony*, London: Sage.
Marty, M.E. and Scott Appleby, R. (1993) 'Introduction', in M.E. Marty and R. Scott Appleby (eds) *Fundamentalism and the State: Remaking Polities, Economies, and Militance*, Chicago, IL: University of Chicago Press.
Massey, D. (1994) *Space, Place and Gender*, Cambridge: Polity.
Mayer, A. (1981) *The Persistence of the Ancient Regime*, London: Croom Helm.
Mayoux, L. (1993) 'A development success story? Low caste entrepreneurship and inequality: An Indian case study', *Development & Change* 24: 541–68.
Mazrui, A. (1990) *Cultural Forces in World Politics*, London: James Currey.
Mbilinyi, M. (1990) ' 'Structural Ajustment', Agribusiness and Rural Women in Tanzania', in H. Bernstein *et al.* (eds) *The Food Question*, London: Earthscan.
Mbiti, J. (1969) *African Religions and Philosophy*, London: Heinemann.
Medhurst, K. (1989) 'Brazil', in S. Mews (ed.) *Religion in Politics. A World Guide*, Harlow: Longman.
Mehmet, O. (1995) *Westernizing the Third World: The Eurocentricity of Economic Development Theories*, London: Routledge.
Middleton, N., O'Keefe, P. and Moyo, S. (1993) *The Tears of the Crocodile: From Rio to Reality in the Developing World*, London: Pluto Press.
Mishra, V. (1985) 'Towards a theoretical critique of Bombay cinema', *Screen* 26, 3: 133–44.
—— (1989) 'The texts of "Mother India" ', in S. Slemon and H. Tiffin (eds) *After Europe*, Australia: Dangaroo Press.
Mississippi Masala (1991) Mira Nair, US.
Mitchell, J.C. (1956) *The Kalela Dance: Aspects of Social Relationship Among Urban Africans in Northern Rhodesia*, Manchester: Manchester University Press.

Mitchell, T. (1995) 'The object of development: America's Egypt', in J. Crush (ed.) *Power of Development*, London: Routledge.
Miyamoto, M. (1982) *The Book of Five Rings: The Real Art of Japanese Management*, London: Bantam Books.
Mohamud, O. A. (1991) 'Female circumcision and child mortality in urban Somalia', *Genus* 67: 203–23.
Montague, A. (1988) 'The changing face of charity', *New Society*, 13 May: 22–4.
Moreland, W. H. and Chatterjee, A. C. (1936) *A Short History of India*, London and New York: Longman and Green and Co.
Morris, J. (1992) *Heaven's Command: An Imperial Progress*, London: Folio Society.
Moore, H. (1988) *Feminism and Anthropology*, Cambridge: Polity.
Morishima, M. (1981) *Why Japan has 'Succeeded': Western Technology and Japanese Ethos*, Cambridge: Cambridge University Press.
Morokvasic, M. (1993) ' "In and out" of the labour market: Immigrant and minority women in Europe', *New Community* 19, 3.
Moser, C. (1993) *Gender Planning and Development: Theory, Practice and Training*, London: Routledge
'Movie Opinion Poll '95' (1995) *Movie International*, March: 49–60.
'Movie Opinion Poll '96' (1996) *Movie International*, March: 46–74.
'Movie Opinion Poll '97' (1997) *Movie International*, April: 48–69.
Moviewatch (1996) Channel 4, UK.
Moynihan, D. (1993) *Pandaemonium: Ethnicity in International Politics*, Oxford: Oxford University Press.
Muller, T. and Espenshade, T. (1985) *The Fourth Wave: California's Newest Immigrants*, Washington DC: Urban Institute.
Murakami, Y. (1984) 'Ie Society as a Pattern of Civilization', *Journal of Japanese Studies* 10, 2.
Myers, W. (1988) 'Alternative services for street children: The Brazilian approach', in A. Bequele and J. Boyden (eds) *Combating Child Labour*, Geneva: International Labour Office.
Nairn, T. (1977) *The Break-up of Britain: Crisis and Neo-Colonialism*, London: New Left Books.
Nakane, C. (1970) *Japanese Society*, Harmondsworth: Penguin.
Nash, C. and Van Der Gaag, N. (1987) *Images of Africa – The UK report*, Oxford: Oxfam.
Nettl, J.P. and Robertson, R. (1968) *International Systems and the Modernization of Societies: The Formation of National Goals and Attitudes*, New York: Basic Books.
New Internationalist (1992) 'Changing Charity: 50 years of Oxfam', 228.
New Internationalist (1993) special issue on Tourism, 19.
NGO-EC Liaison Committee (1989) *Codes of Conduct: Images and Messages Relating to the Third World*, Brussells: NGO-EC Liaison Committee.
Nichols, B. (1981) *Ideology and the Image: Social Representation in the Cinema and other Media*, Bloomington, IA: Indiana University Press.
—— (1991) *Representing Reality*, Bloomington, IA: Indiana University Press.
Nieuwenhuys, O. (1993) 'Street children, NGOs and Social Welfare', paper for the joint seminar on 'Children, NGOs and the State', Open University and London School of Economics, 1 July.
—— (1994) *Children's Lifeworlds: Gender, Welfare and Labour in the Developing World*, London: Routledge.
'No Indian summer in sight for local distributors' (1996) *Screen International*, 31 May: 25.
North, D. (1990) *Institutions, Institutional Change and Economic Performance*, New York: Cambridge University Press.
Nove, A. (1989) '*Glasnost' in Action*, London: Unwin Hyman.
Nugent, S. and Shore, C. (eds) (1997) *Anthropology and Cultural Studies*, London: Pluto Press.
Nyoni, S. (1988–9) 'Images of poverty: A view from Zimbabwe', *Poverty* 71 (Winter): 6–10.

Oberg, K. (1940) 'The Kingdom of Ankole in Uganda', in M. Fortes and E.E. Evans-Pritchard (eds) *African Political Systems*, Oxford: Oxford University Press.

Observer (1996) 'They cooked my brother's heart and ate it', 14 April.

Ogbini, F.E. (1990) 'Linchpin projects as a basis for establishing effective production systems in Third World nations', *International Journal of Technology Management* 5, 5: 499–511.

October (1928) Sergei Eisenstein, USSR.

Olsen, W. (1993) 'Competition and Power in Rural Markets', *IDS Bulletin*, 24, 3 (July).

Olson, M. (1982) *The Rise and Decline of Nations*, New Haven, CT: Yale University Press.

Ortíz, M. (1995) 'Cultura migratoria en Tizapán el Alto, Jalisco', unpublished MA thesis, Universidad Pedagocica de Mexico.

Ortner, S.B. and Whitehead, H. (1981) (eds) *Sexual Meanings: The Cultural Construction of Gender and Sexuality*, Cambridge: Cambridge University Press.

Osman, M.T. (1989) *Malay Folk Beliefs: An Integration of Disparate Elements*, Kuala Lumpur: Dewan Bahasa dan Pustaka.

Ouchi, W. (1981) *Theory Z: How American Business Can Meet the Japanese Challenge*, Reading, MA: Addison-Wesley.

Owen, J. (1996) '"No body, no war": Representations of war and disaster', *Contemporary Politics*, 2, 1 (Spring): 159–67.

Oxfam (1987) *What Makes an Appropriate Picture for Oxfam?*, Oxford: Oxfam.

—— (1991) *Oxfam Review 1990–1*, Oxford: Oxfam.

Oxford Dictionary (1987) *The Oxford Reference Dictionary*, London: Guild Publishing.

Pacey, A. (1990) *Technology in World Civilisation*, Oxford: Basil Blackwell.

Panday, G. (1982) 'Peasant revolt and Indian nationalism: The peasant movement in Awadh, 1919–1922', in R. Guha (ed.) *Subaltern Studies I: Writings on South Asian History and Society*.

—— (1991) 'Hindus and others: The militant Hindu construction', *Economic and Political Weekly of India*, 28 December: 2997–3009.

Pande, B. (1995) '"Bombay", the film (controversial or otherwise) of the decade', *India Link International*, June–July: 41–2.

Paoli, M. (1992) 'Citizenship, inequalities, democracy and rights: The making of a public space in Brazil', *Social & Legal Studies* 1: 143–59.

Parker, M. (1995) 'Rethinking Female Circumcision', *Africa* 65: 506–23.

Parsons, T. (1951) *The Social System*, London: Routledge & Kegan Paul.

Pascale, R.T. and Athos, A.G. (1981) *The Art of Japanese Management*, Harmondsworth: Penguin.

Pasha, M.K. and Samatar, A.I. (1996) 'The resurgence of Islam', in J. Mittelman (ed.) *Globalisation: Critical Reflections*, London: Boulder.

Paterson, R. (1995), 'London life, London television', in D. Petrie and J. Willis (eds) *Television and the Household: Reports from the BFI's Audience Tracking Study*, London: BFI.

Pathak, A. (1994) *Contested Domains. The State, Peasants and Forests in Contemporary India*, New Delhi: Thousand Oaks, and London: Sage Publications in association with The Book Review Literary Trust: New Delhi.

Patterson, O. (1969) 'The ritual of cricket', *Jamaica Journal* 3: 6–15.

Pearce, F. (1992) 'Last chance to save the planet?', *New Scientist*, 30 May: 24–8.

Peires, J.B. (1989) *The Dead Will Arise: Nongqawuse and the Great Xhosa Cattle Killing Movement of 1856–7*, Johannesburg: Raven Press.

Pettman, J. J. (1996) *Worlding Women: A Feminist International Politics*, London: Routledge, and Sydney: Allen & Unwin.

—— (1997) 'Body politics: International sex tourism', *Third World Quarterly* 18, 1: 93–108.

Phillips, M. (1991), 'London: Time machine', in M. Fisher and U. Owen (eds) *Whose Cities?*, London: Penguin.

Philo, G. and Lamb, R. (1986) *Television and the Ethiopian Famine – From Buerk to Band Aid*, London: The Television Trust for the Environment.
Phongpaichit, P. (1982) *From Peasant Girls to Bangkok Masseuses*, Geneva: International Labour Office.
Pieterse, J.N. (1995) 'Globalization as hybridization', in M. Featherstone, S. Lash and R. Robertson (eds) *Global Modernities*, London: Sage.
Pines, J. and Willemen, P. (eds) (1989) *Questions of Third Cinema*, London: BFI.
Pinney, C. (1992) 'The parallel histories of anthropology and photography', in E. Edwards (ed.) *Anthropology and Photography 1860–1920*, London: Yale University Press and Royal Anthropological Institute.
Pires, B. (1994) 'Spot the Bajan: New game at Kensington', *Daily Nation*, 15 April.
Pixote (1981) H.D. Bebenco, Brazil.
Plumwood, V. (1993) *Feminism and the Mastery of Nature*, London: Routledge.
Porter, R. (1994) *A Social History of London*, London: Penguin.
Portes, A. and Borocz, J. (1989) 'Contemporary immigration: Theoretical perspectives on its determinants and modes of incorporation', *International Migration Review* 23, 3: 606–30.
Portes, A. and Zhou, M. (1992) 'Gaining the upper hand: Economic mobility among immigrant and domestic minorities', *Ethnic and Racial Studies* 15, 4: 491–521.
Potter, D. and Thomas, A. (1992) 'Development, capitalism and the nation state', in T. Allen and A. Thomas (eds) *Poverty and Development in the 1990s*, Oxford: Oxford University Press.
Poulton, R. (1988) 'On theories and strategies', in R. Poulton and M. Harris (eds) *Putting People First: Voluntary Organisations and Third World Development*, Basingstoke: Macmillan.
Poulton, R. and Harris, M. (eds) (1988) *Putting People First: Voluntary Organisations and Third World Development*, Basingstoke: Macmillan.
Powell, B. (1996) 'Indian Tinsel Town', *Sunday Telegraph*, 15 September: 17.
Prakash, S. (1984) 'La Musique, la danse et le film populaire' (Music, dance and popular culture), *CinemAction: Les Cinémas indiens*, Paris: Cerf.
Preis, A.-B.S. (1996) 'Human rights as cultural practice: an anthropological critique', *Human Rights Quarterly* 18, 2: 286–315.
Prunier, G. (1995) *The Rwanda Crisis 1959–1994: History of a Genocide*, London: Hurst.
Putnam, R.D. (1993) *Making Democracy Work: Civic Traditions in Modern Italy*, Princeton, NJ: Princeton University Press.
Quingrui and Xiaobo (1991), 'A model of the "secondary innovation" process', Proceedings of the 1991 Portland International Conference on Management of Engineering and Technology, 617–20.
Quinn, M. (1994) *The Swastika: Constructing the Symbol*, London: Routledge.
Radcliffe-Brown, A.R. (1952) *Structure and Function in Primitive Society*, London: Cohen & West.
Radio do Povo (undated) São Miguel, locally distributed booklet.
Rahnema, M. and Bawtree, V. (eds) (1997) *The Post-Development Reader*, London: Zed Books.
Rajadhyaksha, A. (1996) 'Strange attractions', *Sight and Sound*, August: 28–31.
Rajadhyaksha, A. and Willemen, P. (1994) *Encyclopaedia of Indian Cinema*, London and New Delhi: British Film Institute and Oxford University Press.
Rajan, L. (1987) 'Charity statistics 1977–1986: An analysis of trends', in Charities Aid Foundation (ed.) *Charity Trends 1986/7*, Tonbridge: Charities Aid Foundation.
Ranger, T. Somad, Y. and Stuart, O. (eds) (1996) *Culture, Identity and Politics*, Aldershot: Avebury.
Ray, S. (1976) *Our Films Their Films*, Hyderabad, India: Orient Longmans.
—— (1982) 'Under western eyes', *Sight and Sound* 51, 4: 268–74.

Reanda, L. (1991) 'Prostitution as a human rights question: Problems and prospects of United Nations action', *Human Rights Quarterly* 13: 202–28.
Redfield, R. (1960) *The Little Community and Peasant Society and Culture*, Chicago, IL: University of Chicago Press.
Reich, S. (1990) *The Fruits of Fascism*, Cornell: Cornell University Press.
Reuben, B. (1993) *Follywood Flashback*, India: Indus and Harper Collins.
Reuter News Agency (1996) 'Indian leftwing names leader in bid for power', *Guardian*, 15 May: 10.
Reynolds, V., Falger, V. and Vine, L. (eds.) (1987) *The Sociobiology of Ethnocentrism: Evolutionary dimensions of xenophobia, discrimination, racism and nationalism*, London: Croom Helm.
Richards, P. (1996) *Fighting for the Rain Forest: War, Youth and Resources in Sierra Leone*, London: James Currey and Heinemann.
Richman, N. (1994) 'Commissioned review: We're only street children', *ACPP Review & Newsletter* 16, 1: 3–8.
Rickman, H.P. (1979) *William Dilthey: Pioneer of the Human Studies*, New York: Elek.
Ritzer, G. (1995) *Sociological theory*, New York: McGraw Hill.
Robertson, R. (1992) *Globalization: Social Theory and Global Culture*, London: Sage.
Robins, K. (1991) 'Tradition and translation: National culture in its global context', in J. Corner and S. Harvey (eds) *Enterprise and Heritage*, London: Routledge.
Robinson, V. (1986) *Transients, Settlers and Refugees: Asians in Britain*, Oxford: Clarendon.
Rocha, J. (1991) 'Introduction', in G. Dimenstein (ed.) *Brazil: War on Children*, London: Latin American Bureau.
Rock Dancer (1995) Bappi Lahiri, India.
Roebuck, P. (1996) 'W.I. cricket must watch cultural yorker' *Sunday Sun*, 10 April.
Rose, C. (1993) 'Beyond the struggle for proof: Factors changing the environmental movement', *Environmental Values* 2: 285–98.
Rostow, W.W. (1960) *The Stages of Economic Growth: A Non-Communist Manifesto*, Cambridge: Cambridge University Press.
Roszak, T. (1970) *The Making of a Counter-Culture: Reflections on the Technocratic Society and its Youthful Opposition*, New York: Faber & Faber.
Rubinstein, W.D. (1993) *Capitalism and Decline in Britain 1750–1990*, London: Routledge.
Sachs, W. (1992) 'Introduction', in W. Sachs (ed.) *The Development Dictionary*, London: Zed Books.
—— (1992) *The Development Dictionary: A Guide to Knowledge as Power*, London: Zed Books.
Sahlins, M. (1972) *Stone Age Economics*, Chicago, IL: Aldine and Atherton.
Sahliyeh, E. (ed.) (1990) *Religious Resurgence and Politics in the Contemporary World*, Albany, NY: State University of New York Press.
Said, E. (1978) *Orientalism*, New York: Pantheon.
—— (1993) *Culture and Imperialism*, London: Chatto & Windus.
Saifullah Khan, V. (1979) *Minority Families in Britain*, London: Tavistock.
Saikal, A. and Maley, W. (eds) (1995) *Russia in Search of its Future*, Cambridge: Cambridge University Press.
Salaam Bombay! (1988) Mira Nair, India, France and UK.
Sami, I.R. (1986) 'Female circumcision with special reference to the Sudan', *Annals of Tropical Paediatrics* 6: 99–115.
Sanjek, R. (1996) 'Race', in A. Barnard and J. Spencer (eds) *Encyclopaedia of Social and Cultural Anthropology*, London and New York: Routledge.
São Paulo (1986) 31 January.
Sarkar, B. (1992) *The World of Satyajit Ray*, New Delhi: UBSPD.

Sarkar, T. (1991) 'The woman as a communal subject: Rashtrasevika and Ram Janamboomi movement', *Economic and Political Weekly of India*, 31 August: 2057–63.

Sassen, S. (1987) 'Growth and informalization at the core: A preliminary report on New York City', in M.P. Smith and J.R. Feagin (eds) *The Capitalist City*, Oxford: Blackwell.

—— (1988) *The Mobility of Capital and Labour: A Study in International Investment and Labour Flow*, Cambridge: Cambridge University Press.

—— (1989) 'New York City's informal economy', in A. Portes, M. Castells and L.A. Benton (eds) *The Informal Economy: Studies in Advanced and Less Developed Countries*, London: Johns Hopkins University Press.

Satellite Wars (1995) Annie Fienburgh, Channel 4, UK.

Save The Children Fund (SCF) (1988) *Impact of Image Guidelines for Writers, Editors, Advertisers and Photographers*, London: SCF.

Sayat Nova (1969) Sergei Paradzanov, USSR.

—— (1991) *Focus on Images*, London: SCF.

Scheper-Hughes, N. (1987) 'Culture, scarcity, and maternal thinking: Mother love and child death in northeast Brazil', in N. Scheper-Hughes (ed.) *Child Survival: Anthropological Perspectives on the Treatment and Maltreatment of Children*, Dordrecht: D. Reidel.

—— (1995) 'Everyday violence: Bodies, death and silence', in S. Corbridge (ed.) *Development Studies: A Reader*, London: Edward Arnold.

Schindler's List (1993) Steven Spielberg, US.

Schlesinger, P. (1991) *Media, State and Nation*, London: Sage.

Schmitz, H. (1995) 'Collective efficiency: growth path for small-scale industry', *Journal of Development Studies* 31, 4: 529–66.

Schuurman, F. (1993) 'Introduction: Development Theory in the 1990s', in F. Schuurman (ed.) *Beyond the Impasse: New Directions in Development Theory*, London and New Jersey: Zed Books.

Scott, A. (1990) *Ideology and the New Social Movements*, London: Unwin Hyman.

Scott, C. (1995) *Gender and Development: Rethinking Modernisation and Dependency Theory*, Colorado: Lynne Reiner.

Scott, J.C. (1976) *The Moral Economy of the Peasant: Subsistence and Rebellion in Southeast Asia*, New Haven, CT: Yale University Press.

—— (1985) *Weapons of the Weak: Everyday Forms of Peasant Resistance*, New Haven, CT: Yale University Press.

—— (1989) 'Everyday Forms of Peasant Resistance', in F.D. Colburn (ed.) *Everyday Forms of Peasant Resistance*, New York and London: M.E. Sharpe.

—— (1990) *Domination and the Arts of Resistance: Hidden Transcripts*, New Haven, CT: Yale University Press.

Scott, J.W. (1991) 'The evidence of experience', *Critical Inquiry* 17, 4: 773–97.

Screen International Film and TV Year Book (1996) London: Emap Media.

Sealy, M. (1994) 'Match hauls in $1 million', *Weekend Nation*, 15 April.

Searle, C. (1990) 'Race before wicket: Cricket and the white rose', *Race and Class* 31: 343–55.

Segerstale, U. (1990) 'The sociobiology of conflict and the conflict about sociobiology', in J. Van der Dennen and V. Falger (eds) *Sociobiology and Conflict*, London: Chapman & Hall.

Sehgal, M. (1991) 'Indian video films and Asian-British identities', *Cultural Studies from Birmingham* 1.

SEJUP (1994) *National Movement of Street Boys and Street Girls: Objectives, Actions and Achievements*, Brazil (via Greennet): Serviço Brasileiro de Justica e Paz.

Sekula, A. (1984) *Photography Against the Grain*, Halifax, NS: Nova Scotia College of Art and Design.

Seneviratne, H.L. (1978) *Rituals of the Kandyan State*, Cambridge: Cambridge University Press.

Sengoopta, C. (1992) 'Ray's History', *Sight and Sound* 2, 7: 63.

Sethi, H. (1993) 'Survival and Democracy: Ecological Struggles in India', in P. Wignaraja (ed.) *New Social Movements in the South. Empowering the People*, London and New Jersey: Zed Books.

Sewell, J. W. (1988) 'Foreword', in J.P. Lewis (ed.) *Strengthening the Poor: What Have We Learned?*, New Brunswick: Transaction Books.

Shandall, A. A. (1967) 'Circumcision and infibulation of females', *Sudan Medical Journal* 5: 178–212.

Sharif, N. (1994) 'Technology change management: imperatives for developing economies', *Technological Forecasting & Social Change* 47: 103–14.

Sharpley, R. (1994) *Tourism, Tourists and Society*, Cambridgeshire: ELM Publications.

Shaw, P. and Y. Wong (1989) *Genetic Seeds of Warfare: Evolution, nationalism and patriotism*, Boston: Unwin Hyman.

Sheehan, E. (1981) 'Victorian clitoridectomy: Isaac Baker Brown and his harmless operative procedure', *Medical Anthropology Newsletter* 12: 10–15.

Sheman, N. (1983) 'Individualism and the objects of psychology', in S. Harding and M.B. Hintikka (eds) *Discovering Reality: Feminist Perspectives on Epistemology, Metaphysics, Methodology and Philosophy of Science*, Boston, MA: Reidel.

Shore, C. (1997) 'Ethnicity, xenophobia and the boundaries of Europe', *International Journal of Minority and Group Rights* 4, 3–4: 247–62.

Shorter Oxford English Dictionary (1965), 3rd edn, London: Oxford University Press.

Shrage, L. (1994) *Moral Dilemmas of Feminism: Prostitution. Adultery and Abortion*, New York: Routledge.

Simpson, A. (1985) 'Charity begins at home', *Ten8* 19: 21–6.

Sinclair, J. (1992) 'The decentring of globalization: Television and globalization', in E. Jacka (ed.) *Continental Shift: Globalisation and Culture*, Double Bay, NSW: Local Consumption Publications.

Skelton, T. (1996a) 'Globalization, culture and land: the case of the Caribbean', in E. Kofman and G. Youngs (eds) *Globalization: Theory and Practice*, London: Pinter.

—— (1996b) '"Cultures of land", in the Caribbean: A contribution to the debate on development and culture', *The European Journal of Development Research* 8, 2: 71–92.

Sklair, L. (1991) *Sociology of the Global System: Social Change in Global Perspective*, New York: Harvester & Wheatsheaf.

Sklar, R. (1993) *Film: An International History of the Medium*, London: Thames & Hudson.

Slack, A.T. (1988) 'Female circumcision: a critical appraisal', *Human Rights Quarterly* 10: 437–86.

Slater, D. (1992) 'On the borders of social theory: learning from other regions', *Environment and Planning D: Society and Space* 10: 307–27.

So, A.Y. (1990) *Social Change and Development: Modernisation. Dependency and World System Theorists*, London: Sage.

Soja, E. (1987) 'Economic restructuring and the internationalization of the Los Angeles region', in M.P. Smith and J. Feagin (eds) *The Capitalist City: Global Restructuring and Community Politics*, Oxford: Blackwell.

Sokolovski, S. and Tishkov, V. (1996) 'Ethnicity', in A. Barnard and J. Spencer (eds) *Encyclopaedia of Social and Cultural Anthropology*, London: Routledge.

Somaya, B. (1995) 'Signature', *g magazine: glamour, glory, grandeur*, September: 3 (on *Bandit Queen*'s clearance by censors).

—— (1996) 'Signature', *g magazine: glamour, glory, grandeur*, February: 3 (on *Bandit Queen*'s box office success).

South Commission (1990) *The Challenge to the South*, Oxford: Oxford University Press.

Soviet Weekly (1991) 8 August.

Speed (1994) Jan De Bont, US.

Spivak, G. (1986) *Nationalist Thought and the Colonial World*, London: Zed Books.

Sreberny-Mohammadi, A. (1996) 'The Many Cultural Faces of Imperialism', in P. Golding and P. Harris (eds) *Beyond Cultural Imperialism*, London: Sage.

Srinivas, M.R. (1952) *Religion and Society among the Coorgs of South India*, Oxford: Clarendon Press.

Stalker, P. (1991) 'Infotainment', *New Internationalist* 222: 8–9.

Stallabrass, J. (1995) 'Empowering technology: The exploration of cyberspace', *New Left Review* 211: 3–32.

Stoddart, B. (1989) 'Cricket and colonialism in the English-speaking Caribbean to 1914: Towards a cultural analysis', in J.A. Mangan (ed.) *Pleasure, Profit, and Proselytism: British Culture and Sport at Home and Abroad*, London: Frank Cass.

Storper, M. and Scott, A.J. (1990) 'Work organisation and local labour markets in an era of flexible production', *International Labour Review* 129, 5: 573–91.

Street, B.V. (1993) 'Culture is a verb: Anthropological aspects of language and cultural process', in D. Graddol, L. Thompson and M. Bryman (eds) *Language and Culture*, London: Multilingual Matters.

Strike (1924) Sergei Eisenstein, USSR.

Stuart, A. (1994) 'Blackpool Illumination', *Sight and Sound*, February: 26–7 (on Gurinder Chadha's *Bhaji on the Beach*).

Sturdevant, S. and Stoltzfus, B. (1993) *Let the Good Times Roll: Prostitution and the US Military in Asia*, New York: The New Press.

Sullivan, B. (1995) 'Rethinking prostitution', in B. Caine and R. Pringle (eds) *Transitions: New Australian Feminisms*, Sydney: Allen & Unwin.

Sunday Sun (1994), 10 April.

Swain, M.B. (1995) 'Gender in tourism', *Annals in Tourism Research* 22, 2: 247–66.

Swift, A. (1991) *Brazil: The Fight for Childhood in the City*, Florence: UNICEF Innocenti Studies.

Tannenbaum, N. (1995) 'Buddhism, prostitution, and sex: Limits on the academic discourse on gender in Thailand', paper presented at the International Conference on Gender and Sexuality in Modern Thailand, Canberra, ANU.

Tasker, P. (1987) *Inside Japan: Wealth, Work and Power in the New Japanese Empire*, Harmondsworth: Penguin.

Taylor, J. (1987) 'The general theory of icebergs or what you can't see in political advertisements', *Ten8* 26: 44–53.

Taylor, M. (1987) *To Strengthen the Poor: A Statement of Commitment adopted by the Board of Christian Aid July 1987 as a Basis for Action and Reflection*, London: Christian Aid.

Tenbruck, F.H. (1990) 'The Dream of a Secular Ecumene: The Meaning and Limits of Policies of Development in Global Culture', in M. Featherstone (ed.) *Nationalism, Globalization and Modernity*, London: Sage.

Teni Zabytykh (1964) Sergei Paradzhanov, USSR.

Thakore, D. (1997) 'Our women are not bimbets in maxis', *g magazine: glamour, glory, grandeur*, January: 26–7 (interview with Mira Nair).

Thapar, R. (1986) *A History of India*, Harmondsworth: Penguin.

Therborn, G. (1995) 'Routes to/through modernity', in M. Featherstone, S. Lash and R. Robertson (eds) *Global Modernities*, London: Sage.

Thomas, A. (1992) 'Introduction', in T. Allen and A. Thomas (eds) *Poverty and Development in the 1990s*, Oxford: Oxford University Press.

—— (1995) *Does Democracy Matter? – Pointers from a Comparison of NGOs' Influence on Environmental Policies in Zimbabwe and Botswana*, DPP Working Paper 31/ GECOU Working Paper 4, Milton Keynes: GECOU Research Group, Open University.

Thomas, R. (1985) 'Indian cinema: Pleasures and popularity, an introduction', *Screen* 26, 3: 116–21.

Thomas-Hope, E.M. (1986) 'Caribbean diaspora – The inheritance of slavery: Migration from the Commonwealth Caribbean', in C. Brock (ed.) *The Caribbean in Europe: Aspects of the West Indian Experience in Britain, France and the Netherlands*, London: Frank Cass.

Thornhill, A. (1994) 'We won't be sidetracked', *Daily Nation*, 13 April.

Thrift, N. (1996) 'Shut up and dance, or, is the world economy knowable?', in P.W. Daniels and W.F. Lever (eds) *The Global Economy in Transition*, Harlow: Longman.

Tiffin, H. (1981) 'Cricket, literature and the politics of decolonisation: The case of C.L.R. James', in R. Cashman and M. McKernam (eds) *Sport: Money, Morality and the Media*, Brisbane: University of Queensland Press.

Tiger, L. (1990) 'The cerebral bridge from family to foe', in J. van der Dennen and V. Falger (eds) *Sociobiology and Conflict*, London: Chapman & Hall.

Todaro, M.P. (1989) *Economic Development in the Third World*, London: Longman.

Toffler, A. (1981) *The Third Wave*, London: Pan.

Tomlinson, J. (1991) *Cultural Imperialism: A Critical Introduction*, London: Pinter.

—— (1995) 'Homogenisation and globalisation', *History of European Ideas* 20, 4–6: 891–7.

—— (1997) 'Internationalism, globalization and cultural imperialism', in K. Thompson (ed.) *Cultural Change and Regulation: Policies and Controversies* (Block Six of D318: *Media, Culture and Identities*), London: Sage and Open University.

Toulmein, S. (1990) *Cosmopolis: The Hidden Agenda of Modernity*, New York: The Free Press.

Trillo, R. (1993) *Kenya: The Rough Guide*, 4th edn, London: Rough Guides.

Truong, T.-D. (1990) *Sex, Money and Morality: Prostitution and Tourism in South-east Asia*, London: Zed Books.

Tucker, V. (ed.) (1997) *Cultural Perspectives on Development*, London: Frank Cass.

—— (in press) *The Myths of Development*, London: Earthscan.

Turner, B. (1991) *Religion and Social Theory*, London: Sage.

Turton, D. (1997) 'War and ethnicity: global connections and local violence in northeast Africa and former Yugoslavia', *Oxford Development Studies* 25, 1: 77–94.

Tyrrell, H. (1995) 'Bollywood utopias: The role of the Indian popular musical film in the lives of Indians in Britain', unpublished M.Litt. dissertation, University of Strathclyde.

United Nations General Assembly (1994) *Elaboration of an International Convention to Combat Desertification in Countries Experiencing Serious Drought and/or Desertification, Particularly in Africa. Final Text of the Convention, A/AC.241/27*, New York: United Nations.

United Nations (1989) *Violence Against Women in the Family*, New York: United Nations.

United Nations Development Programme (1994) *Human Development Report 1994*, Oxford: Oxford University Press.

Vaid, S. and Sangari, K. (1991) 'Institutions, beliefs, ideologies: Widow immolations in contemporary Rajastan', *Economic and Political Weekly of India*, 27 April: 2–18.

van de Laar, A. (1996)*Economic Theory and the Natural Environment: A Historical Overview*, ISS Working Paper Series no. 209, The Hague: Institute of Social Studies.

van der Dennen, J. and V. Falger (eds.) (1990) *Sociobiology and Conflict*, London: Chapman & Hall.

van der Kwaak (1992) 'Female circumsion and gender identity: a questionable alliance' in *Social Science and Medicine* 35:777–87.

van Esterik, P. (1992) 'Thai prostitution and the medical gaze', in P. and J. van Esterik (eds) *Gender and Development in Southeast Asia*, Montreal: Canadian Council for Southeast Asian Studies.

van Gennep (1960) *The Rites of Passage*, London: Routledge & Kegan Paul.

Vasudevan, R. (1989) 'The melodramatic mode and the commercial Hindi cinema: Notes on film history, narrative and performance in the 1950s', *Screen* 30, 3: 29–50.

—— (1995) 'Addressing the spectator of a "third world" national cinema: The Bombay "social" film of the 1940s and 1950s', *Screen* 36, 4: 305–24.

Verhelst, T.G. (1987) *No Life Without Roots. Culture and Development*, London: Zed Books.
Vickers, A. (1989) *Bali: A Paradise Created*, Australia: Penguin.
Vickers, G. (1970) *Freedom in a Rocking Boat*, London: Allen Lane.
Vieira, P.F. and Viola, E.J. (1992) 'From preservationism to sustainable development: A challenge for the environmental movement in Brazil', *International Journal of Sociology and Social Policy* 12, 4–7: 129–50.
'The View from silly point' (1994) *Daily Nation*, 13 April.
Vizedom, M. (1976) *Rites and relationships: rites of passage and contemporary anthropology* Beverley Hills, California: Sage Publications.
Walby, S. (1989) 'Flexibility and the changing sexual division of labour', in S. Wood (ed.) *The Transformation of Work*, London: Unwin Hyman.
Waldinger, R. (1989) 'Immigration and urban change', *Annual Review of Sociology* 15: 211–32.
—— (1994) 'The making of an immigrant niche', *International Migration Review* 28, 1: 3–30.
Wallerstein, I. (1979) *The Capitalist World-Economy*, Cambridge: Cambridge University Press.
—— (1980) *The Modern World System II: Mercantilism and the Consolidation of the Modern World Economy*, New York: Academic Press.
Wallman, S. (ed.) (1979) *Ethnicity At Work*, London: Tavistock.
Wanek, A. (1993) *Fighting Lucifer: The State and its Enemies in Papua New Guinea*, Stockholm: Department of Social Anthropology, Stockholm University.
Watt, M. (1988) *Islamic Fundamentalism and Modernity*, London: Routledge.
Watts, M. (1991) 'Entitlements or Empowerment? Famine and Starvation in Africa', *Review of African Political Economy* 51: 9–26.
Weber, M. (1958) *The Protestant Ethic and the Spirit of Capitalism*, New York: Scribner.
Wekiya, I. (1991) 'Do it herself: women and technological innovation', *Proceedings of Africa Regional Seminar*, Rugby: Intermediate Technology Development Group.
Werbner, P. (1987) 'Enclave economies and family firms', in J.S. Eades (ed.) *Migration, Labour and Social Order*, ASA Monographs 25, London: Tavistock.
—— (1990) *The Migration Process*, New York, Oxford and Munich: Berg.
—— (1996) 'Essentialising the other: A critical response', in T. Ranger, Y. Samad and O. Stuart (eds), *Culture, Identity and Politics*, Aldershot: Avebury.
Werbner, P. and Anwar, M. (eds) (1991) *Black and Ethnic Leadership: The Cultural Dimensions of Political Action*, London: Routledge.
Wertheim, W.F. (1974) *Evolution or Revolution? The Rising Waves of Emancipation*, Harmondsworth: Penguin.
Whitaker, B. (1983) *A Bridge of People*, London: Heinemann.
White, B. (1982) 'Child workers and capitalist development: An introductory note and bibliography', *Development and Change* 13, 4.
—— (1994) 'Children, work and "child labour": Changing responses to the employment of children', Inaugural Address as Professor of Rural Sociology, Institute of Social Studies, The Hague, 16 June.
Whittaker, A. (1987) 'Brazil's Slave-like Practices', *Anti-Slavery Reporter*.
Whittaker, D.H. (1990) 'The end of Japanese-style employment?', *Work, Employment and Society* 4, 3: 321–47.
Wiener, M. (1981) *English Culture and the Decline of the Industrial Spirit*, London: Cambridge University Press.
Wignaraja, P. (ed.) (1993a) *New Social Movements in the South: Empowering the People*, London and New Jersey: Zed Books.
—— (1993b) 'Rethinking development and democracy', in P. Wignaraja (ed.) *New Social Movements in the South: Empowering the People*, London and New Jersey: Zed Books.
Wilkinson, B.A. (1994) 'Lack of support for Windies', *Barbados Advocate*, 15 April.
Williams, R. (undated) *Marxism and Literature*.
—— (1958) *Culture and Society: 1780–1950*, Harmondsworth: Penguin.

—— (1961) *The Long Revolution*, Harmondsworth: Penguin.
—— (1962) *Communications*, Harmondsworth: Penguin.
—— (1974) *Television: Knowledge and Cultural Form*, Harmondsworth: Penguin.
—— (1981) *Culture*, London: Fontana.
—— (1983) *Keywords*, London: Fontana.
Willis, P. (1977) *Learning to Labour: How Working Class Kids Get Working Class Jobs*, New York: Columbia University Press.
Wilson, B. (1982) *Religion in Sociological Perspective*, Oxford: Oxford University Press
Wilson, G.A. (1993) *Technological capability in small-scale development projects supported by UK-based NGOs*, DPP Working Paper 25, Milton Keynes: Open University.
—— (1995a) 'Technological capability, NGOs, and small-scale development projects', *Development in Practice* 5, 2: 128–42.
—— (1995b) 'From day-to-day coping to strategic management: Developing technological capability among small-scale enterprises in Zimbabwe', DPP Working Paper 33, Milton Keynes: Open University.
—— (1996) 'From day-to-day coping to strategic management: Developing technological capability among small-scale enterprises in Zimbabwe', *International Journal of Technology Management* 11, 6.
Wolf, E. R. (1969) *Peasant Wars of the Twentieth Century*, London: Faber & Faber.
—— (1982) *Europe and the People without History*, Berkeley, CA: University of California Press.
Woodhouse, P. (1992a) 'Social and environmental change in sub-Saharan Africa', in H. Bernstein, B. Crow and H. Johnson (eds) *Rural Livelihoods: Crises and Responses*, Oxford: Open University and Oxford University Press.
—— (1992b) 'Environmental degradation and sustainability', in T. Allen and A. Thomas (eds) *Poverty and Development in the 1990s*, Oxford: Oxford University Press.
Wooldridge, I. (1994) 'Cricket: Passport out of Paradise Islands', *The Daily Mail*, 1 April.
World Bank (1993) *World Development Report*, Washington: World Bank.
Worsley, A. (1964) 'Infibulation and female circumcision: a study of a little known custom', *British Journal of Obstetrics and Gynaecology* 45: 686–91.
Worsley, P. (1964) *The Third World: A Vital New Force in World Affairs*, London: Weidenfeld & Nicolson.
—— (1968) *The Trumpet Shall Sound: A Study of 'Cargo' Cults in Melanesia*, 2nd edn, New York: Schocken Books.
—— (1980) 'One world or three? A critique of the world-system theory of Immanuel Wallerstein', in R. Miliband and J. Saville (eds) *Socialist Register*, London: Merlin Press.
—— (1984) *The Three Worlds: Culture and World Development*, London: Weidenfeld & Nicolson.
—— (1997) *Knowledges*, London: Profile Books, New York: New Press.
Wright, S. (1995) 'Anthropology: Still the uncomfortable discipline?', in A. Ahmed and C. Shore (eds) *The Future of Anthropology: Its Relevance to the Contemporary World*, London: Athlone Press.
Yearley, S. (1994) 'Social movements and environmental change', in M. Redclift and T. Benton (eds) *Social Theory and the Global Environment*, London: Routledge.
York, P. (1995) 'Peter York on Ads, No 99: Diesel Jeans', *Independent on Sunday, Sunday Review*, 1 October: 28.
Young, I.M. (1990) 'The ideal of community and the politics of difference', in L.J. Nicholson, *Feminism/Postmodernism*, London: Routledge.
Young, R. (1990) *White Mythologies: Writing History and the West*, London: Routledge.
Zubaida, S. (1989) *Islam, the People and the State: Essays on Political Ideas and Movements in the Middle East*, London: Routledge.

INDEX

Abbott, D. 242
Abegglen, J. 131
Abu-Lughod, J. 24
Abuladze, T. 278
Adamson, P. 26, 90, 98
Africa 31; images of 104–7; sub-Saharan 63
Agence Free Presse 103
agency 2–3, 16
Agnelli, S. 216
agriculture: and labour migration 181; Tanzania 162; technology, poverty and 63, 64
aid 106, 107–8; imagery and 92–100
Airtours 254
Albrow, M. 153
alienation: from the environment 48; West Indian cricket fans 254–7
All-India films 270
Allen, R. 92
Allen, S. 214
Allen, T. 3, 46, 60, 152, 155, 261, 262; Alice Lakwena 163; Laropi village 59; Lugbara 228; Satyajit Ray 271–2; Somalia 5
Alleyne, G. 257, 259
Alliance for Progress 34
alliances, environmentalist 53–5
alternative communication 167–8
Alvares, C. 53
amenity view of the environment 48
American Anthropological Association 148
American Declaration of Independence 194
American Revolution 198
Amin, M. 91
Amnesty International 217
Anderson, B. 153
Anderson, M. 106
Anderson, M.B. 58, 59–60
Andrei Rublev 278
Angelico, Bishop Dom 169
Anthias, F. 154
anthropology 27–8; culture in 3, 13–15, 17; and development 5; and ethnicity 148–9, 151–2; and ethnicity in industrialised countries 153–5; research on female circumcision 208–9
Anwar, M. 154, 182
Appadurai, A. 24, 153
appeasement policy 140
Appleton, H. 63, 64
Apthorpe, R. 40
archaeology 17
Archer, M. 13–14, 16, 30, 32
Arendt, H. 15
Arnold, S. 90
ascribed trust 61
Asia: East Asia 33, 235; migrants from South Asia 181–2
Askoldov, A. 278
Athos, A.G. 127
Atkinson, P. 105
Atsumi, R. 131
Australia 111
awareness of global interdependence 26

Babel, R. 277
Bachchan, A. 264, 266
Bailey, D.A. 93
Bajran Dal 241
Bali 36
Ballard, R. 154
Band Aid 92, 102
Bandit Queen 263, 271
Bandyopadhyay, J. 52
Bangladeshi women 160
Banks, M. 149, 154, 155
Barbados 10, 251–9
Barbados Cricket Association (BCA) 255
Barbados Tourism Authority (BTA) 254
Barclay, H. 208
Bardhan, P. 160
Baring Brothers Bank 27
Barker, D. 214
Barker, M. 147
Barnard, A. 3
Barnett, C. 140
Barnouw, E. 262, 263, 266

Barry, K. 110
Barsaat 270
Barth, F. 151
Barthes, R. 89, 94, 95, 96, 97
Basic Christian Communities (BCCs) 169, 170–1, 229
Basu, T. 244
Battleship Potemkin 276
Bauman, Z. 16, 25, 28–9
Baumann, G. 3, 4
Bawtree, V. 39
BBC News 91
Beckles, H. 251
beggars 123
Bell, D. 17, 32
Ben-Ari, E. 133
Benedict, R. 14–15
Benhabib, S. 197
Benson, S. 154
Benthall, J. 102, 103, 104, 107
Bernbaum, G. 141
Beyer, P. 237, 238
Bhachu, P. 154
Bhagavan, M.R. 65
Bhaji on the Beach 269
Bhalla, A.S. 64, 66
Bharat Mata (Mother India) 241
Bill of Rights 194, 195, 196
Binford, M.R. 262
biomedicine: and global health 72–3; research on female circumcision 205–8
Bismarck, O. von 20
BJP 241, 247
Black, M. 93
Black Consciousness 237
'Black Roots' group 176
Blackbourn, D. 142
Blades, W. 256
Blair, T. 123, 143
Blanchard, S. 271–2
Blenman, R. 254
Blom Hansen, T. 241–2, 243, 244
Boas, F. 14
Boddy, J. 208, 209
Bollywood 10, 260–73; vs Hollywood 263–5; international film industry 266–9; oppositions to 269–72; popular discourses of Hollywood/Bollywood opposition 265–6; and Third Cinema 261–3
Bollywood or Bust! 269
Bombay 265, 266–7, 270
Bombay riots 245
Bonacich, E. 186
bonded solidarity 186
Border 271
Borocz, J. 182, 185
Boswell, J. 120
Botswana 51, 197; SOIWDP 54–5
bourgeois art 276
box-office takings 263–4
Boyd-Barrett, O. 23
Boyle, C.M. 163
Bozhovich, V. 279
BPL 266
Bracero programme 181
Brake, M. 211
Bramble, B. 53
Brathwaite, L. 258
Brazil 229; local forms of resistance 163, 165–6; people's radio 8–9, 167–79; street children 9, 212–19
Brezhnev, L. 277
Britain 235; cultural chauvinism 140–1; cultural complacency 139–40; development NGOs and the regulation of imagery 7, 87–101; ethnicity in 154; Indian films 267, 268; industrial decline 8, 137–44; Latin American migrants in 187–8; post-war migration to 181–2; Thatcherite diagnosis of cultural malaise 141–2; *see also* English cricket fans
British National Party (BNP) 122
Brockett, L. 111
Brohman, J. 112
Bromley, Y. 148, 149
Brown, J. 214
Buddhism 114, 225
Buerk, M. 91
Burawoy, M. 182
Burgin, V. 89, 94
Burma 114, 115, 225
Burnell, P. 92, 93
Burnt by the Sun 278
Burundi 37, 108, 226
Bustamante, J. 184

cable television 264–5, 268–9
Cain, P.J. 142
Caldeira, T. 168, 169
Calhoun, C. 148, 149
Calvinist Protestant ethics 234–5
Camacho, S. 256
Campbell, A. 152
Campbell, B. 210
CAMPFIRE projects 53, 54
capital-labour contracts 138, 143
capital punishment 196
capitalism 15, 137; gentlemanly 142–3; globalisation and 27, 34; secularisation 236
Caplan, P. 211
Cardoso, R. 169
Caribbean: cricket 10, 251–9; migrants to Britain 181–2

Carlson, S. 108
Cashman, R. 251
Casper 264
Castells, M. 145, 147
Castles, S. 180, 181, 182
Castro, F. 261
catering 188
Catholic Church 18–19; Basic Christian Communities 169, 170–1, 229; people's radio 172, 178; and political transformation 224, 229
censorship 271
Central Bank of Barbados 252
Centre for Contemporary Cultural Studies (CCCS) 154
Centre for Popular Education and Communication of São Miguel (CEMI) 173
Chakravaty, R. 242
Chamberlain, A. 139
Chambers, R. 64, 91
Chaney, D. 100
change: archaeology and 17; development as instigator of religious change 236–7; long-term and weapons of the weak 163–6; religion and unchangingness 235
Chant, S. 67
Charmley, J. 140
Chatterjee, B. 272
Chattopadhyay, 35
Chaui, M. 168
chauvinism, cultural 140–1
Chechnya 37
Cheziya Co-operative Bakery 62–3, 66
Chicago School 17–18, 20
child abuse 209
Child and Adolescent Statute 1990 217
child labour 115, 212–13
child prostitution 115
childhood 214
children: socialisation in Japan 131–2; street children in Brazil 9, 212–19
Chile 163
Chipasula, F. 49
Chipko movement 164–5
Christian Aid 87, 89, 94–100
Christianity: conversion 193, 236; India 234; *see also* Catholic Church, Protestantism
cinema attendance 279–80
cinema distribution 267
circumcision, female *see* female circumcision
citizenship 252–5
city: and identity 8, 117–23; melting-pot theory 117–18
Clark, R. 134
class: culture, hegemony and 18–20; West Indies cricket 255–7
cleaning work 187–8
Cleary, D. 165–6
Cleasby, A. 26
Cliffhanger 263, 265
clitoridectomy 210; *see also* female circumcision
CNN 107
Cohen, A. 150
Cohen, A.P. 154
Cohen, S. 120
Colby, M.E. 47, 49, 55–6
Colchester, M. 165
Coleman, J. 186
Colombian migrants 187–8
colonialism 251; cultural exchanges 234; Other 193; post-colonialism and West Indies cricket 10, 251–9; religious syncretism 228; sexual politics 113–14
Come and See 278–9
Comic Relief 102
commercialism 261–2
communication: alternative 167–8; popular 169–70
communism 36, 275–8
community 197; healers and social health of 78– 80, 82; health and 75–8, 82
community-orientated religious groups 227, 229, 230
company-as-family model 127–36; development 134–5
company unionism 127–8, 133, 135
complacency, cultural 139–40
Comte, A. 16
conflict 40–1, 226; ethnic 37, 39, 104, 145; Japanese company system 133; representations in the Western media 7, 102–8
Confucianism 132, 137, 235
connotation 94–100
Conrad, J. 122–3
conservationism 49–50, 51–2
conservatism 277–8
Conservative Party 138, 141; *see also* Thatcherism
Constantinides, P. 208, 209
consumerism 35
contract cleaning 187–8
control, ownership and 21
Convention to Combat Desertification (CCD) 52
Convention on International Trade in Endangered Species (CITES) 54
Convention for the Suppression of the Traffic in Persons and of the Exploitation of the Prostitution of Others 1949 115
conversion, Christian 193, 236
Corcoran-Nantes, Y. 163

core-periphery distinctions 132–3
Cornia, G.A. 38
Cörvers, R. 47
counter-culture 17–18, 20
counter-hegemony 177–8
Counts, D. 214
Cozier, T. 256
Crang, M. 3
Crawcour, S. 134–5
creativity 277–8
cricket 6, 10, 251–9; alienation of West Indies fans 254–7; and the Caribbean 251–2
Crush, J. 5, 39, 234
Cuba 167
cultural chauvinism 140–1
cultural complacency 139–40
cultural frames 119–21
cultural functionalism 15–16, 17–18, 147
cultural imperialism 23–5
cultural malaise 8, 137–44
cultural materialism 20–1
cultural rights 252–5
cultural studies 20–1
culturalist religious groups 227, 230
culture 2–4; in anthropology 3, 13–15, 17; classic conceptions 6, 13–21; everyday resistance, organisation and 166; factor in development 34–8; and global change 4–5; sex and 114–15
Cummings, C. 251
Cunha, U.D. 272
Curran, J. 102

Dahlman, C.J. 66
Daily Nation 257
Daly, M. 208
Dankelman, I. 165
Dark Nights on the Black Sea 278
David, R. 111
Davidson, J. 165
Davies, M. 214
Davis, N. 110
Daw, S. 259
De Gaulle, C. 139
De Swaan, A. 215, 216
De Waal, A. 155
death squads 217–18
decadence, Western 23–4
decision-making 214–15
Declaration of the Rights of Man 1789 194
deep ecology 49
Defoe, D. 193
de-industrialisation 8, 137–44
demonstrations 174
Deniozos, D. 59
denotation 94–100
dependency theory 33
Desai, M. 270
Descartes, R. 194
desertification 52
development 2, 5, 6, 46; dynamic relationships with religion 237–8; image of 87–90; instigator of religious change 236–7; militarisation, sex tourism and 113–14; religion and 9–10, 232–9
development discourse 234
development education 88, 90
development NGOs 7, 87–101
development theory 6, 30–41
Devereux, G. 209
Devi, P. 263, 271
deviancy theory 17–18
Dimenstein, G. 216, 217
Disaster Emergency Committee appeals 88, 91
disease 207
disguised resistance 162–3, 164
distribution: films 266–9; food aid 106
Dixit, M. 266
documentary/realist representations 99–100
Doi, T. 132
Dolgiye Provody (Long Farewells) 278
Dondurel, D. 280, 281
Dore, R. 132
Dow v. *State of Botswana* 197
Dragadze, T. 149
Dubos, R. 73
Duffield, M. 103, 106, 107, 147
Duffy, J. 210
Dunn, C. 110
Dunston Checks In 263–4
During, S. 3, 21
Durkheim, E. 232
Dutta, J.P. 271

Eade, J. 154, 155
earned trust 61
Eccleston, B. 48, 53–4, 55
eco-development 55–6
ecology movements 52
economic growth: development as 235–6; in LDCs 183
economics 15; pre-dominance in development theory 30–2
education: development education 88, 90; popular 169–70; public school education 141–2; street children 218
Education Means Protection of Women Engaged in Recreation (EMPOWER) 116
Edwards, E. 89, 98
Eisenstein, S. 276
El Dareer, A. 202, 206–8
El Salvador 120

Eley, G. 142
Eliot, T.S. 18
élitism 18
'Elizabeth', image of 89, 95–9
embeddedness, cultural 160
emotional responses: to female circumcision 206, 208, 209–11; to street children 215–16
empires 71, 139
employment: migrants 182, 183–4, 184–5, 186, 187–8; schemes for street children 218; system in Japan 127–8, 132–3
empowerment 38, 237; radio and 167–8; street children 216, 219
End Child Prostitution in Asian Tourism campaign (ECPAT) 115, 116
enforceable trust 186
English cricket fans 252–9
Enloe, C. 113, 114
Ennew, J. 212, 216
entertainment 188; *see also* film industry
entrepreneurship: medical 80–1; migrants 189
Environment 2000 51
Environment Watch Botswana 51
environmentalism 6–7, 45–57, 237; 'modern' view of the environment 47–9; new conception of progress 55–6; new social movements 164–6; north–south alliances 53–5; northern 49–51; southern 51–3
epidemics 71, 75–6
equity 78, 237
Eriksen, T.H. 117–18, 146, 151, 152
escapism 262–3, 270–1
Escobar, A. 4, 39, 40, 232, 238
Espenshade, T. 186
essentialism 148–9, 154
Esteva, G. 53
Ethiopian famine 1985–6 90–3, 102
'ethnic cleansing' 149, 226
ethnic conflict 37, 39, 104, 145
ethnic minorities 17–18, 154, 181
ethnicity 8, 145–55, 209, 225, 226, 230; as an analytical construct 145–8; anthropology and in industrialised countries 153–5; essentialist approaches 148–9; instrumentalist approaches 149–51; local–global relations 153; relationalist approaches 151–2
ethnos theory 148–9
Europe 138, 140
European Court of Human Rights 196
European Union 122, 143
Evans-Pritchard, E.E. 232
Evening Standard 8, 117–23; cultural frames 119–21; poverty in London 122–3; representing London and Londoners 118–19; and transport system 121–2
everyday resistance 8, 159–66
exchanges, cultural 233–4
exploitation 110
export manufacturing 183, 184
export processing zones (EPZs) 67–8, 183

Falger, V. 148
Falklands War 143
family 148; company-as-family model 127–36; health and 77–8, 82; networks 60–1, 185–6; place of safety 214; socialisation and 215; street children and 212–16; ties and migration 185–6
famine 161; Ethiopian 1984–5 90–3, 102
Fanon, F. 251
fear 216, 218
Fei, J.C.H. 181
female circumcision 9, 199, 201–11; anthropological research 208–9; bio-medical research 205–8
femininity 209
feminism 36–7, 110, 196–7, 208, 210
Ferguson, J. 3, 5, 39, 40
Fernandez-Kelly, M.P. 185
festivals 176
film industry: India 6, 10, 260–73; Russia 10, 274–81
film press 267
film stars 264
finance capital 142–3
finance, insurance and real estate (FIRE) 185
Firestone, S. 214
First Blood 263
Fisk, R. 104
Fitch, R. 184
flexibility, new migrants and 9, 180–9
Flintstones, The 264
Foner, N. 182
food aid distribution 106
force, and fraud 19–20
Fordist production techniques 65–6
forest resources 164–5
former Yugoslavia 37, 152, 226
FORWARD 209
Fowler, R. 120
Fox, S. 264
foyers 60
France 139, 140, 195, 198
Francis, P. 253, 254
Frank, A.G. 33
fraud, force and 19–20
Fred Rumsey Travel Group 254
Freire, P. 169
French Revolution 198
Friedman, J. 23
Friends of the Earth 50
friendship networks 185–6

frontier economics 47
Frost, C. 122
Fukuyama, F. 3, 17, 41, 137
Fuller, C.J. 35
functionalism, cultural 15–16, 17–18, 147
functionalist Marxism 15
fundamentalism, religious 225–6, 227, 228–9, 230
fundraising 88, 90
Furnam v. *Georgia* 196
Fyfe, A. 219

Gabayet, L. 184
GABRIELA 116
Gaete, R. 196
Galheigo, S.M. 217
Galichenko, N. 278
Galilei, Galileo 194
Galtung, J. 120
Gamble, A. 143
Gandhi, I. 240, 244–5
Gandhi, M.K. 243
Gandhi, R. 240
Garabwe tailoring co-operative 67, 68
Garcia Canclini, N. 24
Garner, J. 258
Geertz, C. 148
Geisteswissenschaften 14
Geldof, B. 92
Gellner, E. 232
gender: imagery and sex industry 113; and technology 59–60; *see also* women
generational groups 185–6
genocide 37, 197–8
gentlemanly capitalism 142–3
Germany 139, 140, 142, 143
Gettino, O. 261
Giddens, A. 25, 26–8, 153
Giggles, 266
Gill, P. 90, 91
Gillespie, M. 117
Glasbergen, P. 47
Glasgow University Media Group 120
glasnost 276, 278–9
Glazer, N. 149–51, 151–2
Gledhill, J. 168–9
Glendon, M.A. 197
global change 1–2; culture and 4–5
global environmental issues 50–1, 52
global environmentalism 6–7, 45–6; possibilities and dangers 53–5
globalisation 1–2, 117; and ethnicity 153; and health 70–3; religion, development and 237–8
globalised culture 6, 22–9; 'decline of the West' 25–9; global and 22–3; as Westernised culture 23–5
Gluckman, M. 150
Goddard, V. 153
Godelier, M. 15
Goldenberg, S. 241, 242
Goodwin, N. 45
Gorbachev, M.S. 276, 278, 279
Gordon, D. 207
Gordon Drabek, A. 90, 91
Graham-Brown, S. 97
Gramsci, A. 18–20, 195
Grandison, T. 256
Grant, C. 193
Great Traditions 35
Greater London Council (GLC) 119
green revolution 58
Greenpeace 50, 54–5
Greer, G. 210
Greg v. *Georgia* 196
Grigsby, E. 121
group *see* community
Gruenbaum, E. 207, 208
Guest, J. 45
Guevara, Che 32, 34
Gulf War 1991 103
Gumilev, T. 149

Habermas, J. 226
Hacking, I. 99
Haeri, S. 24
Hall, S. 21, 94, 120, 153, 253–4
Hall, W. 259
Hane, M. 134
Hannerz, U. 23, 24, 117, 118, 153
Hantrakul, S. 114, 115
Haraway, D. 238
Harrell-Bond, B. 108
Harris, M. 20, 91
Harris, N. 33, 34, 183
Harrison, P. 91, 92
Hart, A. 92, 93, 99, 100
Harvey, D. 153
Hayes, R.O. 208, 209
Haynes, J. 225, 227, 233
healers 80–1, 82; and the social health of groups 78–80
health 7, 59, 70–83; 'health for all by the year 2000' 70–3; images of 74; national medical culture 80–1, 82–3; putting the 'public' back into public health 81–3; social health as traditional public health 75–8; vs wellbeing 73–5
Hebdige, D. 92
Hegde, P. 54
hegemony 18–21, 176–7
Heidegger, M. 197
Held, D. 153
Helmsing, A.H.J. 65

Henderson, J. 34
Hettne, B. 224, 226
Hewitt, T. 218
hierarchical relations 131
high-tech industry 184
Hill, C. 112, 216
Hindi language 263, 270
Hinduism 270; women and the new Hinduism 10, 240–7
Hirst, P. 143
history: end of 195; *Evening Standard* and 121
Hite, S. 210
Hobbes, T. 194
Hodgson, D. 115
Hoffman La Roche 53
Hollywood 10, 263–6
Holmström, M. 61
Holy Spirit Movement 163
homelessness 123
Honduras 161–2
Honey, J. 144
Hong Kong 34
Hopkins, A.G. 142
Horgan, J. 92
Hosken, F.P. 208
household, Japanese 132
Hum Aapke Hain Koun ...! 264, 268
Human Development Report 38–9
human rights 6, 9; weapons of the weak and 161–3; the West, its Other and 9, 193–200
humanitarian aid *see* aid
Huntington, S. 3, 40–1, 137
Hutton, W. 138
Hutus 37, 106
hybridised culture 24, 36

ideas 14–15
identity 4; city and 8, 117–23; ethnic *see* ethnicity; and politics 226; post-colonial in West Indies 252–5; redefining national identity in India 240–7; sexuality and 211; social and culture 147
ideology 15; Japanese company system 8, 127–36
Ignatieff, M. 145
illegal immigrants 183, 189
imagery: image of development 87–90; negative 88–90, 90–3, 104–5, 108; positive 93–100, 104–5; regulation of 7, 87–101
Images of Africa survey 93
immigrants *see* migrants
imperialism 71, 139
inclusive fitness 148
India 31, 35, 234; Chipko movement 164–5; film industry 6, 10, 260–73; green revolution 58; women and the new Hinduism 10, 240–7
Indian restaurants 120
individualism 234
individuals 48
industrial capitalism 34
industrial decline 8, 137–44
industrialisation 46, 53, 183
infection 207
infertility 208
infibulation 201–5, 206–7
information society 17
innocence 215
institutionalisation of street children 217
institutions, media 103–5
instrumentalism 149–51
integration 1–2; medical culture 80–1
intellectuals 19
interdependence, global 25–6
intermediate circumcision 206, 206–7
International Convention for the Suppression of the White Slave Trade 1910 115
International Film Festival, India's 272
international migration *see* migrants
International Monetary Fund (IMF) 38
international tourism *see* tourism
Internet 108
Iran 36
Iranian Revolution 233
Ishida, T. 131
Islam 36, 224–5, 238; Muslims and Bollywood 270; unchangingness 235; violence against Muslims 245
island mentality 131
ius cogens 198

Jackman, O. 255
Jackson, J.A. 181
Jackson, P. 3, 14
Jafferlot, C. 243
James, C.L.R. 251–2
James, D.D. 64
Jann, A. 53
Jansi ki Rani, Queen 244
Japan 34, 111, 138, 235
Japanese company culture 8, 127–36
Jayawardena, K. 242
Jefferson, T. 21
Jeffrey, P. 154
Jenkins, R. 154, 155
Jenkins, S. 121
Jenks, C. 3
Jennet, C. 164
Jenson, J. 188
Jews 235
Jha, S.K. 263, 272

John Paul II, Pope 237
Johnson, H.E. 162
Joint Forest Management Programmes (JFMP) 53, 54
'joke culture' 163
Jolly, R. 38
Jubilee Line extension 121
Jurassic Park 263, 263–4
justice 78, 194–5

Kabeer, N. 160
Kaida-Hozumi, 92
Kama Sutra 271
Kamata, S. 133
Kamrava, M. 226
Kant, I. 14
Kanwar, R. 245, 246
Kaplan, R. 105
Kaplun, M. 167
Kapoor, R. 271
Kapoor, S. 271
Kayanja, R. 108
Kenez, P. 276
Kennedy, P. 140
Kensington Oval test match 1994 252–9
Kenya 105
Kenyon, S.M. 208
Kerr, C. 16–17
kersevaks (defenders of the faith) 241–2
kersevika movement 243–4, 247
Khan, S.R. 261, 264, 266
Khomeini, Ayatollah 36
Khrushchev, N. 277
Kim, K.-D. 235
Kinjikitili 228
Kinzley, W.D. 134, 135
Kishwar, M. 243, 245
Kissinger, H. 41
Kitchener, Lord 258
Kitching, G. 46
Klimov, E. 278
Kluckhohn, C. 13, 14
knowledge, local *see* local knowledge
Komissar (The Commissar) 278
Korea, South 183, 235
Kouba, L.J. 206
Kouchner, B. 108
Krishnaswamy, S. 262, 263, 266
Kroeber, A.L. 13, 14
Kumar, K.J. 264
Kuper, A. 2, 4
Kyōchōkai (Harmonisation Society) 134
Kyong-Dong, K. 139

labour: capital-labour contracts 138, 143; child 115, 212–13; demand and migration 181–2, 188–9
labour disputes 129, 133
Labour party 138, 143
Lailson, S. 184
Lakwena, A. 163, 228
Lall, B. 264
Lam, T. 188
Lamb, R. 92
landed aristocracy 141–2
language 270
Laropi village, Uganda 59
Las Casas, Fr 196–7
Last, M. 207
Latin America 32; Liberation Theology 169, 170, 229, 237; migrants in Britain 187–8
Lawton, A. 277, 278
lay-out, office 128–30
Le Vine, R. 209
Le Vine, S. 209
Leach, E. 18
Lebra, T.S. 132
Lee-Wright, P. 212
Lees, C. 263
Left, political 138, 143
Legassik, M. 182
Legendre, P. 199
Leheny, D. 112–13
Lenin, V.I. 276–7
Lenshina, A. 228
Lent, J.A. 271
Lerner, D. 30, 31
less developed countries (LDCs), migration from 181–4
Levinas, E. 195
Lewis, A. 31, 181
Lewis, E. 255–6
Leyda, J. 276
Liberation Theology 169, 170, 229, 237
Liberia 105–6
Lidchi, H.J. 94, 95
life-time employment 127
Lindenbaum, S. 209
Lindquist, S. 103, 107
Lines, J. 122
Lipset, S.M. 17
Lissner, J. 88–9, 90
Little Traditions 35
Live Aid 92
livelihoods 60–3
Llobera, J. 153
local forms of resistance 8, 159–66
local–global interaction 153
local knowledge 7, 58–69; social organisation and livelihoods 60–3; technology and poverty 63–8
Locke, J. 194

London: city and identity 8, 117–23; immigrants 119, 187–8
London Underground 121–2
long-term change 163–6
Los Angeles 184, 186
lower-caste Hindus 242, 244
Lucie Clayton's Young Londoner Grooming Course 120
Lugbara 228
Lukes, S. 160
Lull, J. 24
Lumière company 275
lynchpin projects 64
Lyons, H. 209
Lyotard, J.-F. 21

McCormick, D. 61, 63
MacGrane, B. 193, 195, 197
McGrew, A. 153
McGrew, T. 26
Machado, 167
Machiavelli, N. 19–20
Macías, J. 185
McIlwaine, C. 67
McLelland, D. 47
McQuail, D. 23
Macrae, J. 106
Macridis, R.C. 140
Mactan Export Processing Zone 67–8
Magri, 167
Maji-Maji rebellion 228
Malcolm, D. 263
Malenkaya Vera (Little Vera) 278
Maley, W. 279
Malinowski, B. 14
Mamdani, M. 151
Mandelshtam, O. 277
Mandle, W.F. 251
Manning, F. 251
Manu Smitri 242
Manuel, P. 264, 270, 271
Manzo, K. 237
Maquet, J.J.P. 31
marches 242
marginalisation 71–2
market 72–3
Marquand, D. 143
Marshall, M. 255
martial training 241–2, 243–4
Martin, C.J. 188
Martin-Barbero, J. 24
Marty, M.E. 226
Marxism 15, 138, 236
Masagao, 167
Masala films 262, 263
Masao, M. 127
masculinity 211
Massey, D. 26, 29
mastery over nature 47
Mattheson, M. 141
Matthews, B. 256
Mayer, A. 142
Mayoux, L. 61
Mazrui, A. 251
Mbilinyi, M. 162
Mbiti, J. 236
Mead, G.H. 18
meanings: people's radio 176–7; positive imagery 94–100
Medhurst, K. 229
media 19, 21, 26; representations of conflict in Western 7, 102–8; West Indies cricket match 1994 257–9
media institutions 103–5
mediation 21
medical communities 79, 80
medical culture, national 80–1, 82–3
Mehmet, O. 232
melting-pot theory 117–18
mental illness 71
metaphysical subjectivism 196–7
Mexico 183; Mexican-US migration 181, 184–6
Meyerhold, V. E. 277
middle class 141–2
Middleton, N. 48, 49, 52
migrants 9, 180–9, 225; Brazil 168; in Britain 119, 187–8; co-operation and competition among 186–7; international migration in the age of uncertainty 182–4; Mexican-US migration 181, 184–6; post World War II international migration 181–2
Mikhalkov, N. 278
militarisation 113–14
military dictatorships 32
Mill, J.S. 50
Miller, M. 180, 181, 182
Milne, B. 212
Minakshi Temple 35
ministries of health 82–3
Minors' Code 217
Mishra, V. 261
Miss World Contest 242
Mitchell, J.C. 150
Mitchell, T. 5
Mobil Oil Company 51
Modell, J. 186
modernisation 6–7, 45–57; idea of 46–7; 'modern' view of the environment 47–9; northern environmentalist opposition to 49–51; religion and 224–6, 230, 235; southern environmentalists and 51–3
modernisation theory 30–2
Mohamud, O.A. 210

Monanieba (Repentance) 278
monarchy 143
Montague, A. 100
Moore, H. 214
Morelli, Bishop M. 218
Morishima, M. 132
Morley, 21
Morokvasic, M. 188
mortality 210
Moser, C. 232
motherhood 243–4
Movimento de Moradia (Housing Movement) 169
Movimento Sem Terra (Landless Movement) 169
Moynihan, D. 145, 149–51, 151–2
Moyo, S. 48, 49, 52
Mozambique National Resistance (Renamo) 228
Muasher, J. 206
Muller, T. 186
multiculturalism 106, 147
Murakami, Y. 132
Muratova, K. 278
Murdoch, R. 141, 264
Murray, A. 111
music 18, 35, 264
Muslims *see* Islam
Myanmar (Burma) 114, 115, 225
Myers, W. 212, 218
mythical signifier 97

Nair, M. 271
Nakane, C. 131
napramas 228
Nargis 264
Nash, C. 92, 93
national conference of street children 1986 219
National Foundation for the Well-being of the Minor (FUNABEM) 217
national identities: redefining 240–7
national medical culture 80–1, 82–3
National Movement of Street Boys and Girls (MNMMR) 219
National Policy for the Well-being of the Minor 217
Nativity 120–1
natural rights 195
Naturwissenschaften 14
Ndebele 228
negative imagery 88–90, 90–3, 104–5, 108
Nehru, J. 270
Neil, A. 141
neo-liberalism 138, 142–3
Neto, V. de Oliveira 218
networks 60–3, 185–6
New Barbarism thesis 7, 105–7
New Indian Cinema 272
new institution economics 38
New Left Review 138
New Racism 106, 107
New Right 138, 141–2
new world information and communications order 108
New York 149–50, 150–1, 184
news reporting 91–2, 103–4; *Evening Standard* 119–21
Nicaragua 237
Nichols, B. 89, 94, 99
Nieuwenhuys, O. 215, 218–19
Nigeria 207
nkejji fish 64–5
Non-Aligned Powers 33
non-governmental organisations (NGOs) 116; British development NGOs and the regulation of imagery 7, 87–101; environmentalist 49–56; and street children 216, 218–19
North, D. 3, 38
northern environmentalism 49–51; alliance with southern environmentalism 53–5
Northern Line 121
Nove, A. 276–7
Nugent, S. 3
nyangas 79
Nyoni, S. 92, 93

O São Paulo 172
Oberg, K. 37
objectification 90, 93
October 276
Ogbini, F.E. 64
O'Keefe, P. 48, 49, 52
Olsen, W. 160
Olson, M. 139
Omdurman aj Jadida, Sudan 201–5
one nation Toryism 138, 141
OPEC 34
oppositional film-makers 271–2
'organic' intellectuals 19
organisation *see* social organisation
Organiser, The 243
Ortiz, M. 185, 186
Ortner, S.B. 209
Osman, M.T. 35
Other 9, 193–200; voice of and universal rights of the relative 197–9; the West and 193–4
Ouchi, W. 127
outsider's dilemma 198–9
Owen, J. 120
ownership, control and 21
Oxfam 87, 91

Pacey, A. 58, 64
Pakistan 183; Pakistani business community in Britain 187
Palmer, R. 91, 92
Panday, G. 243, 251
PANOS 108
Paoli, M. 217
Paradzhanov, S. 277–8
Parallel Cinema 272
Pariwar 240–7
Park, R. 117, 118
Parker, M. 199
Parsons, T. 15–16, 17
particularism 238
Pascale, R.T. 127
Pasha, M.K. 238
Passage House 218
Paterson, R. 118
Pathak, A. 165
Pathe 275
patriarchy 242
Patterson, O. 251
Peace Corps 34
Pearce, F. 64
peasant producers 161–2
pedagogy of the oppressed 169–70
people's radio 8–9, 167–79; as resistance 176–8; Vila Aparecida and 170–6
perestroika 279–80
Pettman, J.J. 109, 113, 116
petty trading 65
pharaonic circumcision 201–5, 206, 206–7
Philippines 111, 114, 115
Phillips, M. 118–19, 123
Philo, G. 92
Phongpaichit, P. 115
photography 89–90, 99–100
Pichul, V. 278, 279
Pieterse, J.N. 24
Pines, J. 261
Pinochet, General 163
Pires, B. 255
pity 215–16
Pixote 215
planning, state 31–2
Plumwood, V. 47
pluralism, medical 80
Point Four 34
police 217–18
politics 9–10; identity and 226; religion and political transformation 9, 223–31; typology of political religion 226–9
popular communication 173
popular education 169–70
Porter, G. 53
Porter, R. 118
Portes, A. 182, 185, 186, 187
positive imagery 93–100, 104–5
post-colonialism 10, 251–9
post-development school 2, 39–40
post-war consensus 138, 141
Potter, D. 47
Poulton, R. 90, 91
poverty 77–8, 161; in London 122–3; technology, knowledge and 63–8
power: development and 39–40; ethnicity 150–1; female circumcision 208–9; photography and 89–90; policy analysis and power relations in conflict 106; sex tourism 109–16; weakness and 159–60
powerlessness 51
Prakash, S. 264
praxis 16
Preis, A.–B.S. 197
Pre-Raphaelites 141
preventative journalism 108
primordialism 37
prior knowledge 64–5
privacy 198
Proconel (Project for Non-Written Communication of the Eastern Zone) 173
production: of culture 21; films 266–9, 280–1: Fordist techniques 65–6; peasant producers 161–2; small-scale 62–3, 65–8
proletarian art 276
promotion, seniority 127
prostitution: sex tourism 7–8, 109–16; sexual politics of 109–10
Protestantism 137, 234–5
Prunier, G. 37
public health: putting the 'public' back into 81–3; 'traditional' 75–8
public school education 141–2
Punjab riots 244–5
purdah 160

Quakers 236
quality of life 68, 177
Quingrui, 66

race 113, 147, 154; and West Indies cricket 255–7; *see also* ethnicity
racism 154–5
Radcliffe-Brown, A.R. 14
radio 108; and empowerment 167–8; the people's radio 8–9, 167–79
Radio Rebelde 167
Rahnema, M. 39
Rajadhyaksha, A. 263, 270
Rake, D. 256
Ranis, G. 181
Rashtriya Sevika Samiti (Patriotic Association of Voluntary Women) 243–4, 247

rationality 194, 234–5, 238
Ratnam, M. 266–7
Ray, S. 271–2
realism: documentary photography 99–100; socialist 276–7
Reanda, L. 115
Recife 218
recruitment policies 181
Redfield, R. 35
regulation of imagery 7, 87–101
Reich, S. 140
Rekha 271
relationalism 151–2
relativism 193, 194, 196–9
Relator, Lord 258
religion 9–10, 35–6, 198; and development 9–10, 232–9; development as instigator of religious change 236–7; dynamic relationships with development 237–8; health and 59, 78, 79–80; Indian film and 270; and modernisation 224–6, 230, 235; and political transformation 9, 223–31; and resistance 163; as seedbed 234–6; typology of political religion 226–9
representations 7–8; conflict in the Western media 7, 102–8; *Evening Standard* and 118–19; West Indies cricket 257–9
resistance 8–9, 35–6; cultures of 168–9; local forms of 8, 156–66; people's radio as 176–8; sex tourism 116
resource management 48, 53
responsibility, chain of 77
restaurants 188; Indian 120
Reuben, B. 260
revolution 33–4, 198, 276
Reynolds, V. 148
Richards, P. 105, 108
Richman, N. 213
Rickman, H.P. 14
rights: cultural 252–5; human *see* human rights
'rights of the relative' 197–9
Rithambra 244
Ritzer, G. 235
Robertson, R. 26, 235
Robins, K. 24
Robinson, V. 182
Rocha, J. 213, 216, 219
Rock Dancer 264
Roebuck, P. 259
Rose, C. 50, 51
Rostow, W.W. 30–1
Roszak, T. 20
Routemaster bus 121
Royal College of Obstetricians 206
RSS 241, 243, 244, 247
Rubinstein, W.D. 142
Ruge, M. 120
Russia 37; film industry 10, 274–81; *see also* Soviet Union
Russian Revolution 274, 276
Rwanda 37, 106, 226

Sachs, W. 39, 40
safety 214
Sahlins, M. 15
Sahliyeh, E. 227
Said, E. 3, 251
Saifullah Khan, V. 154
Saikal, A. 279
sakhas 241–2, 243–4, 247
Salaam Bombay! 215, 269, 271
Salesian Centre for Minors 218
Samatar, A.I. 238
Sami, I.R. 207
Sangari, K. 245
Sanjek, R. 154, 155
São Miguel Paulista 168; people's radio 170–6; social movements 169–70
São Paulo 163, 167–79
Sarkar, B. 272
Sarkar, T. 244, 245–7
Sassen, S. 182, 183, 184
satellite television 264–5, 268–9
Sauer, C. 14
Save the Children 87, 108
Sayat Nova (The Colour of Pomegranates) 277
Scheper-Hughes, N. 161, 214
Schindler's List 264
Schlesinger, P. 23
Schmitz, H. 60–1
Scott, A. 164
Scott, A.J. 184, 188
Scott, J.C. 159, 162, 164, 168, 182
Scott Appleby, R. 226
Sealy, M. 256
Searle, C. 251
Seaton, J. 155
secularisation/secularism 195, 224, 230, 236
Segerstale, U. 148
Sekula, A. 99
self-advocacy 219
Self-Employed Women's Association (SEWA) 242, 247
Sen, M. 272
Sendero Luminoso 228
seniority promotion 127
Serb ethnicity 152
Serebriany, S. 279
services sector 184–5, 187–8
Sethi, H. 165
Sewell, B. 123
Sewell, J.W. 91

sex 121; international politics of 110–11
sex tourism 7–8, 109–16; international tourism 111–12; militarisation, development and 113–14; resistance 116; sexual politics of prostitution 109–10
sexual revolution 210
sexuality: cultures of and female circumcision 9, 201–11
Shandall, A.A. 206
Sharif, N. 58, 59, 68
Sharpley, R. 112
Shaw, P. 148
Sheehan, E. 210
Shiva, V. 52
Shona 228
Shore, C. 3, 153, 154
short-termism 129
Shrage, L. 110, 113
Sierra Leone 105, 108
signification 94
Sikhs 244–5
Simpson, A. 93
Sinclair, J. 23
Singh, V.P. 240
Sita 244
Skelton, T. 3
Sklair, L. 34
Sklar, R. 277, 279
Slack, A.T. 209
Slater, D. 232
small-scale enterprises 62–3, 65–8
Smart, K. 120
So, A.Y. 235
social capital 186
social conscience 236
social health 81–3; healers and social health of groups 78–80; as 'traditional' public health 75–8
social movements: local resistance 163–6; in São Miguel Paulista 169–70
social organisation: everyday resistance 166; Japanese 127–31; knowledge, livelihoods and 60–3
social structure 16–17
social system 15–16
socialisation 131–2, 215; enforced 130–1
socialist realism 276–7
sociobiology 2, 148
socio-dramas 174, 175
sociology 15–17, 17–18
Soja, E. 184
Sokolovski, S. 149
Solanas, F. 261
Solzhenitsyn, A. 277
Somalia 38
Sontag, S. 90
South Africa 105
South Commission 45
southern environmentalism 51–3; alliance with northern environmentalism 53–5
Southern Okavango Integrated Water Development Project (SOIWDP) 54–5
Sovexportfilm 280
Soviet Union 31, 32, 34; cinema 10, 275–9; *see also* Russia
specialist journalists 104
speech, freedom of 198
Speed 263–4
Spencer, J. 3
Spielberg, S. 263
Spivak, G. 251
Spring Offensive 133
Sreberny-Mohammadi, A. 234
Srinivas, M.R. 35
Stalin, J. 277
Stalker, P. 92
Stallone, S. 263
Star network 264–5
state 195–6; planning 31–2; and street children 217; subsidies 281
status, social 235
Stewart, F. 38
Stewart, R.G. 164
stigma 74
Stoddart, B. 251
Stoltzfus, B. 114
Storper, M. 184, 188
Street, B. 135
street children 9, 212–19; and the family 212–16; responses to 216–19
Strike 276
strikes 129, 133
structural adjustment 38
structure, social 16–17
student protest 36–7
Sturdevant, S. 114
sub-culture 17–18, 20
sub-Saharan Africa 63
subsistence health 80
Sudan 201–5, 206–7, 208, 209
Sullivan, B. 109, 110
Sun 92, 141
sunna circumcision 206, 206–7
Supreme Court 196
survival mechanisms 161–2
Swain, M.B. 112, 113
sweatshop employment 184
Sweden 143
syncretism, religious 227–8, 230, 233–4, 235

Takura Milling Company 66–7
Tannenbaum, N. 114
Tanzania 162
tapestries 163

Tarkovsky, A. 277, 278
Taylor, J. 99
Taylor, M. 98
technology 26, 48; knowledge and the poor 63–8; local knowledges and changing technology 7, 58–69
technology blending 64
television 104, 256–7; film and 264–5, 268–9, 280
temples 35–6
Tenbruck, F.H. 232
Teni Zabytykh (Shadows of Forgotten Ancestors) 277
test-match cricket 252–9
textile industry 134
Thailand 225; sex tourism 110, 111, 112, 113, 114–15
Thatcherism 138, 141–2, 142–3
Third Cinema 261–3, 270
Third World Movement Against the Exploitation of Women (TH-MAE-W) 116
'Third-Worldism' 33–4
Thomas, A. 45, 47, 51
Thomas, R. 266
Thomas-Hope, E.M. 182
Thompson v. *Oklahoma* 196
Thornhill, A. 258
thought, freedom of 198
three jewels system 127–8, 132–3
Thrift, N. 234
Thrupp, L.A. 64
tickets, test-match 254–5
Tiffin, H. 252
Tiger, L. 148
time 17, 236
Tishkov, V. 149
Todaro, M.P. 181
Toffler, A. 17
Tomlinson, J. 23, 24, 27
Toppin, J. 259
torture 198
Touraine, A. 164
tourism: growth in international 111–12; sex tourism 7–8, 109–16; West Indies cricket and 252, 253–9
Toyota 133
trade unions 134–5; company unionism 127–8, 133, 135
'traditional' intellectuals 19
'traditional' public health 75–8
trafficking in women 115–16
transport 121–2
Trillo, R. 201
Trotsky, L. 276
Truong, T.-D. 111
trust 60–1
trust culture 144
Tshomorelo Okavango Conservation Trust (TOCT) 54–5
'Tube Talk' column 121–2
Tucker, V. 39
Turner, B. 198, 199, 234, 238
Turton, D. 152
Tutsis 37
Tyrrell, H. 270

Uganda 64–5, 163
'ummah' 238
UNESCO 108
United Kingdom *see* Britain
United Nations (UN) 39; Special Initiative on Africa 108
United Nations Conference on Environment and Development (UNCED) (Earth Summit) 45, 52
United Nations Convention on the Rights of the Child 115
United States (US) 138, 140, 195–6; ethnicity in New York 149–50, 150–1; films 10, 263–6, 279–80; human rights 194, 195, 196; Mexican migration to 181, 184–6; military bases 114
Universal Declaration of Human Rights 1948 194
universal rationalism 234–5
universalism 28; human rights 197–9; and particularism 237–8

vagabonds 216
Vaid, S. 245
values 14–15, 176–7; and new concept of progress 56; Western and Japanese 127, 128
van de Laar, A. 47, 49, 52
van der Dennen, J. 148
Van Der Gaag, N. 92, 93
van der Kwaak, A. 209
van Esterik, P. 115, 116
van Gennep, A. 209
Vanita, R. 245
Vasconcelos, A. 218
Vasudevan, R. 262, 270
Verhelst, T.G. 164
Vickers, A. 36
Vickers, G. 48
Victoria, Lake 64–5
video 267–8, 280
Vieira, P.F. 165
Vietnam War 114
Vietnamese migrants 188
Vila Aparecida 8, 167–79; and the people's radio 170–6
Vine, I. 148

Viola, E.J. 165
violence 161; India 244–5; and street children 217–18
virtual neighbourhoods 153
Vishwa Hindu Parishad (VHP) 241
Vizedom, M. 209

Walby, S. 188
Waldinger, R. 186
Wallerstein, I. 33, 232
Wallman, S. 154
Watt, M. 235
Watts, M. 161,
weakness 8, 159–66; and power 159–60; weapons of the weak and human rights 161–3; weapons of the weak and long-term change 163–6
Weber, M.16, 137, 234, 235
Wekiya, I. 64–5
welfare 236
wellbeing 73–5
Werbner, P. 154, 187
Wertheim, W.F. 20
West: its Other and human rights 9, 193–200
West Indies 10, 251–9
West Indies Cricket Board of Control (WICBC) 255, 256, 257
Western media: representations of conflict 7, 102–8
Western values 128
Westernisation 225; culture 6, 23–5; decline of the West and globalisation 25–9
Whitaker, B. 90
White, B. 212, 214
Whitehead, H. 209
Whittaker, A. 212
Whittaker, D.H. 128
Wickham, J. 258
Wiener, M. 141–2
Wignaraja, P. 164, 165
wilderness 49–50
Wilkinson, B.A. 255, 257
Willemen, P. 261, 270
Williams, R. 2, 20–1, 176
Willis, P. 21
Wilson, B. 236
Wilson, G.A. 62–3, 64, 67
Wolf, E.R. 35, 39, 181
Wolpe, H. 182
women: Bangladeshi in *purdah* 160; and the new Hinduism 10, 240–7; migration 184; and people's radio 178; self-help groups 60; Tanzanian and agriculture 162; technology, poverty and 63–4; trafficking in 115–16; *see also* female circumcision, sex tourism
women's movement 36–7; *see also* feminism
Wong, Y. 148
Woodhouse, P. 47, 60, 63
Wooldridge, I. 258
Working Women's Forum (WWF) 242, 247
World Bank 38, 40, 70, 71, 73
World Conservation Union (IUCN) 55
World Health Organisation (WHO) 70, 73, 81
world music 35
world-system model 33
World War II 139–40; migration after 181–2
Worsley, A. 206
Worsley, P. 15, 21, 33, 152
Wright, S. 135

Xiaobo, 66
'Xuxa' 178

Yakan, cult of 228
Yearley, S. 50
Yerkes, T. 121
Young, I.M. 197
Young, J. 120
Young, R. 90
Yugoslavia, former 37, 152, 226

Zambezi Society 51
Zee TV 264–5, 268–9
Zensanren (All-Japan Producers Union) 134
Zhou, M. 185, 186, 187
Zimbabwe 51, 228
Zubaida, S. 36
Zumbi 178